T0186233

Surgical Oncology

David N. Krag, M.D.
University of Vermont Medical School
Burlington, Vermont, U.S.A.

LANDES
BIOSCIENCE

GEORGETOWN, TEXAS
U.S.A.

VADEMECUM
Surgical Oncology
LANDES BIOSCIENCE
Georgetown, Texas U.S.A.

Please address all inquiries to the Publisher:
Landes Bioscience, 810 S. Church Street, Georgetown, Texas, U.S.A. 78626
Phone: 512/ 863 7762; FAX: 512/ 863 0081

ISBN: 1-57059-633-6

Library of Congress Cataloging-in-Publication Data

Surgical oncology / [edited by] David N. Krag.
    p. ; cm. -- (Vademecum)
    Includes bibliographical references and index.
    ISBN 1-57059-633-6 (spiral)
    1. Cancer--Surgery. I. Krag, David N. II. Series.
    [DNLM: 1. Neoplasms--surgery. 2. Medical Oncology. QZ 268 S96105 2000]
    RD651.S8832 2000
616.99'4059--dc21   00-057733

Dedication

This book is dedicated to Paul Gross, Ph.D. at the Unviersity of the Pacific. It is a small gesture to recognize his influence as an outstanding educator whose powerful messages have stood the test of time.

Contents

Editor

David N. Krag, M.D.
SD Ireland Professor of Surgical Oncology
University of Vermont Medical School
Burlington, Vermont, U.S.A.
david.krag@uvm.edu

Contributors

James C. Alex
Division of Facial Plastic
    and Reconstructive Surgery
Yale University School of Medicine
New Haven, Connecticut, U.S.A.
Chapter 5

David A. August
Division of Surgical Oncology
UMDNJ-Robert Wood Johnson Medical
    School
The Cancer Institute of New Jersey
New Brunswick, New Jersey, U.S.A.
Chapter 14

Paul L. Baron
Department of Surgery
Medical University of South Carolina
Charleston, South Carolina, U.S.A.
Chapter 3

Mary Sue Brady
Department of Surgery
Gastric and Mixed Tumor Service
Memorial Sloan-Kettering Cancer Center
New York, New York, U.S.A.
Chapter 4

Laurence H. Brinckerhoff
Department of Surgery
University of Virginia
Charlottesville, Virginia, U.S.A.
Chapter 2

Peter Cataldo
University of Vermont College of
    Medicine
Burlington, Vermont, U.S.A.
Chapter 9

Robert M. Goldstein
Department of Surgery
Baylor University Medical Center
Dallas, Texas, U.S.A.
Chapter 10

Frederic W. Grannis, Jr.
Section of Thoracic Surgery
City of Hope National Medical Center
Duarte, California, U.S.A.
Chapter 12

Jay K. Harness
University of California, Davis–East Bay
Alameda County Medical Center
Oakland, California, U.S.A.
Chapter 17

James C. Hebert
University of Vermont College of Medicine
Burlington, Vermont, U.S.A.
Chapter 11

Neil Hyman
University of Vermont College of Medicine
Burlington, Vermont, U.S.A.
Chapter 8

Thomas J. Kearney
Division of Surgical Oncology
UMDNJ-Robert Wood Johnson
    Medical School
The Cancer Institute of New Jersey
New Brunswick, New Jersey, U.S.A.
Chapter 14

Jason Klenoff
Section of Otolaryngology
Yale University School of Medicine
New Haven, Connecticut, U.S.A.
Chapter 5

Joseph A. Kuhn
Department of Surgery
Baylor University Medical Center
Dallas, Texas, U.S.A.
Chapter 10

Bruce Jason Leavitt
University of Vermont College of Medicine
Burlington, Vermont, U.S.A.
Chapter 13

Brian W. Loggie
Division of Surgical Oncology
University of Texas Southwestern
  Medical Center at Dallas
Dallas, Texas, U.S.A.
Chapter 7

Todd M. McCarty
Department of Surgery
Baylor University Medical Center
Dallas, Texas, U.S.A.
Chapter 10

Frederick L. Moffat
Division of Surgical Oncology
Daughtry Family Department of Surgery
University of Miami School of Medicine
Miami, Florida, U.S.A.
Chapter 18

Thomas A. Roland
University of Vermont College of Medicine
Burlington, Vermont, U.S.A.
Chapter 16

Philip D. Schneider
Department of Surgery
University of California
  Davis Medical Center
Sacramento, California, U.S.A.
Chapter 6

Stephen A. Shiver
Wake Forest University
Baptist Medical Center
Winston-Salem, North Carolina, U.S.A.
Chapter 7

Craig L. Slingluff, Jr.
Department of Surgery
University of Virginia
Charlottesville, Virginia, U.S.A.
Chapter 2

Lorraine Tafra
The Breast Center at Anne Arundel
  Medical Center
Annapolis, Maryland, U.S.A.
Chapter 1

Lee W. Thompson
Department of Surgery
University of Virginia
Charlottesville, Virginia, U.S.A.
Chapter 2

Preface

This book distinguishes itself from other major oncology texts in that it is not designed to be encyclopedic but distills the most important information that should be known by those interested in surgical oncology. It is therefore approachable and readable. It should be able to be read during the length of a rotation on a surgical oncology service.

Chapters are arranged by organ of involvement. The chapters are designed with a common theme arranged in a practical fashion. Each chapter begins with defining the scope of the particular problem and includes incidence, risk factors for development of disease, and methods for screening. This is followed by methods of diagnosis, preoperative evaluation, and staging. Treatment options are defined for the most common, treatable presentation of the disease. Clearly stated treatment outcomes are presented including survival and important immediate and long term side effects. Guidelines for posttreatment surveillance are also presented. Important aspects of Radiation and Systemic therapy are covered in separate chapters.

The last two chapters cover two areas of clinical science which are expanding in the field of surgical oncology in a dramatic manner: Chapter 17 covers diagnostic and interventional ultrasound and chapter 18 covers sentinel node surgery.

Over the past ten years a new generation of surgical oncologists have entered the surgical work force of the United States. Few training programs are now without a service devoted to surgical oncology. This book is written to aid that group of surgical oncologists in transmitting essential information in a concise, up-to-date, and readable manner. This book meets the need of the medical student or surgical resident rotating on a surgical service that emphasizes oncology. It will also be useful to the busy practicing general surgeon for reviewing the most current information on organ specific cancer therapy.

The chapters have been written by outstanding clinical educators. All are active teaching clinicians that day-to-day manage and teach the issues outlined in each chapter.

David N. Krag

Acknowledgments

It is important to acknowledge the support of the S.D. Ireland family. They have faced the threat of cancer and fought back by supporting research in a concrete and meaningful manner.

CHAPTER 1

Breast Cancer

Lorraine Tafra

Scope of the Problem: Incidence

Breast cancer is the most frequently diagnosed malignancy of woman in the United States. In 1998 the incidence was found to be approximately 180,000 with 44,000 deaths attributed to breast cancer. For women, the lifetime risk for developing breast cancer is 1 in 8. Until 1998, mortality from breast cancer had remained unchanged but now appears to be slowly decreasing. More universal application of screening mammography has resulted in gradual decrease in the size of the primary tumor at time of diagnosis. Mortality reduction appears to be at least in part due to the impact of screening mammography in finding tumors at an earlier stage.

Worldwide

In general, less industrialized nations tend to have lower rates of breast cancer, with Japan being an exception to this rule. The highest rates are seen in Europe and North America.

Race

The overall incidence of breast cancer is higher among white women than African-American woman, but the mortality is higher for the latter. Research has shown that most African-American woman will present with later-stage disease, a fact that may be secondary to cultural factors.

Risk Factors

Gender

Breast cancer is predominantly a disease occurring in women. Only about 0.5% of all breast cancer cases occur in men.

Age

For most women, this is the single most important risk factor. The majority of breast cancers are diagnosed after the age of 50.

Family History

A major breakthrough providing insights into breast cancer risk occurred in 1990 with the identification of the BRCA1 gene. This large gene is located on chromosome 17q. Breast cancer in BRCA1 families is transmitted as a classic Mendelian autosomal dominant trait with high but incomplete penetrance. BRCA1 mutations are associated with an extremely high risk of breast cancer development (50% chance

Surgical Oncology, edited by David N. Krag. ©2000 Landes Bioscience.

1

by age 45, and 85% lifetime risk). This inherited form of breast cancer, though, occurs in only 5-10% of patients with breast cancer. The percentage of patients with breast cancer due to known hereditary factors varies with age and is higher in the younger age group. In patients less than 30 years of age the risk of an identifiable abnormal gene is 33% and in patients between ages of 40 and 49, 13% will have the abnormal gene. Genetically transmitted breast cancer should be suspected in women with multiple relatives with the disease, particularly when the family history includes premenopausal or bilateral breast cancer. The risk of developing breast cancer is increased 1.5-3.0 times if a mother or sister has the disease, and risk may be greater if a sibling is affected. For most women with a family history, the lifetime risk of developing the disease does not exceed 30%. Genetic conditions associated with an increased risk of developing breast cancer include Li-Fraumeni, Cowden disease, Muir Torre, Peutz-Jeghers, and ataxia-telangiectasia.

Menstrual and Reproductive Factors

Factors that have been linked to breast cancer include early age of menarche (age < 12), late age of menopause (> 55), and late age at first pregnancy (> 30). It appears that increased duration of exposure to endogenous estrogens is a common factor.

Another interesting risk factor may be related to the lifetime number of menstrual cycles. In less industrialized countries women may have only about 100 menstrual cycles in their lifetime due primarily to childbearing and nursing. Women living in industrialized countries may have an average lifetime number of menstrual cycles exceeding 400. During each menstrual cycle the breast undergoes cell growth stimulation. Increased cell growth stimulation may contribute to an increased rate of cancer.

Weight and Diet

These two variables have been suggested as a cultural explanation for the differences in incidence rates between countries. National per capita fat consumption correlates with incidence and mortality from breast cancer. Although Japanese women have a lower incidence of breast cancer, after moving to the United States, their incidence approaches that of Caucasian women. However, numerous epidemiological studies have found no correlation of fat consumption to an increased incidence of breast cancer. Weight and diet remain inconclusive risk factors for developing breast cancer.

Radiation Exposure

Exposure to radioactivity from therapeutic irradiation, medical diagnostics, or nuclear explosions has been shown to increase the risk of breast cancer. The risk is greatest when the radiation was received before the age of 40.

Lactation

Prolonged lactation in recent studies has been shown to reduce the incidence of breast cancer in premenopausal women. This may also be related to a lower frequency of menstrual cycles.

LCIS and Atypical Hyperplasia

These pathological findings found at the time of biopsy place the patient at higher risk for subsequent development of breast cancer. This risk is increased further if associated with a positive family history.

Exogenous Estrogen

The effect on risk of developing breast cancer by exogenous hormones for oral contraception and hormone replacement therapy is controversial. It appears that there is some increased risk of developing cancer in the overall population but this is relatively small. An important area of investigation is to define which groups of women may be at exceptionally higher risk due to exogenous estrogens.

Prevention

Medical Prevention

During the 1990s, Tamoxifen was evaluated by the National Surgical Adjuvant Breast and Bowel Project (NSABP) for prevention of breast cancer. Women were randomized to receive Tamoxifen versus a placebo.[1] This antiestrogen agent was chosen since a prior study showed that when Tamoxifen was used as adjuvant therapy in patients with breast cancer, it reduced the rate of contralateral breast cancers by approximately 40%. The results of the breast cancer prevention trial were released in 1999 and showed a 49% reduction in the incidence of breast cancer in women taking tamoxifen.

Tamoxifen has both short and long term side effects. Although it has been demonstrated to reduce the incidence of breast cancer in a defined patient population over a defined period of time, the indications for use in the general population remain to be determined.

Surgical Prevention

In high risk patients, bilateral prophylactic mastectomy can decrease the incidence of breast cancer by at least 90%. The physical and psychological price for this method of prevention is very high and should be undertaken only after considerable evaluation.

Screening

The elements of screening for breast cancer are self breast examination (SBE), clinical breast examination (CBE), and mammography. Ultrasound is not currently part of standard screening methods but may have a role in women with dense breasts and multiple cysts. Prospective randomized trials have established that breast cancer mortality is reduced in woman ages 50-74 by approximately 26% if patients are screened annually with mammography. It has been less well established for women ages 40-50. The American Cancer Society and other groups continue to recommend mammography and breast exams every 1-2 years for these women.

Methods of Diagnosis

In order to make a definitive diagnosis of breast cancer, tissue must be acquired and analyzed by a pathologist. It is simplest to categorize the various methods of tissue acquisition according to:

1. the method of guidance to the target tissue and
2. the type of tissue acquisition device.

There are three types of guidance systems: radiographic, ultrasound, and palpation (the hand). There are also three general types of tissue acquisition devices: fine needle aspiration (FNA), core (specialized cutting needles), and a scalpel. Any of the guidance systems may be linked with any of the tissue acquisition devices. For instance a stereotactic biopsy simply means that the guidance system is radiographic. There are a variety of clever minimally invasive tissue acquisition devices being developed but common to all are methods of accurate guidance to the target tissue and methods of removing a sample of tissue.

Fine needle aspiration involves passage of a needle (about 25 gauge) into the lesion. The needle is passed several times back and forth in the lesion while the syringe barrel maintains negative pressure. The contents of the needle are then placed on a glass slide, smeared, and fixed in a variety of manners. This type of sample preparation does not provide microscopic architecture and interpretation requires an experienced cytopathologist. This method may not allow differentiation between DCIS and an invasive lesion. These problems have prevented wide acceptance of FNA as the preferred method of tissue sampling.

Core biopsies are performed using specialized cutting needles, usually 14-16 gauge. The simplest types of core biopsy devices are spring loaded and collect a single core of tissue. The physician places the tip of the core needle at the edge of the lesion and presses a button to "fire" the device. The inner part of the needle advances exposing a trough. The outer part of the device then advances over the trough cutting a small portion of tissue that remains in the trough. This firing action is very rapid and makes a snapping noise as the springs rapidly advance the needles. The device is then removed from the breast and the core biopsy specimen collected. Typically 3-6 specimens are collected. More complex and expensive core biopsy devices allow repeated removal of small specimens without removing the entire device from the breast. The needle is advanced into the lesion. A side window is opened along the shaft of the needle near the sharp tip. A vacuum is applied pulling a small sample of tissue into the hollow shaft. An inner cutting shaft is then advanced forward to cut the sample of tissue. The inner cutting shaft containing the tissue sample is removed and the sample collected. The inner cutting shaft is then passed back through the needle and the process can be repeated several times. This method can piecemeal remove blocks of tissue.

A core biopsy provides more tissue than FNA. It can be processed and evaluated by standard histopathological techniques. Importantly, this method provides architecture and can reliably distinguish invasive from noninvasive cancer. However, it is still a sampling technique and does not typically remove the entire lesion in question. The surgeon must take the responsibility to insure that an incompletely removed lesion is not cancer.

Excisional biopsy is a surgical procedure that removes the entire lesion within a defined block of tissue. If the lesion is palpable the guidance system is the surgeon's hand. If the lesion is not palpable, mammography is typically used to aid in localization of the lesion. Just before surgery a bent-tipped wire is mammographically guided into the breast. This wire is then used to guide the surgeon to the location of the nonpalpable lesion.

An advantage to excisional biopsy is that the entire lesion can be excised and the most definitive diagnosis obtained. If the lesion is benign, in most cases the patient needs no further follow up for that lesion because it has been removed. A disadvantage is that if the lesion is malignant, a re-excision is generally required to obtain therapeutically sufficient margins of normal tissue. Re-excision usually leads to unnecessary tissue loss and less optimal cosmesis.

Types of Breast Cancer

There are essentially two types of breast cancer: invasive and noninvasive. Invasive breast cancers have invaded the basement membrane and have the potential for metastases. Noninvasive breast cancers do not have the ability to metastasize.

Ductal Carcinoma In Situ (DCIS)

This type of breast cancer can be characterized by a number of features, but the most prognostically significant characteristics include the presence of comedo features, nuclear grade and size. The form of DCIS that carries the greatest chance for recurrence is comedo type, large size, and high nuclear grade. If this is obtained on core biopsy there is a 25% chance that an invasive component will be found.[3] Others forms of DCIS include micropapillary and cribiform, which also carry a better prognosis than comedo. DCIS is most frequently found as a mammographically-detected abnormality and usually appears as a cluster of pleomorphic calcifications. Its incidence has increased as screening mammography has increased.

Lobular Carcinoma In Situ (LCIS)

LCIS is typically not identified by prebiopsy imaging studies. It is found by the pathologist as an incidental finding in a biopsy specimen in which another benign or malignant lesion was targeted. LCIS is considered more to be a marker of risk for future development of cancer than an actual cancer. The cancer type that may subsequently develop can be ductal or lobular carcinoma.

Invasive Breast Cancer

Infiltrating ductal carcinoma is the most common form of invasive breast cancer and accounts for about 75% of all invasive breast cancers. It arises from ductal cells lining milk ducts. Histologic subtypes which have a more favorable prognosis include tubular, colloid or mucinous, and medullary carcinoma.

Infiltrating lobular carcinoma arises from cells of the lobular unit. Lobular carcinoma has a higher frequency of bilaterality. It also tends to be more infiltrative than ductal carcinoma and its size is frequently underestimated by mammography. Tumors of low frequency include mesenchymal tumors or sarcomas of the breast. These include fibrosarcoma, malignant fibrous histiocytoma, liposarcoma, leimyosarcoma, osteogenic and chondrosarcoma, and angiosarcoma. Malignant cystosarcoma phylloides are fibrous tumors that can be difficult to distinguish from benign phylloides tumors.

Paget's Disease of the Nipple

This is an uncommon form of breast cancer that is clinically characterized by an eczematoid reaction of the nipple and symptoms of itching, erythema and nipple discharge. In 45% of patients there is an associated underlying breast mass which is an infiltrating ductal carcinoma or ductal carcinoma in situ. It is often treated with mastectomy but may be treated by central breast lumpectomy if the cancer is well localized. Prognosis is similar to infiltrating ductal carcinoma and is determined similarly by the stage of the disease.

Preoperative Evaluation

Patients must be evaluated for the ability to withstand the stress of the proposed surgical procedure. Most patients tolerate surgery for breast cancer since a body cavity is not entered and critical organs are not manipulated. Very frail patients and those of advanced age must have their therapy customized to that which they can acceptably tolerate.

Preoperative Staging

The staging system for breast cancer is based on the recommendation of the American Joint Committee on Cancer (AJCC). This uses the TNM system where T refers to tumor, N to lymph nodes and M to metastases. TNM definitions are displayed in Table 1.1. Stage groupings are in Table 1.2. Preoperatively, histologic staging of the tumor is limited to knowing only the type of tumor. The final T and N status designation is determined after pathological analysis of the resected specimen. Preoperative serum laboratory studies are generally not useful for detecting systemic metastases. A baseline chest roentgenogram is obtained. Additional studies are obtained to evaluate symptoms or to establish baseline conditions if a patient is entering a clinical trial.

Treatment Options

Surgical Therapy

Surgical therapy for breast cancer remains the most successful single form of therapy. Decisions on which surgical procedure is best for any particular patient are determined by a number of factors and include: size of the tumor, size of the breast, resectability (is it fixed to underlying muscle), comorbid factors, and patient preference. It is important to understand both the indications for each procedure as well as the definitions of those procedures. An excellent diagrammatic overview of the treatment of breast cancer by stage of disease was recently published.[4]

Definition and Indications of Breast Surgery Procedures

Partial Mastectomy (PM), Lumpectomy, Wide Excision, Telectomy, and Quadrantectomy

These terms all refer to breast sparing surgery or resection of a portion of the breast which contains the primary tumor. The goal of this procedure is to remove

Table 1.1. TNM system of breast cancer staging

Primary Tumor	
Tis	Ductal carcinoma in situ, Paget's disease of the breast with no tumor
T1	Tumor equal or less than 2 cm
T2	Tumor greater than 2 cm but no greater than 5 cm
T3	Tumor more than 5 cm
T4	Tumor of any size with direct extension to chest wall or skin (includes peau d'orange, skin ulceration, inflammatory carcinoma
Lymph Node Involvement	
N0	No regional lymph node metastases
N1	Metastases to movable ipsilateral axillary nodes
N2	Metastases to fixed ipsilateral axillary nodes
N3	Metastases to internal mammary lymph nodes
Distant Metastases	
M0	No distant metastases
M1	Distant metastases

Table 1.2. Stage grouping of breast cancer

Stage	T	N	M
I	T1	N0	M0
IIA	T0	N1	M0
	T1	N1	M0
	T2	N0	M0
IIB	T2	N1	M0
	T3	N0	M0
IIIA	T0	N2	M0
	T1	N2	M0
	T2	N2	M0
	T3	N1	M0
	T3	N2	M0
IIIB	T4	Any N	M0
	Any T	N3	M0
IV	Any T	Any N	M1

the tumor with a sufficient rim of normal breast tissue. It is unproven how large the margin of normal tissue should be in order to achieve optimal local control. A partial mastectomy is typically combined with whole breast irradiation (RT). The combination of PM and RT is termed breast conserving therapy. It is usually indicated in any woman who has a malignancy measuring less than or equal to 4 cm. and in which the cosmetic result will be considered satisfactory. Contraindications include: first or second trimester of pregnancy, two or more gross tumors in separate quadrants of the breast (multicentric disease), diffuse, indeterminate, or malignant-appearing microcalcifications, and a history of prior therapeutic irradiation to the breast. A history of collagen vascular (connective tissue) disease has also been relative contraindication based on a number of anecdotal reports of severe fibrotic reactions to irradiation in patients with lupus and scleroderma.

Patients positive for BRCA1 or BRCA2 gene mutations may have a higher local recurrence rate. Retrospective data is being gathered to determine whether patients with strong family histories in fact have higher recurrence rates with breast sparing surgery. Positive margins on a lumpectomy specimen indicate that the tumor was incompletely excised. Except for very frail or patients of advanced age, pathologically positive margins are not acceptable. Re-resection is required and this may result in the need for a mastectomy.

Since 1970, there have been six prospective randomized trials using modern radiation techniques in which conservative surgery and radiation therapy have been compared with mastectomy. These trials differ somewhat in patients selection, the methods of surgery and radiation therapy, and the length of follow-up. Nevertheless, all of these trials show equivalent survival between the two treatment options and are summarized in Table 1.3. A recent overview of these randomized trials, as well as a number of other unpublished results, showed equivalence in mortality.[5] It should be remembered that patients have a choice of mastectomy versus breast conserving surgery and should be offered such a choice. There is frequently no single "right" choice for a woman with breast cancer and it is helpful to arrange for visits with the radiation oncologist, the medical oncologist and to encourage her to not rush making a decision.

Total Mastectomy

This refers to removal of the breast. Total mastectomy may be performed as the only operative procedure in which case regional lymph nodes are not removed. An older term for total mastectomy without removal of regional lymph nodes is simple mastectomy. The indication for this procedure includes DCIS that cannot be encompassed by a partial mastectomy. DCIS may not be encompassed effectively by a partial mastectomy because it is very extensive or because it is present simultaneously in diverse locations of the breast. Other indications include patient preference for management of DCIS even though eligible for breast conserving therapy. Mastectomy is also indicated for local recurrence in a breast previously treated for cancer with partial mastectomy and breast irradiation. When patients with invasive cancer are treated with total mastectomy, an axillary dissection is usually included. However, frail patients or those of advanced age are frequently considered for lessor surgical procedures that may include total mastectomy without a regional node resection.

Total Mastectomy and Axillary Lymphadenectomy (TM and AD)

This involves removal of the breast and removal of the regional axillary lymph nodes. TM and AD replaces the older term "modified radical mastectomy". In order to explain to a patient the meaning of modified radical mastectomy, the meaning of radical mastectomy must be explained. Since radical mastectomy is rarely performed and is an alarming procedure to patients, avoiding this reference is desired. The indications for TM and AD are multifocal cancer, large tumor to breast ratio, positive margins following partial mastectomy, and patient preference.

Radical Mastectomy

Removal of the breast, the skin of the breast, axillary lymph nodes, and the pectoralis major muscle. This procedure was introduced by Halsted[6] in the late 1800s. It is rarely used now. Even tumors that are locally invading the muscle are usually treated with limited excision of the pectoralis muscle, thus obtaining negative margins but not removing the muscle in its entirety. Clinical trials have demonstrated no observable survival difference between total mastectomy and axillary resection versus radical mastectomy.

Axillary Dissection or Axillary Lymphadenectomy

This procedure usually means removal of level I and II axillary lymph nodes (lateral and posterior to the pectoralis minor muscle). More thorough node resection includes level III lymph nodes and interpectoral lymph nodes. The rationale for surgically removing axillary nodes are:

1. staging and prognosis,
2. regional control, and
3. possible improved survival.

Regional node resection usually is performed when the primary breast tumor is an invasive cancer. The extent of resection of axillary nodes and the value of axillary lymphadenectomy is controversial. Clinical trial data indicate that the status of the regional nodes is an independent prognostic indicator. The status of the nodes is an integral element of the TNM staging system. It is also clear that surgical resection of the axillary nodes results in excellent (nearly 100%) long term regional control. It carries significant morbidity including edema, sensory (and uncommonly motor) neurological complication, and increased susceptibility to serious infection of the upper extremity. Associated psychological distress is noted in a high percentage of women and this side effect is frequently underestimated. Since about 60-70% of patients do not have regional metastases, there is significant motivation to find alternatives to axillary lymphadenectomy. With the introduction of sentinel node biopsy for breast cancer (discussed below), it has become clear that the role of axillary dissection is in transition. Axillary dissection can probably be eliminated for patients with localized DCIS, DCIS with small areas of microinvasion, and for pure tubular carcinoma less than 1 cm in size because the risk of nodal involvement is extremely low.

The NSABP-06 clinical trial randomized patients to prophylactic regional lymphadenectomy at the time of surgical treatment of the primary breast cancer or to delayed regional lymphadenectomy only if regional lymph nodes subsequently became clinically involved with tumor. This did not show a statistically meaningful difference in survival between the two groups. Three other prospective randomized European studies however did show a survival advantage to performing prophylactic axillary node dissection. Although in the United States it is generally assumed that axillary lymphadenectomy has no bearing on survival, it is possible that there regional lymphadenectomy results in improved survival by as much as a 5-10%.

Sentinel Node Biopsy

This is a new technique that maps the drainage pattern of a tumor to the first set of lymph nodes most likely to contain metastatic disease. As applied to breast cancer it appears that a limited removal of the set of lymph nodes (usually 1-3 nodes) allows determination of whether regional metastases have occurred. Validation of the accuracy of this method of identifying the nodes most likely to contain metastases has been performed in a multi-center study for breast cancer.[7] This study showed that injection of a radioactive tracer (technetium sulfur colloid) into normal breast tissue immediately surrounding the tumor resulted in labeling those nodes most likely to contain cancer. Intraoperatively a hand held gamma detector was used to identify the radiolabeled sentinel nodes. Interestingly, a small percentage of patients had drainage to pathologically positive lymph nodes which were located outside the axillary area. If those patients had only a conventional axillary lymph node resection, nodes containing cancer would have been left behind. Although it appears that the status of sentinel nodes predicts whether regional metastases have occurred, there is no data as to whether this will result in long term survival or long term regional control as good as conventional regional lymphadenectomy. This technique is now being evaluated in prospective randomized trials and is considered experimental until such data is available.

Whole Breast Radiation Therapy

The goal of radiation therapy is to reduce the rate of local or regional tumor recurrence by treating residual microscopic disease that may have spread beyond the margin of surgical resection. It is an integral part of breast conserving therapy and is therefore routinely used in patients that have had a partial mastectomy. Irradiation of the breast following partial mastectomy reduces local recurrence in the breast from almost 50 to about 10%.

It is controversial whether all patients with DCIS treated with lumpectomy need to receive radiation therapy. There are certain groups of patients with DCIS that have a very low risk of local recurrence where the benefits of radiation therapy are less clear.[3] These include noncomedo DCIS, clear margins greater than 1 cm, or DCIS less than 1 cm in size.

Chest Wall and Regional Node Radiation Therapy

Recent studies have shown improved regional control and increased survival when higher risk patients that have had a total mastectomy and regional node dissection are also treated with irradiation to the chest wall and regional nodes. This data remains controversial, but a consensus statement has been released supporting the role of radiation therapy postmastectomy for patients with > 4 involved lymph nodes.[8]

Timing of Chemotherapy vs. Radiation Therapy

Theoretically, delaying radiation therapy in women receiving lumpectomy could increase the chance of developing a local recurrence, and delaying chemotherapy in patients with positive lymph nodes could increase the chance of developing distant metastases. However, giving both therapies at the same time increases morbidity. Radiation therapy typically follows completion of systemic therapy based on the rationale that systemic metastases are the most critical element of the disease to treat.

Systemic Therapy

It has long been hypothesized that at the initial presentation of breast cancer, it is already a systemic disease. Meta-analysis of clinical trials involving tens of thousands of women have demonstrated that systemic adjuvant therapy improves survival of breast cancer patients. It is this data which most convincingly supports the notion that the systemic disease is a component of breast cancer and that treating this component improves survival. Systemic therapy is not without complications and careful evaluation of the risks versus benefits must be performed and presented clearly to the patient. There may be categories of patients that do not benefit from systemic adjuvant therapy either because their cancer is more advanced and not responsive to drugs or because they are at such low risk of systemic disease that the risks of systemic therapy outweigh the benefits.

The most well studied regimen of chemotherapy for breast cancer patients in the adjuvant setting is CMF (cyclophosphamide, methotrexate, and 5-fluorouracil). This usually consists of IV administration of methotrexate and 5-flourouracil on days 1 and 8, and oral cyclophosphamide on days 1-14, with 14 days rest without chemotherapy – for a total cycle length of 28 days. This regimen improves disease free and overall survival. The usual length of therapy is 6 months or 6 cycles and studies have shown no benefit to more prolonged therapy. There is data to suggest that the addition of doxorubicin to patients at high risk of relapse (greater than four positive nodes) may enjoy improved survival over standard CMF. For patients with a poor prognosis, stem cell transplant is currently under evaluation. These patients receive higher doses of chemotherapy then they could otherwise tolerate because their bone marrow is harvested prior to administering the chemotherapy and is then transfused back after the chemotherapy. Although there was tremendous enthusiasm for this relatively morbid regimen, recent data has not shown that this improves survival.

The decision regarding which chemotherapeutic agents to use, the doses, timing of the doses and coordination with radiation therapy is complex. Many factors must be considered including patient age, general health, risk of recurrence and patient preference. Patients with a risk of recurrence less than 10-15% may be spared adjuvant therapy because the absolute risk reduction is so small that the toxicity and expense of treatment are not as well justified as for higher risk patients.

Tamoxifen

This antiestrogen is one of the most studied drugs in oncology and is used both in the adjuvant setting and for metastatic disease. It has a relatively low toxicity profile and is administered orally at a dose of 20 mg/day. It improves survival in postmenopausal and premenopausal patients whose tumors are estrogen receptor positive. Tamoxifen has also been shown to decrease further the recurrence rate in patients receiving chemotherapy. There may be a small benefit for women with estrogen receptor negative tumors. Tamoxifen should be given for at least 2 years and is usually given for 5 years. Studies indicate no survival benefit for use of Tamoxifen beyond 5 years.

Bilateral Prophylactic Mastectomy

This therapy is usually reserved for women with very high risk of developing breast cancer. Women positive for BRCA1 or BRCA2 may choose this prophylactic treatment since the lifetime risk of developing cancer approaches 90%. Women with less well characterized risk factors including LCIS, family history of breast cancer, or previous personal history of breast cancer are often less inclined to choose this method of prophylaxis. In assisting a women make this choice the surgeon must determine the likelihood of being able to detect a breast cancer at an early stage. If a women has breasts with dense glandular tissue, mammography may not provide a good screening tool. In addition if the patient has large breasts with irregular glandular tissue, CBE may not provide good screening for early stage breast cancer. These factors should be taken into consideration when evaluating a patient for possible bilateral prophylactic mastectomies. This allows the patient to fully compare the risks of getting breast cancer and the risks of dying from that breast cancer to the total morbidity of bilateral prophylactic mastectomy.

Treatment Outcomes

Survival Rates

The overall survival rate of all stages of breast cancer ranges from 60-80% at 5 years. Patients without metastases to lymph nodes have a higher survival rate than patients with positive lymph nodes. For patients with distant metastases, the 5-year survival drops to 18%. The best survival rate is seen in patients with DCIS where only 1-2% of patients will die of their disease.

Only about 50% of patients who have developed a local recurrence following mastectomy will survive 5 years, and this drops to 26% at 10 years. The longer the interval of time from surgery to local recurrence the greater is the survival rate. The prognosis is much better following a recurrence after breast conserving surgery then after mastectomy, with the majority (60-75%) of patients alive and disease free after 5 years.

Prognostic Parameters

These are factors that are associated with disease-free and overall survival. In addition to providing the patient information as to their risk of dying from breast cancer, they guide recommendations for adjuvant therapy. It is important to realize that there is a difference between a prognostic factor, which is a factor shown to correlate with survival independent of therapy, and a predictive factor, which is any measurement associated with response or lack of response to a particular therapy.

Lymph Node Status

One of the most important prognostic parameters is the number of positive axillary lymph nodes found on axillary dissection. As the number of involved lymph nodes increases, the survival decreases (Table 1.4).

Tumor Size

Next to axillary nodal status, the most important prognostic factor is tumor size. Even with negative axillary nodes, patients are found to have decreased survival rates with increasing tumor size. The number of axillary nodes containing metastases varies directly with increasing size of tumor (Table 1.5).

Lymphatic and Vascular Invasion, Histologic Grade, Nuclear Grade

All reflect potential aggressiveness of the tumor. The presence of lymphatic and/or vascular invasion and the presence of a poorly differentiated tumor (high histological and nuclear grade) all correlate with a poorer survival.

Estrogen and Progesterone Receptor Status

ER (estrogen receptor) and PR (progesterone receptor) positivity may be associated with a slightly better survival. It is, however, highly predictive for responsiveness to hormonal therapy such as Tamoxifen..

Ploidy

DNA ploidy measures the average amount of DNA per cell, with diploid being normal and aneuploid being abnormal. In general, the greater the amount of the aneuploid population, the worse the prognosis.

S Phase Fraction

Refers to the percentage of cells preparing for mitosis by their active synthesis of DNA . High S phase in general correlates with poor survival rate.

Mitotic Index

Refers to the number of mitoses seen on routine H & E; the higher the mitotic index the poorer the survival rate.

Oncogene Products Like Her2/*neu*

Overexpression of this growth factor receptor is believed to be associated with a poorer prognosis, but of more importance is that it is predictive of response to treatment regimens which include the new monoclonal antibody Herceptin (anti-Her2/*neu*).

Surgical Complications

Complications related to surgery of the breast include psychological distress, infection, hematoma, and seroma of the lumpectomy cavity or mastectomy site. Complications from axillary dissection include damage to the long thoracic nerve, thoracodorsal nerve, axillary vein, and sensory nerves. Motor nerve injury is uncommon but sensory nerve injury is very common and is permanent. A serious

1

Table 1.3. Randomized trials comparing breast conserving surgery + XRT to mastectomy

Trial	Years	Patients	Duration years	Survival (%) Mastectomy	Survival (%) BCS & XRT
NSABP Trial B-06	1976-1984	2105	12	62	62
NCI-Milan	1973-1980	701	13	69	71
NCI-US	1979-1987	237	10	75	77
EORTC	1980-1986	903	7	75	75
Institut Gustave-Roussy	1972-1979	179	10	79	78
Danish Group	1983-1987	905	6	82	79

Table 1.4. Nodal status and 10 year survival

Number of Positive Nodes	10 Year Survival (%)
0	80
1 – 3	63
4 +	29
4 – 6	41
7 – 12	31
13 +	13

Table 1.5. Relationship between tumor size and axillary node status

Tumor Diameter (cm)	Axillary Node-Positivity (%)
0.1 – 0.5	28.6
0.6 – 1.0	24.7
1.1 – 2.0	34.1
2.1 – 3.0	42.1
3.1 – 4.0	50.1
4.1 – 5.0	56.5
> 5.0	64.5

and common complication from axillary dissection is lymphedema of the upper extremity. Lymphedema typically presents months to years following the surgical procedure. Lymphedema can be very difficult to control and can result in significant functional and psychological distress. Education and early intervention are important in effectively managing upper extremity edema.

Local Recurrence and Treatment

Local recurrence after mastectomy will usually present as one or a few subcutaneous nodules and typically are located near the mastectomy incision. The majority of local recurrences following mastectomy occur within the first 5 years postoperatively, however local recurrences after 15-50 years have been reported. Larger tumors and a higher numbers of involved lymph nodes are important factors that increase the risk of developing local recurrence following mastectomy. Local recurrences following mastectomy should be treated with wide excision and chest wall irradiation. The role of systemic therapy is not well defined.

The majority of recurrences following breast conserving surgery occur near the incision or site of the previous breast tumor. Treatment of an ipsilateral breast tumor recurrence is mastectomy. Although frequently administered, the benefit of systemic adjuvant therapy following a recurrence is unknown.

Posttreatment Surveillance

There is no rigidly established monitoring scheme. Typically the breast cancer patient is evaluated every 4 months for the first year 3 years after diagnosis, and then every 6 months for life. If the patient has been treated with breast conserving surgery, a mammogram of the affected breast is taken 6 months postoperatively to establish a new mammographic baseline. Mammograms are then performed yearly although some physicians obtain mammograms of the affected breast at six month intervals. Although frequently performed, chest roentgenogram and blood analyses have not been demonstrated to improved survival. Symptoms suggesting recurrence, for example focal bone pain should be thoroughly evaluated. Patients receiving Tamoxifen should be routinely evaluated by a gynecologist because of the increased risk of developing endometrial cancer.

Special Considerations

Breast Cancer in the Male Breast

Male breast cancer is unusual, and accounts for about 0.5% of all breast cancer. Risk factors are similar to those for women. The prognosis is similarly related to the stage of presentation. Male breast cancer usually presents as a subcutaneous mass beneath the nipple-areola complex. It usually presents in men between the ages of 60-70—approximately 10 years older than the mean age of presentation of women. The only pathological entity not seen in the male breast is lobular carcinoma in situ, and infiltrating lobular carcinoma is rare. Treatment is total mastectomy and axillary lymphadenectomy. Since randomized trials of adjuvant therapy have not been performed with male breast cancer patients, decisions for systemic adjuvant therapy are guided by the same factors as are used in female breast cancers.

1

Breast Cancer During Pregnancy

Fortunately this occurs in less than 5% of breast cancer patients. It usually presents as a breast mass, and delays in diagnosis are common. This is due to the considerable enlargement of the breast associated with pregnancy precluding effective clinical breast examination. Mammography is generally not utilized during pregnancy because the proliferative changes in the breast result in a "white out" of the mammogram. Although surgical biopsy or US guided core biopsy are generally well tolerated in pregnant women, delays in performing biopsy are common. US is usually performed in place of mammography for diagnostic purposes. Because of the inability to shield the fetus from radiation, breast conserving therapy must be carefully considered. There has not been any documented advantage to aborting the fetus. Although it has been thought that breast cancer presenting during pregnancy carries a worse prognosis, the decreased survival rate is probably attributable to the later stage of presentation, i.e., double the rate (61%) of pregnant patients presenting with positive nodes than nonpregnant patients (38%). Chemotherapy is controversial during pregnancy and can usually be delayed so as to not threaten the fetus.

Occult Primary Breast Cancer with Axillary Metastases

Infrequently a woman will present with an enlarged axillary lymph node which, upon biopsy, reveals metastatic breast cancer with no identifiable breast cancer. When a mastectomy is performed in these women, 64% will have an identifiable occult breast cancer not detected by imaging studies. Magnetic resonance imaging is emerging as a useful imaging study for this clinical situation.

Selected Readings

1. Fisher B, Constantin JP, Wickerham DL et al. Tamoxifen for prevention of breast cancer: Report of the National Surgical Adjuvant Breast and Bowel Project P-1 study. J Natl Cancer Inst 1998; 90:1371-1386.
    Landmark article that reports the results of the first prospective randomized prevention trial that showed Tamoxifen could prevent about 50% of breast cancers.
2. Stereotactic Biopsy of Breast Cancer. Pass H. Principles and Practice of Oncology Updates. 1998; 12:1-7.
    This is an excellent overview of the indications and limitations of core biopsy as well as the results of large clinical trials.
3. Silverstein, MJ. Ductal Carcinoma in Situ of the Breast. Baltimore, Williams & Wilkins 1997.
    This text is edited by one of the foremost experts on DCIS and scans the history of the disease to the current controversies surrounding surgical and radiation management. It also provides extensive data to support conclusions and identifies areas that require further investigation.
4. Update: NCCN Practice guidelines for the treatment of breast cancer. NCCN Proceedings. Oncology May 1999; 15 (5A):41-66.
    This is an excellent diagrammatic representation of the treatment and work up of breast cancer at all stages of the disease.

5.   Early Breast Cancer Trialists' Collaborative Group. Effects of radiotherapy and surgery in early breast cancer: an overview of the randomized trials. N Engl J Med 1995; 333-1444.
This is a very good overview of the results of the prospective randomized trials that compared breast sparing surgery plus radiation therapy to mastectomy showing no difference in survival.

6.   Wagner, FB. History of Breast Disease and its Treatment. In: Bland KI and Copeland EM, eds. The Breast. Philadelphia: W.B. Saunders Company. 1991:1-16
This is a wonderful history of the management of breast disease and breast cancer.

7.   Krag D, Weaver D, Ashikaga T et al. The sentinel node in breast cancer—a multicenter validation study. N Engl J Med 339:13:941-6.
Results of the first multicenter trial that evaluated this new surgical technique for use in breast cancer patients.

8.   Consensus Statement of Postmastectomy Radiation Therapy. Harris JR, Halpin-Murphy P, McNeese M et al. Int J Rad Onc Biol Phys 1999; 44(15):989-990.
This is a review of the multi-center randomized trials that showed improved survival with radiation therapy after mastectomy.

1

CHAPTER 2

Melanoma

Lee W. Thompson, Laurence H. Brinckerhoff, and Craig L. Slingluff, Jr.

Scope of the Problem

Melanomas represent a minority of skin cancers (3-5%), but melanomas cause 65% of the deaths from skin cancer. Although possibly now beginning to plateau, the incidence of melanoma has been increasing faster than any other cancer.

Melanoma is a malignancy of melanocytes. Melanocytes originate from neural crest cells; therefore, melanoma can occur anywhere neural crest cells migrate in the embryo. Most commonly primary melanoma occurs in the skin. The eye and mucous membranes are less common but well described areas primary melanomas can occur.

Risk Factors for the Development of Melanoma

Ultraviolet Radiation

Intermittent unaccustomed sun exposure has the best correlation with melanoma. An example of this is the strong relationship between sunburns and the development of melanoma. Either the people most susceptible to sunburn are also predisposed to melanoma, or the intermittent exposure does not allow the body time to protect itself from the UV light.

Direct evidence of the relationship between UV light and melanoma has been difficult to demonstrate in humans; however, nonhuman animal models support the contribution of UV light to melanoma. Direct DNA damage, promotion of growth, and suppression of immune systems response to melanoma are three ways UV light has been hypothesized to contribute to melanoma development.

Ultraviolet light is characterized into three groups (Table 2.1). The UVB group is thought to have the most direct role in carcinogenesis.

Because of the relationship between melanoma and ultraviolet light, it is logical that a typical melanoma patient has a fair complexion (celtic complexion), has a history of sunburning easily, and may be younger than other types of cancer patients (80% of melanoma patients are from ages 25-65).

Other Risk Factors

A few diseases and conditions other than UV light exposure have been associated with increased risk of melanoma:

Patients with dysplastic nevi (sporadic) have a 10% lifetime risk of developing melanoma, and a 20-fold increased risk over the general population. These melanomas may arise in dysplastic nevi or de novo.

Surgical Oncology, edited by David N. Krag. ©2000 Landes Bioscience.

Table 2.1. Characteristics of ultraviolet light

	Wavelength	Characteristics
UVA	320-400 nm	penetrates deeper (dermis) and may be responsible for changes of aging; may play a role in carcinogenesis
UVB	290-320 nm	responsible for sunburn and melanin production; probably the most important for carcinogenesis
UVC	200-290 nm	should be completely absorbed by the ozone

Dysplastic nevus syndrome (B-K mole syndrome, or atypical mole syndrome) is often associated with a family history of melanoma, and when that does occur, the patient has approximately a 100% lifetime risk of melanoma.

Congenital nevi can be characterized as small (< 1.5 cm), medium (1.5-20 cm), or large (> 20 cm). The large congenital nevi have the strongest correlation with increased risk of melanoma, 5-20% lifetime risk. These nevi have an irregular surface, variation in color especially brown, and hypertrichosis. Malignant transformation of these lesions has been reported during childhood.

A familial predisposition is recognized and is thought to be a genetic risk that is modulated by environmental factors. Patients with a family predisposition have a 8-12% chance of cutaneous melanoma. Some have reported evidence of autosomal dominant inheritance with variable penetrance. Patients with a history of previous melanoma have at least a 10-fold increase risk of a subsequent melanoma.

Screening for Melanoma

Physical Exam of the Skin

Self-examination of the skin by the educated patient and by routine physician's exam are the best and cheapest way to screen for melanoma. Because melanoma can be successfully treated if recognized at an early stage and because it usually begins in the skin, the appearance of melanoma is very important. The following characteristics aid in the identification of melanomas, and many of them can be remembered with the ABCDE mnemonic:

1. Asymmetry
2. Border irregularity–Borders that blend, are difficult to see, and nondiscrete are more likely melanoma. Smooth, discrete, uniform borders are more likely benign.
3. Color variation–Variations of black and brown with possible shades of white, red, or blue. The variation of color across a lesion is more worrisome than any particular color in a uniform lesion.
4. Diameter greater than 6 mm–a melanoma can be smaller than 6 mm, but lesions larger than 6 mm should be regarded with high suspicion.

5. Elevation–The surface of the lesion should be smooth and uniform. Any nodular area, indentation, or nonuniformity in elevation should raise the suspicion of melanoma.

Furthermore, ulceration, itching, bleeding, or rapid changes are signs of melanoma. Melanomas usually occur in the sun-exposed areas of the skin, but may occur anywhere.

Histologic Classification

There are four main types of melanoma. The categories are based on location and histologic criteria (Table 2.2).

Methods of Diagnosis

All suspicious lesions should be biopsied if possible. A biopsy of primary melanoma must include subcutaneous tissue because prognosis and further treatment will depend on the Breslow thickness as will be described below. Incisional biopsy (full thickness but only partial excision) and total excisional biopsy (full thickness removal of the entire lesion) are preferred methods of biopsy. A small 2-4 mm punch biopsy of the most suspicious area is an alternative form of "incisional" biopsy that provides full thickness specimens. For small lesions, total excisional biopsy is satisfactory. With all lesions full thickness including subcutaneous tissue must be obtained. Shave biopsies are to be avoided because frequently there is inadequate information regarding depth of the melanoma. The wound should be closed primarily, but can be allowed to heal by secondary intention if it is very small (for example a small punch biopsy). If the primary melanoma is over a lymph node basin the incision should be made so that it can be included in a lymph node dissection if later indicated.

Preoperative Evaluation

The preoperative evaluation should begin, as with any other medical problem, with a good history and physical exam. The history should include symptoms of the lesion of concern such as itching; bleeding; changes in color, texture, or size; and duration of the lesion. Other suspicious lesions or lesions that have regressed are important to investigate. Risk factors such as sun exposure, sun burns, and previous skin cancers are important. Any masses or nodules near or in the area of draining lymph nodes will need to be evaluated. A good review of systems will be an important first step to ruling out distant metastasis. In particular symptoms that should raise concern about possible metastases include bone pain, headaches, changes in mental status, weight loss, change in bowel function, and nausea. The physical exam should include most importantly a good skin exam noting skin type, complexion, and other suspicious lesions. The draining lymph node beds should be thoroughly scrutinized for any masses, or enlarged nodes. The skin exam should be head to toe. This thorough exam should cover the entire cutaneous surface including nonsun exposed areas. The skin between the lesion and nodes should be evaluated for in transit metastasis.

Melanoma in situ or less than 0.75 mm thick may need little more than history and physical exam. For melanoma greater than 0.75 mm in thickness a more extensive preoperative evaluation will be required. Laboratory tests should include an LDH and liver panel, but may also include baseline electrolyte panel and complete

2

Table 2.2. Classification

Four major histologic subtypes of melanoma	Characteristics
1) Lentigo Malignant Melanoma	10-15% of melanomas Least aggressive Typically on sun exposed areas of head, neck, and dorsum of hand More common in females Median age is 70 Large, flat, brown with areas of darker and lighter pigmentation Radial growth of abnormal melanocytes in epidermis with minimal invasion of papillary dermis. The radial growth phase may last for years before vertical growth is seen. Vertical growth is associated with raised areas on the previous flat lesion, and this vertical growth is the progression to invasive phase. If only the radial growth is seen it is called a Hutchinson's Freckle or Lentigo Maligna. Radial growth is not associated with malignancy.
2) Superficial Spreading Melanoma	70% of cutaneous melanomas Intermediate aggressiveness Median age is fifth decade of life Males equal females Should see both radial and vertical growth phases Vertical growth is associated with the nodular areas of the lesion, and is associated with greater potential to metastasize. These lesions have the characteristic variation in color, irregular borders, and irregular surface
3) Nodular Melanoma	15-30% of melanomas Most aggressive type Peak is fifth decade of life Twice more common in males than females Common to occur in normal skin and not from pre-existing nevi Usually bluish-black, but may be more uniform in color and borders Almost all vertical growth
4) Acral Lentiginous Melanoma	2-8% of whites; but 35-60% of dark skinned people Occurs on palms, soles, and subungual locations Long radial growth phase followed by the vertical growth phase with its usual increase in metastatic potential. More than 3/4 involve the large toe or thumb May be confused with subungual hematoma

blood count. Initial radiologic evaluation will include a chest x-ray; but for clinically positive nodes or evidence, by history and physical exam, of more extensive disease a CT scan of the chest, abdomen, and/or pelvis may be needed. PET scanning may be useful for staging patients with signs of advanced disease, as well.

Staging

Depth of invasion, nodal status, and presence of distant metastasis are the three best predictors of prognosis for cutaneous melanoma. Depth of invasion has the greatest prognostic significance.

Depth of Invasion

Two classification systems based on depth of invasion are used for melanoma:

Clark Levels of Invasion

Five levels are defined by histology of invasion (Table 2.3). The levels are based on the idea that as tumors invade through barriers such as the basement membrane into areas of richer lymphatics and blood vessels the potential for distant metastasis increases.

Breslow Thickness

Many studies demonstrate a continuous inverse relationship between tumor thickness and mortality. Breslow and colleagues designed a staging method determined by thickness alone. Even within a single Clark's level, gradations of thickness have independent prognostic information. Clark's levels correlate somewhat with thickness relative to the thickness of the involved skin, whereas Breslow thickness is an absolute measurement. Thickness is more reproducible and more objective than determining level of invasion. Measurements are made from the top of the granular layer to the base of the tumor.

Nodal Status

Survival drops significantly with positive lymph nodes. As more nodes are positive, the prognosis worsens. With 1-2 positive nodes 5 year survival is 30-55%, and with multiple nodes the 5 year survival is 8-26%.

Distant Metastasis

Another important prognostic factor is the presence of distant metastasis. Current therapy for distant metastasis has minimal effect on survival; thus, long-term survival of patients with distant metastases is rare. The most common sites of distant metastasis are lung, liver, brain, bone, gastrointestinal tract, and distant skin. However, melanoma has the potential to metastasize to any tissue, even to unusual sites such as the heart and the spleen. The first site of visceral involvement is commonly lung or liver. Brain metastases are commonly the ultimate cause of death. Metastatic melanoma also has the potential not to become clinically apparent until up to 30 years after the treatment of the primary lesion.

Because these three factors are so critical to prognosis, the current staging scheme is based primarily on them. The staging system defined by the American Joint Committee on Cancer is a 4-stage TNM system and is described in Table 2.4. The T stage is based on the tumor thickness (T1: < 0.76 mm or Clark II; T2: 0.76-1.5 mm

Table 2.3. Clark levels of invasion

Level	Definition
I	All tumor cells confined to the epidermis with no invasion through the basement membrane (melanoma in situ)
II	Tumor cells penetrate through basement membrane into the papillary dermis, but not to the reticular dermis
III	Tumor cells fill the papillary dermis and abut the reticular dermis but do not invade it.
IV	Tumor invades reticular dermis
V	Tumor cells invade the subcutaneous tissue

or Clark III; T3: 1.5-4 mm or Clark IV; T4: > 4 mm or Clark V). Stage I includes T1N0M0 and T2N0M0. Stage II includes T3N0M0 and T4N0M0. Nodal metastases or intransit metastases qualify for N1 status. In transit metastases are metastatic lesions that appear in the skin or subcutaneous tissue between the primary site and the draining nodes, but greater than 2 cm from the primary site. Bulky nodes are graded as N2. Stage III includes $T_{any}N1M0$ and $T_{any}N2M0$, and Stage IV includes $T_{any}N_{any}M1$. There was some inconsistency in the AJCC staging manual in the mid-1990s that listed T4N0M0 lesions as stage III. However, recommendations in 1999 for changing the AJCC staging criteria are being formally considered, and they likely will be adopted by the time this text is in print. The proposed changes include incorporation of ulceration of the primary lesion as a part of staging the primary lesion, and the N staging will also take into account the number of positive nodes, which is well-documented as the single most important prognostic factor in patients with nodal metastases.

Other factors that are important in evaluating prognosis in patients with clinically localized melanoma have been defined. Even though they are not included in the formal staging systems, they should be noted. These include:

1. Anatomic location–from best to worse four categories of sites are independent variables that affect survival.
    a) Extremities (excluding areas of worse prognosis)
    b) Hands and feet
    c) Trunk
    d) Head and neck
2. Women have better prognosis than men
3. Ulceration
4. Radial growth characteristics have better prognosis than vertical growth characteristics.
5. Blue coloration appears to have a worse prognosis.

Mucosal melanomas have a much more aggressive and more malignant behavior than cutaneous melanomas. The worse prognosis is possibly due to lack of early diagnosis and lack of immune response. Ocular melanomas have a different clinical course and management than cutaneous melanomas or mucosal melanomas. They are staged by different criteria.

Table 2.4. Abbreviated table for staging of melanoma

Stage 0	Tis N0 M0	Melanoma in situ
Stage IA	T1 N0 M0	Less than 0.75 mm thick
Stage IB	T2 N0 M0	0.75-1.5 mm thick
Stage IIA	T3 N0 M0	1.5- 4 mm thick
Stage IIB	T4 N0 M0	Greater than 4 mm thick
Stage III	T_{any} N1 M0	Lymph node metastasis or in-transit metastases
Stage IV	$T_{any} N_{any}$ M1	Distant Metastasis

Treatment

Excision of Primary

The primary melanoma is usually diagnosed with an incisional or excisional biopsy; therefore, the original incision made for the biopsy and all remaining tumor must be excised to fascia with a wide margin. The size of the margin required depends on the depth of the primary melanoma. A 0.5 cm margin is recommended for melanoma in situ. Historically melanoma was excised with wide margins that often required split thickness skin grafting. The wide margins caused significant morbidity, so recently trials were performed to address appropriate margin size. A 1 cm margin is appropriate for melanoma that is less than or equal to 1mm thick. A 2 cm margin is considered by most to be adequate for 1 mm to 4 mm thick melanomas; and a 2-3 cm margin for melanomas that are greater than 4 mm thick. The site of the primary melanoma, such as on the head and neck, may influence the extent of resection. Attempts should be made to follow the above guidelines while minimizing morbidity.

Lymph Node Metastases

Patients with clinically positive lymph nodes should have a complete nodal dissection. The diagnosis is generally confirmed by fine needle aspiration. Overall 5 year survival of patients who have positive nodes resected is between 30 and 50%. Without resection of nodal metastasis patients certainly die; therefore, lymph node resection for positive nodes is a potentially curative procedure. Edema of the extremity and wound infection are the most common complications. Occasionally patients present with nodal disease and no identifiable primary lesion, also known as unknown primary. These patients should have complete nodal dissection. The prognosis for patients with unknown primary is similar to that of other patients with nodal disease.

The management of patients with clinically negative nodes is controversial. In the past, surgeons have either waited until lymph nodes become clinically positive or performed an elective lymph node dissection (ELND) early. Older literature suggests that patients with thin (less than 1 mm thick) or thick (greater than 4 mm thick) melanomas, who do not have clinically positive nodes, are the least likely to benefit from elective lymph node dissections. The patients with intermediate thickness melanomas had been considered the group most likely to benefit. However, three randomized prospective trials failed to show a survival benefit to ELND when compared to observation and subsequent resection of palpable nodes. A subgroup of patients with 1-2 mm thick lesions were found in one study to have a survival improvement with elective lymph node dissection but the role of ELND is not established.

Melanoma appears to metastasize over an identifiable route. Cutaneous areas usually drain first to a limited set of lymph nodes known as sentinel lymph nodes. If the sentinel lymph nodes do not have melanoma, then the other lymph nodes are unlikely to be positive. Radionuclide or blue dye can be injected into the area of the primary melanoma, and the marker traced to the first or sentinel lymph nodes. The sentinel lymph nodes can then be biopsied without removing the entire drainage basin. These lymph node biopsies are now routinely performed for melanomas greater than 1 mm thick. Sentinel lymph node biopsies are relatively small procedures with minimal morbidity and small number of risks.

Sentinel lymph node biopsy provides prognostic information that is of some direct value to patients and that affects decisions about adjuvant therapy. In addition, completion lymph node dissection is generally recommended when the sentinel node biopsy is positive, though the long-term survival advantage of that surgery is not defined. Adjuvant therapy with interferon alfa-2b originally demonstrated benefit to patients with bulky nodal disease, and sentinel lymph node biopsies were designed to identify a group of patients with nodal disease (stage III) so that they could receive adjuvant therapy. The benefit of interferon alfa-2b has not been confirmed with follow up studies. Identification of patients who have stage III disease also permits their enrollment in some trials of experimental therapy.

Distant Metastasis

Surgery

Metastatic melanoma has remained a deadly disease with no good treatment. Rare patients will develop a complete response on current experimental protocols or after surgical resection. Resection of some metastases is warranted for palliative reasons. In addition, for some patients with isolated pulmonary metastases or subcutaneous recurrences, resection of metastasis has provided prolonged disease-free survival. Surgical excision or gamma-knife irradiation can effectively palliate patients with solitary brain metastases. Other resections of single lesions that will require relatively low risk resections are supported by some groups. Until more effective systemic therapy is available, surgical resection remains a reasonable option for some patients with distant metastases. The risks and benefits must be carefully considered and discussed with each patient, because surgery for metastatic disease in most patients is unlikely to provide survival benefit.

Chemotherapy and Biologic Agents

Dacarbazine, platinum analogues, nitrosoureas, and tubular toxins have all shown some response, but the response is almost universally short and partial. Patients with skin, subcutaneous, lymph node, and occasionally lung are more likely to have a response. These patients are also the group most likely to benefit from aggressive surgical resections. Some of the agents are now available as outpatient regimens.

No combination chemotherapy has been proven more advantageous than to DTIC alone. DTIC has a 20% response rate but, as with other chemotherapeutic regimens, this has not been durable. One combination known as the "Dartmouth Regimen" or CBDT (carmustine (BCNU), cisplatin, DTIC, and tamoxifen) had some initial favorable results and is favored by many oncologists. Phase II and Phase III studies are ongoing to determine if there is an advantage of CBDT to DTIC alone.

Interleukin-2 does not have direct antitumor effects. Interleukin-2 directly affects T lymphocytes, causing their proliferation. Because of the importance of T lymphocytes in animal models of melanoma rejection, interleukin-2 is widely used in melanoma research for the proliferation of antimelanoma lymphocytes. The use of IL-2 has spread into human trials, and this cytokine is a part of many ongoing immunotherapy trials. Interleukin-2 therapy alone has a 7% complete response rate that is sustained in 3/4 of patients, but use of interleukin-2 is limited by severe systemic toxicity. Biologic agents such as tumor vaccines and cytokines are now being combined. These clinical trials are ongoing and results are too preliminary to make recommendations for therapy.

Isolated Limb Perfusion

In patients with isolated but extensive limb involvement, isolated limb perfusion has been performed for 40 years. In this procedure a tourniquet is placed on the limb and with an extracorporeal bypass circuit the limb is perfused with melphalan and/or tumor necrosis factor for 60-90 minutes. Hyperthermia of 39-40°C is often added. This procedure obviously avoids systemic toxicity while delivering high doses of antitumor agents to the affected area. Prospective well-controlled trials have not been conducted; therefore, the benefit is not certain.

Adjuvant Therapy

Clinical responses to systemic therapy of melanoma with existing agents do occur, but generally do not provide cure. Thus, efforts to use them in the adjuvant setting, after resection of high-risk disease, are being explored.

Interferon alpha-2b

This drug was approved in 1995 by the FDA for postsurgical adjuvant therapy for high risk patients. A trial by the Eastern Cooperative Oncology Group (ECOG) showed a significant survival benefit with high-dose interferon alfa-2b. However, a repeat study with the same regimen failed to show a survival benefit. The balance of the treatment and control arms of the trial has been criticized. The current regimen requires a full year of high dose therapy, and the toxicity of the regimen is high enough that many patients choose not to take it. Indomethacin may help with some of the side effects, but the fatigue syndrome experienced by many on interferon alfa does not have an effective treatment. Patients with resected stage IV and stage IIB may also be candidates for interferon alpha but clinical trial results are not yet available to demonstrate benefit in these patients. Interferon alpha has theoretical promise in combination with tumor vaccines or defined melanoma antigen vaccines.

Experimental Therapy

Immunotherapy

The immune response to melanoma has been extensively studied. With occasional clinical spontaneous regressions, tumor infiltrating lymphocytes capable of destroying melanoma cells, anecdotal responses of patients to immunotherapy, and the ability of cytokines to affect the growth of melanoma without any obvious direct antitumor effects there is promise in therapies that modulate the immune response to cancer. Antibodies specific for melanoma have been well characterized and are used routinely in immunohistochemistry to diagnose melanoma. The role of these antibodies in therapy has yet to be demonstrated. Interferon alfa-2b, interleukin-2, and tumor necrosis factor were discussed above as mainstays in therapy. Each of these as well as many other are included in many ongoing clinical trials. During the last decade the importance of the lymphocytic response to melanoma has been shown. In animal models the rejection of melanoma is a T-helper dependent (CD4+), cytotoxic T-lymphocyte (CD8+) mediated response. The human cytotoxic T-lymphocyte (CTL) antigens on the surface of melanoma cells are now being identified. With these new targets and the increased understanding of how to deliver these antigens, many new melanoma immunotherapy trials are now underway. Early results have shown promise, but no uniform success has been shown.

Posttreatment Surveillance

The incidence of melanoma is increased in patients with a history of prior melanoma. Therefore the goals of surveillance are both to detect recurrence of the previously treated melanoma and to detect new primary melanomas. Surveillance should be life-long. A program which includes education as to the appearance of melanoma is vitally important to patients with a history of melanoma as well as the general public.

The interval duration between examinations is somewhat arbitrary but evaluation every 6 months is logical. The dual goal of each visit is detection of treatable recurrence and early detection of new primary melanoma. Detection of treatable recurrence includes careful examination of the primary resection site, regional lymph nodes, and the skin and soft tissues between the primary tumor site and regional node basin(s). Any and all suspicious nevi or skin nodules should be biopsied. Patients with very high risk lesions may require more frequent or more in-depth evaluation. This may include LDH, liver panel, chest x-ray and possibly chest/abdomen/and pelvis CT scans. Although the merits of detecting systemic disease can be debated, it may allow patients to be directed to appropriate clinical trials early and also to optimize planning for palliation.

Special Considerations

Ocular Melanoma

The eye is the second most common site of primary melanoma after the skin. Melanocytes in the conjunctiva and uveal tract are considered the precursor cells of ocular melanomas. Ocular melanoma can be divided into uveal melanoma and conjuctival melanoma.

Conjunctival melanomas usually arise at the limbus (or on the bulbar conjunctiva), but may also present on the palpebral or forniceal conjunctiva. Conjunctival melanomas account for only 2% of all ocular melanomas. Conjunctival melanomas are treated by local excision with supplemental cryotherapy to the surrounding conjunctiva. Enucleation is not recommended because the procedure does not completely remove the conjunctiva. Exenteration (removal of the eyelids, eye, mucous membranes, and orbital contents), has not been clearly proven to improve overall survival, and is performed by some for patients with massive invasive melanomas or tumors that arise in unfavorable locations such as the palpebral or forniceal conjunctiva. Metastasis tends to occur through the lymphatics, like melanomas of the skin. A good nodal exam should be performed in all patients and include palpation of the preauricular, submandibular, and cervical nodes. Five and 10-year survival rates for conjunctival melanomas are approximately 85% and 70%, respectively.

Uveal melanomas are the most common melanomas of the eye. They can occur in the iris, ciliary body, and choroid, with choroidal melanomas (85%) being the most common. Uveal melanomas have a 10:1 predilection for whites over non-whites. The uveal tract is a highly vascular and lacks lymphatic channels; thus, uveal melanomas do not spread by the usual lymphatic route typical of most melanomas but tend to spread hematogenously. The liver is the most common site of metastasis. Melanoma of the iris is classified as an anterior melanoma, and has a better prognosis than the posterior melanomas, ciliary body melanoma and choroidal melanoma. Histologic grade is also important in the prognosis of uveal melanomas.

Iris melanoma can present with painless glaucoma, heterochromia, sector cataracts, spontaneous hyphemas, or be seen on routine eye exams. Small stable iris melanomas can be treated by observation alone. Small resections such as iridectomy, iridotrabeculectomy, and iridocyclectomy can also be performed. Tumors that are more extensive, that cause intractable glaucoma, or that cause unsalvageable loss of vision are best treated by enucleation. The mortality for iris melanomas is less than 5%.

Ciliary body melanomas (CBM) are difficult to diagnose and may go unnoticed until they are at a late stage. For small lesions, simple excision can be adequate treatment. Radioactive plaque therapy is the treatment of choice for medium to large tumors. Large tumors that cannot be managed any other way or that produce intractable glaucoma may need treatment by enucleation. The mortality for ciliary body melanomas is near 40%.

Choroidal melanomas may present with decreased vision, floaters, or scotomas (visual field defects), as an unexplained cataract, or may be completely asymptomatic. Small choroidal melanomas can be treated with photocoagulation. Medium to large tumors may be treated with local resection if possible, but enucleation has traditionally been required for a majority of choroidal melanomas. Radiotherapy has gained popularity and radioactive plaques have improved short term survival. Ongoing clinical trials are addressing the role of radiotherapy. Enucleation is usually first line therapy in patients with blind, painful eye or when the tumor exceeds 40% of the ocular volume. Exenteration is reserved for patients with extraocular extension. Mortality for all choroidal melanomas is between 25 and 40%.

Pregnancy and Melanoma

The role of pregnancy in the clinical course of patients with melanoma has been debated without resolution for years. Melanoma has been reported to enlarge during pregnancy then regress between pregnancies. The possible negative impact of pregnancy on melanoma has left many physicians unsure how to advise their patients who are pregnant or who are thinking of becoming pregnant.

Both a decrease in the disease free interval and an increased incidence of nodal metastasis have been reported in melanoma patients who are pregnant. The overall survival has not been shown to be statistically different in pregnant patients versus patients who are not pregnant. The etiology behind these differences has been hypothesized to be hormonal. Estrogen has been most often implicated, although basic science research has not been able to support this hypothesis. Recent reports of the importance of tamoxifen in the Dartmouth Regimen has again raised the issue of effects of female hormones on melanoma growth and metastasis.

The initial treatment for melanoma is surgical excision. Because the surgery can be performed on pregnant patients with minimal increased risk to patient or unborn child, the initial treatment should not be altered in patients who are pregnant. In the past, concern about the possible effects of pregnancy on prognosis caused some practitioners to recommend abortion in some cases. However, survival does not appear to be related to the pregnancy; therefore, if patients are diagnosed with melanoma while they are pregnant, termination of the pregnancy has no role in the treatment of the patient with melanoma. Because of the reported increase in nodal metastasis and decrease in disease free survival, the workup and follow up should be aggressive. Sentinel node biopsy would be a reasonable option because of the higher incidence of nodal metastasis, and this could save the patient an elective node dissection. Follow up should be considered the same as the high-risk melanoma patients. Subsequent pregnancy does not appear to affect prognosis.

Selected Readings

1. Breslow A. "Thickness, cross-sectional areas and depth of invasion in the prognosis of cutaneous melanoma" Ann Surg 1970; Nov;172(5):902-8.
2. Clark WH Jr, From L, Bernardino EA et al. The histogenesis and biologic behavior of primary human malignant melanomas of the skin. Cancer Research 1969 Mar; 29(3):705-27.
    References 1 and 2 are key references in the establishment of tumor thickness as the primary prognostic factor in melanoma therapy. The therapy used today is based on tumor thickness because of these studies.
3. Veronesi U, Cascinelli N, Adamus J et al. Thin stage I primary cutaneous malignant melanoma. Comparison of excision with margins of 1 or 3 cm. N Eng J Med 1988 May 5; 318(18):1159-62.
4. Balch CM, Urist MM, Karakousis CP et al. Efficacy of 2 cm surgical margins for intermediate-thickness melanomas (1-4 mm). Results of a multi-institutional randomized surgical trial" Annals of Surgery 1993; Sep;218(3):262-7; discussion 267-9.
    References 3 and 4 are two of the trials that helped reduce the morbidity of surgical resection of primary melanoma by demonstrating the safety of reduced margins.

5.   Gershenwald JE, Thompson W, Mansfield PF et al. Multi-institutional melanoma
      lymphatic mapping experience: the prognostic value of sentinel lymph node status
      in 612 Stage I and II Melanoma Patients. J Clin Oncol 1999 March; 17(3):976-83.
      Reference 5 is a current update and explains the important concepts of sentinal node
      biopsy for melanoma. The exact role of sentinal node biopsy is now becoming estab-
      lished in melanoma therapy.
6.   Grin JM, Grant-Kels JM, Grin CM et al. Ocular melanomas and melanocytic
      lesions of the eye. J Amer Acad Dermatol 1998 May; 38(5 Pt 1):716-30.
      This review article explains the biology and treatment of ocular melanoma. Ocular
      melanoma is very different than cutaneous melanoma.
7.   Brinckerhoff LH, Thompson LW, Slingluff CL Jr. Melanoma Vaccine. Curr Opin
      Oncol 2000; 12(2): (in press).
      This review article outlines the current status of immunotherapy for melanoma.
8.   Geraghty PJ, Johnson TM, Sondak VK et al. Surgical therapy of primary cutane-
      ous melanoma. Sem Surg Oncol 1996;12(6):386-93.
      In depth review for cutaneous melanoma.
9.   Slingluff CL Jr, Seigler HF. Melanoma. In: Sabiston Textbook of Surgery; The
      Biologic Basis of Modern Surgical Practice (Fifteenth Edition). 1997: 515-28.
      In depth review for cutaneous melanoma.

CHAPTER 3

Nonmelanoma Skin Cancers

Paul L. Baron

Introduction

Nonmelanoma skin cancer (NMSC) is the most common cancer diagnosed in the United States. It is estimated that at least 1 million patients are treated for this disease each year. The most common types are basal cell carcinoma (BCC) and squamous cell carcinoma (SCC). Patients rarely die from these cancers since they rarely metastasize. Most of the morbidity results from local invasion. Treatment frequently includes some form of surgery to remove the tumor with clear margins and thereby prevent recurrence.

Scope of the Problem

The significance of nonmelanoma skin cancers has been under appreciated. This is primarily because the mortality rate is very low. In addition, their incidence is not included in most hospital tumor registries. Registries usually record cases of melanoma or metastatic skin cancer. Despite this, there can be significant functional, cosmetic, and financial morbidity. It has been estimated that the annual cost of treating NMSC in the United States alone is $500 million.[1]

More than 30 different types of skin cancer have been described. Many are indolent, while some can metastasize. The vast majority of these tumors are basal and squamous cell cancers. Table 3.1 lists many of the different types of NMSC. The less common but important ones (keratoacanthoma, Paget's disease, Merkel cell tumor, dermatofibrosarcoma protuberans, and sebaceous carcinoma) will be covered briefly in the section entitled "Unusual Skin Cancers." The primary focus of this Chapter will be on basal and squamous cell cancers.

Risk Factors

The incidence of NMSC has been on the rise since the 1960s. National Cancer Institute surveys showed an increase in frequency of 15-20% from 1972-1978. Exposure to ultraviolet radiation, the most important risk factor for developing NMSC, has probably been on an increase during this time period. This would result from an increased sun exposure in those trying to obtain a tan from sunbathing, tanning booths, or the increase in outdoor sports. Although the mortality rate is generally low from NMSC, it has actually been on the rise in men since 1980. It is estimated that 2000 people die each year from NMSC. The most common cause of death is metastatic squamous cell cancer.[1]

Ultraviolet radiation is composed of UVA (320-400nm), UVB (290-320 nm), and UVC (200-280 nm). Both UVA and UVB cause mutations in cellular DNA.

Surgical Oncology, edited by David N. Krag. ©2000 Landes Bioscience.

Table 3.1. Nonmelanoma skin cancers

Basal cell carcinoma
Squamous cell carcinoma
Keratoacanthoma
Paget's disease of the nipple
Extramammary Paget's disease
Merkel cell tumor
Dermatofibrosarcoma protuberans (DFSP)
Sebaceous carcinoma
Atypical fibroxanthoma
Malignant fibrous histiocytoma (MFH)
Angiosarcoma of the head and neck
Microcystic adnexal carcinoma
Mucinous carcinoma
Kaposi's sarcoma

Failure to adequately repair these defects leads to unrestrained cell growth and development of cancers. In addition, ultraviolet radiation has been implicated in causing immunosuppression and thereby preventing tumor rejection.[2]

An increased disposition to develop NMSC occurs in those who have fair complexion and a tendency to sunburn rather than tan when exposed to the sun. Members of this higher risk group who live near the Equator, work outdoors, or participate in recreational tanning with insufficient sunscreen are at particularly high risk.

Gender, race and geography also seem to play a role in the risk for developing skin cancer. Males are more prone to developing NMSC than females. The Asian population has a decreased risk of dying from skin cancer than whites. Males living in England and Wales have a slightly increased risk of dying from these cancers when compared with those living in Japan. Australian males have an almost three-fold increase in mortality rate when compared to Japanese males. These differences are not as apparent when the females are compared.[3]

Recent studies have suggested that loss of ozone from the atmosphere has resulted in decreased protection from the sun. The ozone layer prevents most of the ultraviolet radiation from reaching ground level. It has been claimed that the upward spread of chlorofluorocarbons used in spray cans, plastic foam, refrigerators and air conditioners are depleting the ozone layer. Mathematical calculations predict that the incidence of skin cancer increases by 2-4% for each 1% reduction in the ozone layer.[1] Despite the fact that the ozone layer has decreased the last 15 years, actual measurements have failed to show any increase in biologically effective ultraviolet radiation at the ground level.

Some patients also inherit rare conditions that predispose them to NMSC. Albinism, an autosomal recessive disorder characterized by the lack of melanin, increases the risk of developing skin cancer at a younger age than the rest of the population. This can be diminished if these individuals limit the amount of exposure to the sun. Patients with xeroderma pigmentosum have an autosomal recessive defect that inhibits their tissue from repairing damaged DNA. They also tend to develop skin cancers at a younger age than the normal population. It has been estimated that their risk for developing skin cancer is increased by a factor of 2000. Epidermodysplasia verruciformis is a rare autosomal recessive skin disease that starts early in life and is linked to papilloma viruses. Various types of skin cancer occur at an early age in these patients on sun-exposed skin.

Immunosuppressed patients are at increased risk for developing skin cancer. This immunocompromised state may be due to medication for patients undergoing organ transplantation, chemotherapy for cancer, cancer itself, or HIV infection. Kidney transplant recipients have been shown to be at increased risk for developing squamous cell carcinoma within a few years after their transplant. These individuals are at increased risk for metastatic disease.

Scarred or traumatized skin due to ulceration, burns, frostbite, or exposure to arsenic and shale oil is also at an increased risk for developing NMSC. Marjolin's ulcers are aggressive squamous cell carcinomas that arise from old burns or traumatic scars. This has been reported to occur as late as 30 years after the initial injury. About 20% of these patients develop lymph node metastases.

Exposure to excessive radiation was found to cause chronic dermatitis and subsequent squamous cell cancer. Basal cell carcinoma has also been described in relation to radiodermatitis. The usual victims were those who worked with x-rays. Currently, radiation-induced skin cancers are rare because of the rigorous safeguards employed.

A number of different precursor lesions can develop into invasive SCC. At this point, there are no known precursors for BCC. Actinic keratoses are the most common premalignant lesions. They occur on chronically sun-exposed skin. The presence of these lesions identifies patients who have been exposed to excessive ultraviolet radiation and need close supervision. It is estimated that about 20% of these will develop into SCC.

Bowen's disease occurs on both sun-exposed and unexposed areas of the body. This form of SCC in situ has been associated with arsenic ingestion, radiation therapy, viral agents, and sun exposure. It is estimated that only 1 patient per 1000 per year develops invasive SCC.[1]

Screening

Any patient with an increased risk for developing NMSC should be screened. These risk factors have been noted in the previous section and include: excessive sun exposure, fair skin, various inherited syndromes, immunosuppression, history of significant skin trauma from burns or radiation, or the presence of actinic keratoses.

Patients with a history of NMSC are at risk of developing further NMSC. Patients with at least one NMSC have a 17% risk of developing a second one within 1 year, a 35% risk in 3 years, and a 50% risk in 5 years. These patients need to be examined at regular intervals after treatment of the initial NMSC. It should also be noted that patients with a history of NMSC have a relative risk of 3-17% of developing melanoma.

In screening these patients, it is important to examine the entire skin surface. Thus, the patient needs to be completely undressed so that the whole body including genitalia can be examined. All lymph node basins need to be palpated to rule out metastatic disease.

Diagnosis

The most important step in managing these patients is to establish a correct diagnosis. The methods of obtaining tissue include shave, punch, incisional, or excisional biopsy. The location of the practice and type of referrals determine whether a general surgeon finds him/herself performing biopsies of skin lesions. Often a patient is referred after a primary care physician or dermatologist has performed the biopsy and obtained a diagnosis. The type of cancer will dictate whether further excision is needed, how extensive the procedure should be, and whether lymph nodes need to be surgically excised. It is important that an excisional biopsy be planned and performed in such a manner to allow optimal wide re-excision should this be necessary.

Ideally, all skin cancers need to be biopsied before definitive treatment can be performed. This helps determine the most appropriate method of managing the NMSC. The only problem with this approach is that it increases the cost and inconvenience to the patient. When a biopsy is felt to not be necessary, it is appropriate for the physician to treat the lesion. It is imperative that a specimen be submitted to ensure that the correct treatment was provided. As a result, if the pathology shows that the treatment was inadequate, the patient can return for further therapy. Similarly, it can be reasonable to perform posttreatment biopsies on lesions that were managed by cryosurgery or radiation therapy to ensure adequate treatment.

Basal cell carcinoma is the most common form of NMSC. Four out of five NMSC are BCC. It is rarely diagnosed in patients under 40, and is more common in men. The distribution of these lesions follows the pilosebaceous follicles and sun-exposed areas of skin. The most common location is on the head and neck region. Although they may occur on areas without follicles, they virtually never arise from oral, anal or vaginal mucosa (unlike squamous cell cancers). [4]

There are three main types of BCC: nodular, superficial, and aggressive-growth patterns. [5] The nodular type is the most common, occurring in 50-70% of cases of BCC. These present as raised, well circumscribed, smooth, pearly, translucent papules with telangiectatic blood vessels (Fig. 3.1). The stretching of the skin accentuates the pearly appearance. On occasion, the lesion may ulcerate to form a rodent ulcer. The name stems from the fact that it simulates a rat bite. Superficial BCC accounts for 10% of all BCC. These lesions present as well defined, discrete, slightly raised plaques. At times there can be a pearly rolled border with overlying scale. This can be difficult to distinguish from Paget's disease and Bowen's disease. Both nodular and superficial BCC can be pigmented and resemble melanoma. Aggressive-growth BCC accounts for 10-15% of all BCC and includes those tumors that have poorly circumscribed growth patterns. Under this heading is included micronodular or infiltrative growth patterns. They usually present as ill-defined plaques. These lesions often have extensive sub-clinical spread, which can result in significant surgical defects following excision for negative margins.

Squamous cell cancer is the second most common form of skin cancer. Unlike BCC, these lesions have a real risk of metastasis. As a result, SCC accounts for the majority of deaths due to NMSC. These cancers usually occur in patients after the age of 50. There is a strong correlation between lifetime ultraviolet radiation exposure and SCC. Thus, it occurs more frequently in older individuals who are fair skinned and live closer to the Equator. Except for cases in which there is a predisposing condition (chronic ulceration, burn, etc.), these lesions occur primarily on sun-exposed portions of the body (Figs. 3.2 and 3.3). They primarily occur on the head and neck region, and the dorsum of the hand. SCC rarely is found in skin folds or

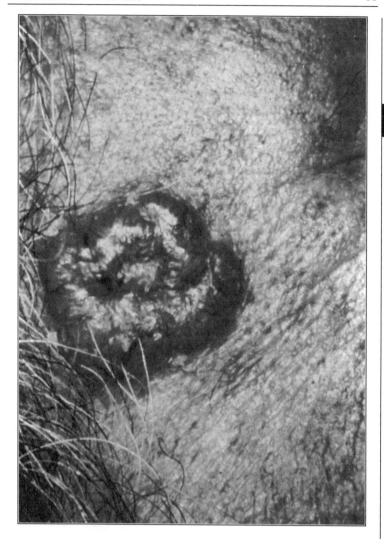

3

Fig. 3.1. Typical nodular basal cell carcinoma of the temple. Note the translucent rolled border. (courtesy of Pearon Lang).

covered areas of the body. It tends to present as a well-defined, red, scaling, plaque, or as a firm, red nodule. At times, there is ulceration with necrotic debris.[4]

The overall metastatic risk ranges from 3% for primary cutaneous lesions, to 11% for mucocutaneous disease, and up to 30% for SCC that arises in inflammatory conditions or chronic wounds.[6] Certain features identify patients at increased risk. These include: location on the ear or lip, large size (> 2 cm), depth greater than

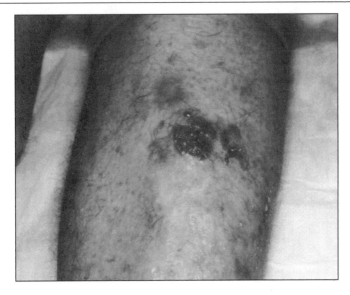

Fig. 3.2. Squamous cell carcinoma arising in an old burn scar (Marjolin's ulcer). The exophytic lesion has rolled borders resembling granulation tissue (courtesy of Pearon Lang).

Fig. 3.3. Squamous car-cinoma on the dorsum of the foot in a patient who had pre-viously received radiation therapy for a plantar wart. Note that it is very vascular and exophytic. (courtesy of Pearon Lang).

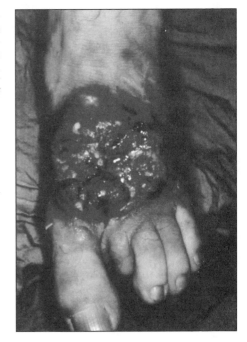

4 mm, poorly differentiated histology, perineural invasion, those arising in a chronic ulcer or previously radiated field, or immunosuppressed patients.

Preoperative Evaluation

The key step to preoperative evaluation is the physical exam focusing on most likely locations of spread based on the histological type of tumor. Although it is very uncommon for NMSC to metastasize, SCC (and other more uncommon tumor types) can metastasize to lymph nodes and lungs. Thus, a careful examination of the nodal basin at risk for metastases is very important. If the nodes are palpable, a fine needle aspiration should be performed of the lymph nodes in question. Ultrasound guidance of the fine needle aspiration may be useful to target the most suspicious area of the lymph node. A chest x-ray should be obtained in any patient with SCC that has an increased risk of metastatic disease. In instances where the lymph nodes are clinically positive, it may be appropriate to also obtain a chest CT scan. If the liver function tests are abnormal, an abdominal CT scan should be obtained. If there is bone pain, a bone scan should be performed. The identification of metastatic disease may alter the approach to the primary tumor. For example, if widely metastatic disease is found, it may be less appropriate to perform an extensive resection of the primary tumor. However, aggressive treatment of symptomatic metastases may be part of an overall treatment plan.

Staging

The American Joint Committee on Cancer (AJCC) has developed a staging system for NMSC (Table 3.2).[7] This system primarily pertains to basal and squamous cell cancers.

Treatment

Squamous cell and basal cell carcinomas are effectively treated by a variety of methods. In most instances, treatment options are similar for both histological types.

The surgical goal in treating patients with nonmelanoma skin cancers is to completely remove the primary tumor to prevent local recurrence while simultaneously preserving function in a cosmetically acceptable manner. The actual method of treatment needs to take into account the tumor type, its risk of recurrence, size, site of the body, and the patient's age and comorbid conditions. In general, surgical methods are very effective for achieving long term local control. However, radiation or cryosurgery are also effective and may be the preferred method when function and cosmesis are at risk.

BCC is most often treated with surgical excision. Cure rates (recurrence-free survival) of 95-99% are achieved with this method as long as the histologic margins are clear. A margin for resection should be 2-5 mm for most lesions less than 2 cm in diameter. Those lesions that have a higher likelihood of leaving residual tumor (aggressive-growth pattern) should be excised with a 1 cm margin. Superficial BCC can be excised to include a portion of subcutaneous tissue. If the subcutaneous tissue is involved with tumor, it may be necessary to excise underlying fascia or muscle.[8] If the margins are still positive and left untreated, the recurrence rate is 16-42%. Most surgeons would recommend performing a re-excision to achieve free margins. However, in select cases, especially with the more indolent histologic types

3

Table 3.2. Staging for nonmelanoma skin cancer[7]

Primary Tumor (T)

Tx	Primary tumor cannot be assessed
T0	No evidence of primary tumor
Tis	Carcinoma in situ
T1	Tumor 2 cm or less in greatest dimension
T2	Tumor more than 2 cm but not more than 5 cm in greatest dimension
T3	Tumor more than 5 cm in greatest dimension
T4	Tumor invades deep extradermal structures, i.e., cartilage, skeletal muscle, or bone

Note: In the case of multiple synchronous tumors, the tumor with the highest T category will be classified and the number of separate tumors will be indicated in parentheses, e.g., T2 (5)

Regional Lymph Nodes (N)

Nx	Regional lymph nodes cannot be assessed
N0	No regional lymph node metastases
N1	Regional lymph node metastases

Distant Metastasis (M)

Mx	Distant metastases cannot be assessed
M0	No distant metastasis
M1	Distant metastasis

Stage Grouping

Stage 0	Tis	N0	M0
Stage I	T1	N0	M0
Stage II	T2	N0	M0
	T3	N0	M0
Stage III	T4	N0	M0
	Any T	N1	M0
Stage IV	Any T	Any N	M1

of BCC (nodular and superficial), it is reasonable to not re-excise and reoperate only when a recurrence is identified.

Primary BCC will recur less than 10% of the time.[9] There are certain features that identify patients at higher risk of recurrence: location on the midface and ears, size greater than 2 cm, micronodular and infiltrative growth patterns, and recurrence of a previously treated BCC. Recurrent BCCs may be treated effectively by Mohs' technique. In this method, the visible tumor is initially removed, and the edges of the specimen examined histologically for any evidence of tumor at the margins. If any margin is identified to harbor tumor cells, then the tumor bed margin corresponding to that surface is further re-excised. This process is continued until all margins are found to be free of tumor. The five-year recurrence rate for excision of primary BCC by Mohs' technique is only 1% and is 5.6% for excision of a recurrent BCC.[9] This compares favorably to a recurrence rate of 17.4% for routine surgical excision of recurrent BCCs.

A common method of initial treatment for small BCC is curettage and electrodesiccation. A curette is used to scrape the lesion away and then electrodesiccation is applied to a 1-2 mm rim of tissue around the defect. This eschar is curetted again, and electrodesiccation applied a second time. The resulting wound heals by granulation. Five-year local control rates of 95% are reported when this technique for small, low risk BCC.

Two other methods of treating BCC include cryosurgery with liquid nitrogen and radiation therapy. These lesions are very sensitive to radiation treatment. A drawback to these two techniques is that unless a small sample has been obtained (for example by a small 2 mm punch biopsy) there is no pathologic specimen to confirm that the tumor has been diagnosed accurately or treated adequately.

SCC, like BCC, can effectively be treated by surgical excision. Lesions that are well-differentiated and smaller that 2 cm can be adequately excised with a 4 mm margin. If the SCC is larger or less differentiated, then it should probably be excised with a 1 cm margin. Recurrence rates of 5-6% have been reported following this technique.[10] Recurrences are more frequent for lesions greater than 1 cm, poorly differentiated cytology, and histologic evidence of invasion into the deep dermis. If the margins are positive, repeat excision is recommended. Lesions that have a positive margin or are at high risk for recurrence, should be considered for treatment by Mohs' technique. Local control obtained by Mohs' surgery for SCC is 97% for primary lesions and 76% for recurrent lesions. Curettage and electrodesiccation, cryosurgery, and radiotherapy are effective options for treating SCC.

The technique of lymphatic mapping and sentinel node biopsy has emerged as the standard of care in evaluating patients with melanoma to determine the presence of microscopic metastatic disease to lymph nodes. The theory behind this is that the first set of draining nodes, the so-called "sentinel nodes", are the lymph nodes most likely to contain metastases from the melanoma. Thus, if histologic exam of the sentinel nodes shows no evidence of tumor, then it is likely that all other nodes in the draining basin are also free of tumor. A variety of techniques employing isosulfan blue dye and/or radiolabeled colloid have been used to localize sentinel nodes.

One of the earliest experiences with identifying sentinel nodes was actually with SCC of the penis, predating the work with melanoma patients. Since BCC virtually never metastasizes, it would only be appropriate to use lymphatic mapping and sentinel node biopsy in SCC at high risk for metastatic disease. This would include lesions that are large, poorly differentiated, deeply invasive, arise in chronic wounds or in immunosuppressed patients. If sentinel nodes are positive for metastatic disease, then the patient should be returned to the operating room for a formal node dissection.

Posttreatment Surveillance

Patients with a history of NMSC are at risk for local recurrence and development of new lesions. In rare instances of squamous cancers, they can also develop metastases. It is important to note that these patients are also at increased risk for melanoma. A reasonable approach is for these patients to undergo examination by their physician every six months the first year, and then annually for life.

It should be emphasized to these patients that they perform self-examination of their skin on a regular basis. An effective approach is for a woman to perform her monthly breast exam at the same time as her skin examination. The best time for this is a few days after her menstrual period. For men, it is reasonable to recommend the first day of the month as a way of reminding them to do this. Skin surveys by the spouse is also an effective way of identifying recurrence.

Prevention of new NMSC is important. One can reduce ultraviolet exposure by a number of methods:

1. Minimize exposure to the sun between 10 am and 3 pm.
2. Wear sunglasses, protective clothing, and a wide brim hat.
3. Use a sunscreen that has both UVA and UVB protection with a sun protection factor (SPF) of at least 15.
4. Avoid artificial tanning devices and booths.

Unusual Tumors

As noted earlier, there are other less common but important types of skin cancer. This section includes a brief description of some of the more significant ones.

Keratoacanthoma arises from the hair follicle. Since it occurs on sun-exposed skin, it is believed that these lesions are due to ultraviolet radiation. Keratoacanthoma goes through three clinical stages. During proliferation, it appears as a firm, smooth papule. During maturation, it becomes dome-shaped with a central umbilication and keratinous core (Fig. 3.4). The lesion then involutes, resulting in a depressed, hypopigmented scar. It is often difficult to distinguish keratoacanthoma from SCC both clinically and histologically. On rare occasions, these lesions have been found to metastasize. As a result, many authors just regard these as low grade well-differentiated SCC. Excision with clear margin is adequate treatment.

Paget's disease is a term used to describe a nonsquamous intraepithelial neoplasm. Mammary Paget's disease presents in the skin of the nipple. The cause is an underlying duct carcinoma in situ that extends up into the epidermis via the mammary ducts. The patient usually presents with burning, pain, pruritus and soiling of the bra. Examination frequently shows an erythematous, scaling patch. Extramammary Paget's can occur on the vulva, anogenital, and axillary regions. One third of these patients have an underlying carcinoma of the adnexa. Histologically, both the mammary and extramammary versions contain Paget's cells. These are large ovoid cells with pale cytoplasm. Since most cases indicate an underlying malignancy, treatment should be directed at removing the primary tumor. Either total mastectomy or partial mastectomy with postoperative radiation therapy is used to treat mammary Paget's. Lymph nodes are only removed if there is evidence of an invasive cancer. Wide local excision is the treatment for extramammary Paget's.

Merkel cell carcinoma is one of the most aggressive NMSC. Merkel cells arise from a pluripotential basal cell in the epithelium. They are felt to function as a receptor of mechanical stimuli and as a target for mechanosensory nerves during fetal development.[4] About 700 cases of Merkel cell cancer are reported in the literature. These tumors usually occur in the head and neck area, especially in the periorbital and eyelid regions. They present as a painless solitary dermal nodule. The surface is usually smooth and can have a slightly erythematous to deeply violaceous with associated telangiectasias. Lymph node metastases occur in 20% of patients at presentation, while 50% of the patients develop them at some time during the course of their disease. Nuclear scans with ^{123}I metaiodobenzylguanidine (MIBG) and ^{111}In octreotide have been described as identifying early evidence of metastatic disease. It is generally recommended that these lesions should be treated by wide local excision and lymph node dissection if the regional nodes are clinically positive. Resection margins of 1 cm on the head and neck and 3 cm on the other areas of the body are appropriate. Since these cancers aggressively spread like melanoma, it has been suggested that these patients should undergo sentinel lymph node biopsy at the time of

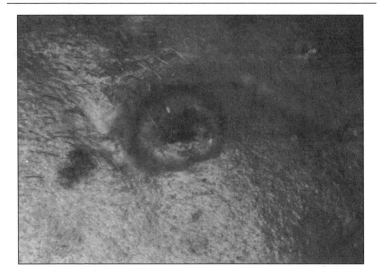

3

Fig. 3.4. This is a keratoacanthoma of the cheek in its mature phase. Note the rolled borders with a central keratotic plug. Also note the adjacent areas of sun-damaged skin. (courtesy of Pearon Lang).

resection of the primary. If the sentinel node is positive, then the patient should have a lymph node dissection. There may also be merit in adding radiation therapy to the tumor bed.[11] These patients have a 5-year survival rate of 50%. Patients who develop systemic disease usually die within 6 months.

Dermatofibrosarcoma protuberans (DFSP) is an uncommon fibroblastic tumor of the dermis. There appears to be an increased incidence of this disease among African-Americans. About 50-60% occur on the trunk, 20-30% on the extremities, and the remaining 10-15% on the head and neck region. These generally present as a slowly growing indurated plaque that can range in color from skin-colored to red to purple. The tumor tends to spread with projections of neoplastic cells beneath clinically normal skin. As a result, recurrence rates as high as 60% are reported after standard surgical excision. Although these lesions rarely metastasize, uncontrolled local disease can cause death from spread into contiguous vital structures. Treatment is either by wide local excision or possibly Mohs' technique. The latter appears to have the lowest local recurrence rates. There does not appear to be any need for lymph node dissection in these patients.

Sebaceous (meibomian) gland carcinoma is a rare malignancy that arises from the adnexal epithelium of the sebaceous gland which is a component of the pilosebaceous unit. Although these are found throughout the hair-bearing areas of skin, there are five types of sebaceous glands on the ocular adnexa. Not surprisingly, 75% of all sebaceous gland carcinomas are located on the eyelid. Clinically these present as small, firm, slowly growing nodules. Surgical excision for clear margins is the recommended treatment. These cancers are aggressive with metastases occurring in 14-25% of cases.[4]

Special Considerations

NMSC is a very common malignancy, but is rarely a cause of death. Ultraviolet radiation is the most common cause of these lesions. Most patients can be adequately treated with simple surgical excision. Those at high risk for recurrence may benefit from Mohs' technique. Close lifelong follow-up of these individuals is needed since they are prone to develop new NMSC.

Selected Readings

1. Strom SS, Yamamura Y. Epidemiology of nonmelanoma skin cancer. Clin Plast Surg 1997; 24:627-36.
    A review of the risk factors for developing NMSC.
2. Grossman D, Leffell DJ. The molecular basis of nonmelanoma skin cancer. Arch Dermatol 1997; 133:1263-1270.
    This article presents some of the changes that take place in malignant trasformation of skin.
3. Lee JAH. Epidemiology of cancers of the skin. In: Friedman RJ, Rigel DS, Kopf AW et al, eds. Cancer of the skin. 1st ed. Philadelphia: W. B. Saunders Company, 1991: 14-24.
    This chapter identifies some of the environmental and inherited predisposing features for the development of NMSC.
4. Skidmore RA, Flowers FP. Nonmelanoma skin cancer. Med Clin N Amer 1998; 82:1309-23.
    A review of the diagnosis and management of patients with NMSC.
5. Lowe L. Histology. In: Miller SJ, Maloney ME, eds. Cutaneous oncology. 1st ed. Malden, Massachusetts: Blackwell Science, Inc 1998:633-645.
    A review of the histologic features of NMSC. This textbook is an excellent source of information about all skin malignancies and their treatment.
6. Demetrius RW, Randle HW. High-risk nonmelanoma skin cancers. Dermatol Surg 1998; 24: 1272-1292.
    This article identifies patients with NMSC who are at high risk for recurrence after treatment.
7. Carcinoma of the skin. In: Fleming ID, Cooper JS, Henson DE, et al, eds. AJCC cancer staging handbook. 5th ed. Philadelphia: Lippincott Williams & Wilkens, 1998:147-151.
    This is the standard textbook for staging cancers.
8. Goldberg DP. Assessment and surgical treatment of basal cell skin cancer. Clinics in Plastic Surgery 1997; 24:673-686.
    This article presents some of the surgical probelms in managing patients with NMSC.
9. Randle HW. Basal cell carcinoma: identification and treatment of the high-risk patient. Dermatol Surg 1996; 22:255-261.
    This article identifies which BCCs have a significant potential to recur following treatment.
10. Roth JJ, Granick MS. Squamous cell and adnexal carcinomas of the skin. Clinics in Plastic Surgery 1997; 24:687-703.
    This discusses some of the surgical methods of managing patients with these cancers.
11. Gruber S, Wilson L. Merkel cell carcinoma. In: Miller SJ, Maloney ME, eds. Cutaneous oncology. 1st ed. Malden, Massachusetts: Blackwell Science, Inc 1998:710-721.
    This is a thorough discussion of the difficulties in managing patients with Merkel cell tumors.

CHAPTER 4

Soft Tissue Sarcomas

Mary Sue Brady

Soft tissue sarcomas are malignant tumors that arise from mesodermally-derived extraskeletal tissue. Surgeons have traditionally held the leadership role in the care of the patient with soft tissue sarcoma (STS). In order to continue in this role, they must maintain a thorough understanding of all aspects of the disease. The purpose of this Chapter is to provide the student and surgeon in training a broad understanding of this disease, from the presentation of the patient with a STS to an overview of the management of patients with metastatic disease.

Scope of the Problem

STSs are relatively uncommon tumors, accounting for 1% of malignancies diagnosed in adults in the United States. In 2000, an estimated 4,600 adults will be diagnosed with a STS.[1] The median age at diagnosis of adults with STS is 56 years,[2] and the incidence in males and females is approximately equal (53% vs. 47%, respectively).[1] STSs are relatively more common in children, accounting for approximately 5% of all childhood malignancies and occurring in about 850 children per year in the United States. Approximately 57% of all adult patients diagnosed with STS will die of their disease.[1] The survival of children with STS is more favorable. Approximately one third of those diagnosed with rhabdomyosarcoma and 10-15% of those with nonrhabdomyosarcoma STS will die of their disease.

Certain histologic subtypes of STS tend to occur in children, the most common of which is rhabdomyosarcoma, usually arising in head and neck or genitourinary sites. Other subtypes of STS associated with pediatric patients include synovial, epithelial, and alveolar STS. Common types of STS in young adults include embryonal rhabdomyosarcoma (ERMS), synovial sarcoma, alveolar sarcoma, and desmoid tumors. The median age at diagnosis of patients with various histologic subtypes of STS is provided in Table 4.1.

The most common anatomic location of adult STS is the extremity, occurring in half of all patients. Lower extremity sites are twice as common as upper extremity and include tumors of the buttocks, groin, leg, and foot. Upper extremity sites include sarcomas arising in the shoulder, axilla, arm, and hand. Visceral sarcomas occur in intraabdominal organs and the gastrointestinal and genitourinary tracts. These account for approximately 15% of cases of STS. Sarcomas which arise in retro- or intraperitoneal sites also account for approximately 15% of cases seen at our center.[3] Within visceral sarcomas, "gynecologic sarcomas" refer to sarcomas of the female genital tract, and "genitourinary sarcomas" are exclusive of these. Sarcomas of the back, chest wall and abdominal wall are considered truncal sarcomas, and account

Surgical Oncology, edited by David N. Krag. ©2000 Landes Bioscience.

Table 4.1. Median age of patients with different histologic subtypes, MSKCC prospective soft tissue sarcoma database, 1982-1992

Median age (yrs)	Histology	No. of patients
21	ERMS	76
30	Alveolar sarcoma	14
33	Desmoid	124
35	Synovial fibrosarcoma	150
37	Rhabdomyosarcoma	178
55	Leiomyosarcoma	29
56	MFH	348
	Liposarcoma	451

ERMS = embryonal rhabdomyosarcoma
MFH = malignant fibrous histiocytoma

for approximately 10% of all cases. Tumors of the cranium, orbit, ear, nose, oral and nasopharyngeal cavities, and neck are considered 'head and neck' sarcomas. Thoracic sarcomas include those that arise in the heart, lung, and mediastinum. Head and neck and thoracic sarcomas account for 5 and 2%, respectively, of all sarcomas seen.[4]

STSs of the gastrointestinal tract usually arise in the stomach and occur less commonly in the small bowel, colon and esophagus. Genitourinary sarcomas are uncommon, accounting for less than 5% of all sarcomas. The most common site of origin is paratesticular, followed by prostate/seminal vesicle, bladder, and kidney.[5] Gynecologic sarcomas are also uncommon, and occur predominately in the uterus.

Risk Factors for Development of the Disease

The vast majority of patients with STS have no known risk factors for the disease. Genetic disorders manifested in several well described clinical syndromes predispose to soft tissue and bone sarcoma in only a small percentage of patients. Therapeutic radiation and extremity lymphedema are clearly risk factors for the development of STS, albeit in a very small percentage of all patients at risk. In addition, several environmental factors have been implicated in the pathogenesis of STS, but the role that these play, if any, is unclear. Well established risk factors for STS are summarized in Table 4.2.

Genetic Risk Factors

STSs occur with increased frequency in individuals with a predisposition to cancer because of mutations in tumor suppressor genes (Li-Fraumeni syndrome, neurofibromatosis, hereditary retinoblastoma, familial adenomatous polyposis/Gardner's syndrome). These patients inherit or develop a germline mutation in the gene so that only one additional genetic "hit" is required for the development of a tumor. Mutations in tumor suppressor genes are usually inherited in an autosomal dominant pattern, and a careful family history commonly reveals other affected first degree relatives. Typically, these patients develop tumors at a significantly younger age than the general population. These syndromes, the genes involved, and the sarcomas found in these patients are listed in Table 4.3.

Table 4.2. Environmental risk factors for STS

Sarcoma	Site	Agent
STS	All	Radiation
Lymphangiosarcoma	Extremity	Lymphedema/Radiation
Angiosarcoma	Liver	Vinyl chloride, Thorium oxide
Desmoid tumors	Abdominal wall	Parturition

Table 4.3. Genetic disorders associated with sarcoma

Disorder	Gene	Chromosome	Sarcomas
Li-Fraumeni syndrome	p53	17p	Soft tissue, bone
Retinoblastoma, hereditary	Rb	13q	Soft tissue, bone
Neurofibromatosis	NF	17q	Soft tissue, MPNT
Gardner's syndrome/FAP	APC	5q	Abdominal desmoids

APC = adenomatous polyposis coli, FAP = familial adenomatous polyposis,
MPNT = malignant peripheral nerve sheath tumor

Li-Fraumeni Syndrome

Patients with Li-Fraumeni syndrome possess a germline mutation in the tumor suppressor gene p53. As a result, they develop malignancy at a younger age than the unaffected population and tend to develop multiple tumors. Tumors associated with this syndrome are listed in Table 4.4. These patients are at increased risk of developing a soft tissue or bone sarcoma—approximately 25-fold higher than the unaffected population.

Neurofibromatosis

Type 1 neurofibromatosis (von Recklinghausen's disease, NF-1) is an autosomal dominant disease that occurs in 1 of 3500 live births. Criteria for diagnosis include two of the following:

1. 6 or more "cafe au lait" macules
2. two neurofibromas of any type or one "plexiform" neurofibroma,
3. freckling in the axillary or inguinal regions,
4. optic glioma
5. two or more iris hamartomas
6. a characteristic osseous lesion (sphenoid dysplasia or thinning of long bone cortex)
7. a first-degree relative with NF-1.

CNS tumors account for at least half of all malignancies diagnosed in patients NF-1, with gliomas the most common histologic subtype. A relatively small percentage of these patients (7-10%) will develop neurofibrosarcomas (malignant peripheral nerve sheath tumors or MPNTs) at some point in their lifetime, a risk over 100 fold higher than that in the unaffected population. Viewed in another way, of patients with MPNTs, approximately half are patients with NF-1. Another uncommon tumor that occurs more commonly in patients with NF-1 is pheochromocytoma, although it occurs in only a small percentage of NF-1 patients.

Table 4.4. Tumors associated with Li-Fraumeni syndrome

Adenocarcinoma of the breast
Brain tumors
Sarcoma
Adrenocortical carcinoma
Leukemia

Familial Adenomatous Polyposis and Gardner's Syndrome

Patients with Gardner's syndrome are a subset of those with familial adenomatous polyposis (FAP). Both groups of patients develop innumerable adenomatous polyps throughout the large bowel at a young age and have a nearly 100% lifetime risk of developing colorectal cancer if the colon is not removed. In addition, they are also at risk of developing intra-abdominal desmoid tumors following colon resection. Desmoid tumors are low grade fibrosarcomas that have a propensity to recur locally but rarely metastasize. In addition, these patients also develop adenomatous polyps and carcinoma of the small bowel and duodenum, often many years after prophylactic colectomy. Patients with Gardner's syndrome have FAP as well as extraenteric manifestations of the disease. Gardner's syndrome patients manifest great variability in phenotypic expression of the disease, but typical features include multiple benign soft tissue tumors (usually epidermoid inclusion cysts) and osteomatosis. The gene that is mutated in individuals with FAP, the "APC" gene (adenomatous polyposis coli), has been identified, but the genetic basis for the extracolonic manifestations of Gardner's syndrome is unclear.

Retinoblastoma

Retinoblastoma is a malignant tumor of the eye that occurs in patients with germline and/or spontaneous mutations in the tumor suppressor gene, retinoblastoma (Rb). The molecular basis for the development of retinoblastoma is a recessive, loss of function mutation in the gene, which is located on chromosome 13q14. Individuals who inherit one mutated allele will develop one or more tumors due to a high rate of somatic mutation of the other allele. Osteosarcomas are the most common nonocular tumors in these patients, although STSs also occur with increased frequency. Patients treated with radiation therapy for retinoblastoma that possess a germline mutation in the Rb gene have a particularly high risk for the development of osteosarcoma of the bone within the radiation field.

Environmental Risk Factors

Radiation

Prior exposure to therapeutic radiation is the most common etiologic factor associated with the development of soft tissue or bone sarcoma. Despite this, patients with radiation-associated sarcoma account for less than 3% of all sarcoma patients seen at MSKCC and account for only a small number of patients treated with therapeutic radiation.[6] Radiation-associated sarcomas account for a significant percentage of certain histologic subtypes of sarcoma, however. In a ten year interval at Memorial Sloan-Kettering Cancer Center (MSKCC), radiation-associated sarcomas accounted for 3.1% of all osteosarcoma, 4.9% of malignant fibrous histiocytoma (MFH), and 20% of cases of angiosarcoma or lymphangiosarcoma seen.[7]

In a recent review of our experience with these tumors, the most common antecedent malignancies in patients who developed radiation-associated sarcoma were breast cancer, lymphoma, and carcinoma of the cervix. The most common radiation-associated sarcomas to occur in soft tissue were MFH and lymphangiosarcoma.[6]

As a group, these tumors are associated with a poor prognosis because most are large and high grade at presentation.

Lymphedema

Chronic lymphedema due to surgery, radiation, or parasitic infection can lead to lymphangiosarcoma of the extremity. The most common clinical setting is the postmastectomy patient with lymphangiosarcoma of the ipsilateral upper extremity (Steward-Treves syndrome). While postoperative lymphedema alone can lead to lymphangiosarcoma, it is much less common than in patients with lymphedema and a history of prior radiation exposure. Modern surgical and radiotherapeutic techniques have led to a decreased incidence of severe lymphedema of the extremity and subsequently, a decreased incidence of the disease.

Chemical Exposure

Dioxins ("Agent Orange") and phenoxy herbicides (phenoxyacetic acid and chlorophenols) have been implicated but not proven to be involved in the development of STS. Thorium dioxide ("Thorotrast") exposure is associated with the development of angiosarcomas of the liver. Chemotherapy has never been shown to predispose to STS in the adult.

Trauma/Parturition

Patients with STS will often give a history of recent trauma to the site. In this setting, it is probable that the tumor preceded the trauma, and that relatively minor trauma led to the discovery of the tumor. The "trauma" of parturition appears to be fairly strongly associated with the development of desmoid tumors of the abdominal wall, however, for they classically occur in postpartum women (Fig. 4.1).

Screening for STS

STSs are uncommon tumors. For this reason, it is not cost effective to screen the general population. Patients who have undergone therapeutic radiation for an antecedent malignancy should certainly be followed at least yearly for their remaining life, however, as the median time to development of a radiation-associated STS is 10 years (range 1 to 43 years)[7]. Patients who are known to possess a germline mutation in one of the tumor suppressor genes mentioned above should undergo periodic medical evaluation, although no specific diagnostic tests for soft tissue sarcoma beyond physical examination are of proven value in these patients.

Methods of Diagnosis and Clinical Presentation

The diagnosis of STS is made by pathologic examination of tissue obtained following biopsy or resection of a soft tissue tumor. A decision as to whether to biopsy or remove a soft tissue tumor is based on the history of the lesion as well as findings on physical examination. The vast majority of patients who present with a soft tissue

4

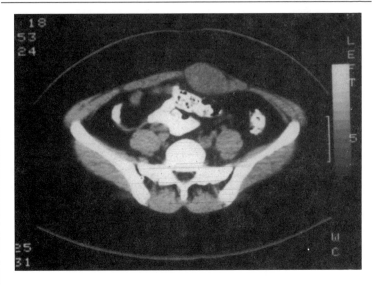

Fig. 4.1. An abdominal CT scan demonstrating a desmoid tumor of the anterior abdominal wall in a 38 year old postpartum patient.

mass of the trunk or extremity will have a benign tumor. A history of recent growth or the appearance of a new lesion of significant size should prompt consideration of a biopsy to determine the nature of the tumor. A preoperative biopsy may not be necessary in patients who present with intra-abdominal or retroperitoneal tumors unless the results would preclude initial operative management, as in patients with testicular carcinoma or lymphoma.

Frequency of Histologic Subtypes

Liposarcoma is the most common histologic subtype of STS in adults, followed by leiomyosarcoma and MFH. Together, these three histologic subtypes account for over half of patients with STS referred to our center.

Certain histologic subtypes occur in certain sites. The most common subtype in the lower extremity is liposarcoma, while in the upper extremity MFH predominates. Visceral sarcomas are largely leiomyosarcomas and retroperitoneal sarcomas are usually either liposarcomas or leiomyosarcomas. The most common histologic subtypes of truncal sarcoma are MFH, fibrosarcoma, and liposarcoma. The three most common head and neck sarcomas are fibrosarcoma, MFH, and embryonal rhabdomyosarcoma.

Extremity Sarcomas

Most patients with STS of the extremity present with a painless mass. Pain may occur due to tumor necrosis or direct compression of a sensory nerve trunk. Proximal extremity tumors tend to be larger at presentation than those arising in the distal extremity (Fig. 4.2). STS of the hand or foot occurs but is uncommon.

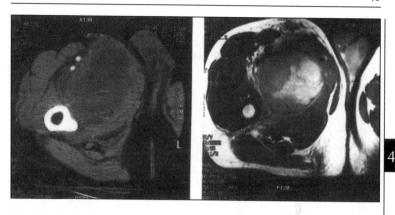

4

Fig. 4.2. MRI of a 45 year old male with a large (10 x 10 x 15 cm), high grade malignant fibrous histiocytoma of the right groin and proximal thigh. T1 weighted image is on the left, T2 weighted image on the right.

Retroperitoneal Sarcoma

Retroperitoneal sarcomas tend to be large at presentation, and occur in close proximity to vital structures (Fig. 4.3). Most patients present with a palpable abdominal mass. Less commonly, patients will have neurologic signs and symptoms due to compression of peripheral nerves or lumbosacral nerve roots. Gastrointestinal symptoms are uncommon, with intestinal obstruction occurring in less than 5% of patients at presentation .

The differential diagnosis of a patient presenting with a retroperitoneal mass includes lymphoma, renal cell carcinoma, and testicular carcinoma. It is particularly important to consider the diagnosis of testicular carcinoma in a male with a retroperitoneal mass. Testicular cancer is the most common malignancy in men between the ages of 15 and 30 years. Serology for tumor markers (beta-HCG, alpha-fetoprotein, and LDH) and a testicular ultrasound should be performed to rule out an occult testicular primary in all male patients with a retroperitoneal tumor of unclear etiology. Imaging characteristics of right-sided testicular tumors with retroperitoneal metastasis are paracaval or inter-aortocaval location. Left-sided testicular tumors with retroperitoneal metastasis have an inter-aortocaval and para-aortic location.

Visceral Sarcomas

Patients with visceral sarcomas present in much the same was as do patients with adenocarcinoma arising at the same anatomic site. When symptomatic, gastric sarcomas cause abdominal pain, ulceration and bleeding, and/or an abdominal mass. A patient with a STS of the cardia of the stomach may present with dysphagia, while nausea and vomiting may be due to a sarcoma of the distal stomach. STSs that arise in the small and large bowel mimic adenocarcinoma of the same location and cause abdominal pain, bleeding, or intra-abdominal perforation.

Patients with genitourinary sarcoma present like those with retroperitoneal sarcoma when the tumor arises in the kidney. Patients with bladder sarcoma most

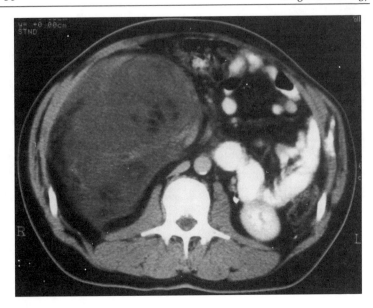

Fig. 4.3. CT scan of a 50 year old male with a large retroperitoneal liposarcoma, low grade, inferior to the right kidney.

commonly present with hematuria, and paratesticular sarcomas are usually discovered due to a palpable mass. Sarcomas of the prostate mimic advanced prostatic carcinoma, causing symptoms of obstruction and stranguria.[5]

The vast majority of gynecologic sarcomas arise in the uterus, and patients most commonly present with abnormal vaginal bleeding.

Head and Neck

The most common sites of STS of the head and neck are the skin, paranasal sinuses, and upper aerodigestive tract. Sarcomas of the nasopharynx or paranasal sinuses may present with nasal obstruction or epistaxis. Neck sarcomas may be mistaken for enlarged lymph nodes.

Pathologic Diagnosis

STSs are often difficult to categorize pathologically, and subsequently require significant expertise and experience on the part of the pathologist. The pathologist provides the surgeon with information that is critical to the appropriate management of the patient. This information includes the histologic grade and subtype of the tumor as well as the status of the clinical margins following resection. By definition, a grossly positive margin is one in which the surgical margin appears involved to the naked eye, and this is confirmed by microscopic examination. A margin should be considered microscopically positive when the tumor is close to or involves the surgical margin on microscopic exam.

The surgeon and pathologist should not hesitate to seek a second opinion when there is a question about the diagnosis or histopathology of a soft tissue tumor. Review at a center with expertise in the diagnosis of STS is frequently necessary and appropriate.

Histology

STSs arise from undifferentiated connective tissue (MFH), fibrous tissue (desmoid, fibrosarcoma), joints (tendosynovial, chordoid, epithelioid, and clear cell sarcoma), muscle (leiomyosarcoma, rhabdomyosarcoma), fat (liposarcoma), vessels (lymphangiosarcoma, hemangiopericytoma) and nerves (malignant peripheral nerve tumor, primitive neuroectodermal tumor). Chondrosarcoma and osteogenic sarcoma can also arise in soft tissue (extraskeletal osteogenic sarcoma). Within histologic subtypes the clinical behavior of a STS varies with size, grade and location.

Grade

Tumor grade is the best indicator of the potential for a STS to metastasize. Management of the patient is simplified by utilization of a grading system that distinguishes between high and low grade tumors only. High grade tumors have a very high likelihood of metastasizing (> 50%) compared to low grade tumors (< 15%). The distinction between high and low grade is made based on cellular differentiation (well vs. poor), cellularity (hypocellular vs. hypercellular), stroma (abundant vs. scant), vascularity (hypo- vs. hypervascular), necrosis (minimal vs. much), and number of mitoses per 10 high power fields (< 5 vs. > 5).

STSs may contain both high and low grade components. This commonly occurs in "de-differentiated liposarcoma" in which a low grade liposarcoma recurs with areas of high grade tumor. When over 20% of the tumor contains high grade elements it should be classified and managed as a high grade sarcoma.

Immunohistochemistry

The technique of immunohistochemistry is based on the propensity of tissues of specific histogenesis to express specific cell surface antigens. Antibodies are used to identify these cell surface proteins. Immunohistochemistry is very useful in the pathologic diagnosis of STS, particularly with regard to histiogenic subtype. It can be used on paraffin-embedded or frozen tissues and is particularly useful in distinguishing between highly undifferentiated melanoma, lymphoma, carcinoma, and sarcoma. A panel of antibodies is used to determine the specific type of sarcoma, as within categories there is considerable variation in antigen expression. A simplified description of commonly used antibodies in the immunohistochemical analysis of STS and other tumors is provided below.

Vimentin is expressed by a variety of tumors of mesodermal and neuroectodermal origin, including many carcinomas, melanoma, and most STSs. Epithelial membranes and cells of epidermal origin will stain positive for cytokeratin, as will most carcinomas. STSs are usually cytokeratin-negative and vimentin positive. The demonstration of cytokeratin and epithelial membrane antigen (EMA), although characteristic of carcinoma, can also be seen with epithelioid and synovial sarcoma. Factor VIII-related antigen is produced by endothelial cells and can be very useful in the diagnosis of poorly differentiated angiosarcoma. Desmin is an extremely useful marker for muscle tumors. The vast majority of rhabdomyosarcomas stain positively, even when poorly differentiated. Myoglobin, though found exclusively in skeletal muscle, is less sensitive than desmin.

Neuron-specific enolase is an antigen found on the surface of cells of neural crest origin. It is expressed in neuroblastoma, APUDomas, ganglioneuromas, and ganglioneuroblastomas. Neurofilament and S-100 are useful in distinguishing between tumors of neuronal origin (neurofilament positive) or nerve sheath origin (S-100 positive). While S-100 immunoreactivity is seen in chondrocytes, adipocytes and Schwann cells, the differential diagnosis of a particular STS rarely involves more than one of these histologic subtypes.

Cytogenetics
Certain chromosomal abnormalities occur in particular subtypes of STS. Cytogenetic analysis can provide the correct pathologic diagnosis in poorly differentiated tumors that lack immunohistochemical markers. These are listed in Table 4.5.

Electron Microscopy
Electron microscopy is a valuable technique for categorizing the small number of tumors that are negative for immunohistochemical markers. Lipid may be appreciated in liposarcomas, while clear cell sarcomas often contain melanosomes.

Radiologic Diagnosis

STS of the Extremity and Trunk
Patients who present with very small or superficial tumors do not necessarily require an imaging study, as excisional biopsy and pathologic evaluation may be the most reasonable diagnostic study. Either a CT scan or MRI is appropriate in the evaluation of all other patients with a soft tissue mass of the extremity or trunk. MRI is increasingly utilized however, if cost or availability are not an issue. While both studies are useful in the measurement of tumor size and assessment of depth, several benign tumors have characteristic MRI appearances that may facilitate the correct preoperative diagnosis. These include lipomas, hemangiomas, nerve sheath tumors and myxomas. A soft tissue mass of the extremity that does not appear characteristic of one of the benign tumors listed above, is heterogeneous in appearance, does not involve bone, and displaces surrounding structures is highly suggestive of a STS.

Retroperitoneal Tumors
A CT scan with oral and intravenous contrast is the diagnostic procedure of choice for patients with a retroperitoneal mass. Most patients who present with a retroperitoneal mass will be found to have a malignant tumor, with lymphoma, testicular carcinoma, renal cell carcinoma, and STS in the differential diagnosis. A CT scan can be very useful in determining which of these is the most likely diagnosis. An isolated retroperitoneal mass without evidence of additional central or peripheral adenopathy is unlikely to represent lymphoma. Tumors that arise in the midline in a male patient should be considered testicular carcinoma until proven otherwise (with a negative testicular ultrasound and negative tumor markers). Direct involvement or destruction of the kidney is more likely due to a primary renal cell

Table 4.5. Nonrandom cytogenetic abnormalities in STS

Histologic Subtype	Cytogenetic Abnormality
Synovial sarcoma	t(X;18)(p11.2;q11.2)
Alveolar rhabdomyosarcoma	t(2;13)(q35-37;q14)
Ewing's sarcoma	t(11;22)(q21-24;q11-14)
Myxoid liposarcoma	t(12;16)(q13;p11)
Extraskeletal myxoid chondrosarcoma	t(9;22)(q22;q11-12)

MPNT=malignant peripheral nerve sheath tumor

4

carcinoma than a STS, which more commonly displaces the kidney and its vasculature.

Preoperative Evaluation

In general, patients with STS > 5 cm in size or deep in location should undergo preoperative imaging with either an MRI or CT scan. Appropriate, high quality imaging will facilitate operative treatment planning. While either a CT scan or MRI will provide approximately equal visualization of adjacent neurovascular structures for patients with STS of the extremity, MRI may be superior to CT scanning in evaluating the anatomic compartment and individual musculature involved. In patients with suspected STS without a histologic diagnosis, CT imaging may be more appropriate as a CT guided needle biopsy can be used to sample areas of the tumor that are likely to be diagnostic, avoiding necrotic areas or benign components of heterogenous tumors.

The most likely site of metastasis in patients with STS of the extremity or trunk is the lungs. A preoperative CT scan of the chest should be performed in any patient with STS at significant risk for distant metastases. These include all patients with high grade sarcoma, particularly if they are over 5 cm in size. Either a CT scan of the chest or a chest radiograph is appropriate for patients with a large low grade sarcoma or small, high grade extremity sarcoma.

It is important to establish the histologic diagnosis of sarcoma and the grade of the tumor preoperatively in patients with large soft tissue tumors of the extremity or trunk. This will allow appropriate treatment planning with the radiation oncologist. Brachytherapy may be considered in patients with high grade sarcoma as it has been shown to decrease local recurrence in patients undergoing complete operative resection.[8] For patients who are found to have low grade sarcoma, preoperative treatment planning for postoperative external beam radiation therapy (EBRT) can be initiated. Preoperative determination of the tumor grade is also important because patients with large (≥ 8 cm) high grade STS may be appropriate candidates for participation in neoadjuvant or adjuvant chemotherapy protocols.

The most efficient technique for obtaining a pathologic diagnosis of STS depends on the size and anatomic site of the tumor. Incisional biopsies or core tissue biopsies are appropriate for large soft tissue tumors. CT guided techniques may be used when necessary, or open operative biopsies if these are nondiagnostic or it is anatomically more appropriate (such as for tumors of the hand or foot). Incisional biopsies of soft tissue tumors of the extremity should be placed axially. When performing an incisional biopsy it is important to ensure that the biopsy scar can be included in the definitive resection without sacrificing cosmetic or functional results. Hemostasis is extremely important, as bleeding with hematoma formation has the potential to spread tumor cells. Excisional biopsies are indicated for small, superficial tumors (generally ≤ 3 cm).

A fine needle aspiration for cytologic diagnosis may be used in patients with a suspected recurrence of STS of the trunk or extremity. If the cytology is positive, this is all that is required to proceed with treatment planning. If nondiagnostic, a core or incisional biopsy must be performed.

STS Staging

AJCC (American Joint Commission on Cancer)

The AJCC staging system utilizes tumor grade, size, regional lymph nodes status, and distant metastasis in the assignment of stage for all adult (over age 16) patients with STS (Table 4.6). It applies to all histologic subtypes of STS except Kaposi's sarcoma, desmoid tumors, and dermatofibrosarcoma protuberans (DFSP), and all tumor sites with the exception of CNS and visceral. Tumor grade and size are the predominant determinants of stage for most patients with STS. Other determinants are whether the tumor is superficial (above the noninvolved muscular fascia) or deep (below or involving the muscular fascia), and whether there are lymph node or distant metastases present. As depicted, lymph node metastases have the same prognostic significance as distant metastases, although these infrequently occur.

Prognostic Variables

Prognostic factors for both recurrence and survival for all patients with STS in our prospective institutional database include size, grade, site, histopathology, and adequacy of resection. In the extremity, high grade and large size tend to occur together, while in the retroperitoneum a very large tumor that has not metastasized suggests that it is low grade (Fig. 4.3). High grade sarcomas are more likely to recur locally than low grade sarcomas, and are much more likely to metastasize. Extremity sarcomas are associated with a more favorable prognosis than tumors of the viscera or retroperitoneum. The overall 5-year survival of patients with retroperitoneal sarcoma is 54%, clearly less favorable than that of patients with extremity STS, who have a 5-year survival of 76%.[8,9] When the more common histopathologic subtypes are considered, patients with primary leiomyosarcoma have a worse prognosis than those with liposarcoma, MFH, or tenosynovial sarcoma.[10] The adequacy of resection has a direct influence on local recurrence rates in all sites; in the retroperitoneum it is also a predictor of survival.

Extremity

Tumor mortality in the great majority of patients with extremity STS is determined by the development of metastatic disease. A local recurrence may increase the morbidity of treatment, but a significant percentage of patients will survive following aggressive local treatment for recurrent disease.

Table 4.6. American Joint Committee on Cancer (AJCC) Staging of STS, 1997

STAGE	G	T	N	M
IA	1-2	1a-1b	0	0
IB	1-2	2a	0	0
IIA	1-2	2b	0	0
IIB	3-4	1a-1b	0	0
IIC	3-4	2a	0	0
III	3-4	2b	0	0
IV	any	any	1	0
			0	1

G1: low grade
G2: intermediate grade
G3: high grade
G4: undifferentiated
T1: tumor ≤ 5 cm
T2: tumor > 5 cm
   a: superficial tumor
   b: deep tumor
N1: regional nodal involvement
M1: distant metastasis

Clinical and pathologic factors that increase the risk of local recurrence are often different from those that affect distant recurrence. In a recent analysis of 1,041 adult patients with extremity STS, significant independent adverse prognostic factors for local recurrence were age greater than 50 years, recurrent disease at presentation, microscopically positive surgical margins, histologic subtypes of fibrosarcoma and malignant peripheral-nerve tumor, and large tumor size. Adverse factors for distant recurrence were intermediate tumor size, high grade, deep location, recurrent disease at presentation, leiomyosarcoma, and nonliposarcoma histology.[8] Tumor grade is by far the greatest predictor of death from sarcoma within the first 2 years. After 2 years grade, size, and depth are equally predictive of outcome.[11] When a large tumor (> 10 cm) is high grade, survival decreases to 20% at 5 years.[12]

While high grade tumors are clearly associated with a poorer survival than low grade, a designation of "low grade" reduces but does not eliminate the possibility that a soft tissue tumor of the extremity will metastasize. Donohue and colleagues found that 14% of 130 patients with low grade extremity STS developed metastases.[13]

Retroperitoneal and Visceral Sites

Adequate surgical resection of STS is an essential component of management of STS arising in any site, but particularly for those arising in the retroperitoneum and visceral sites. While a significant percentage of patients with extremity STS can be cured with aggressive management of locally recurrent disease, this is much less likely to occur in patients with visceral and retroperitoneal disease, where local recurrence is a much more common cause of death. Isolated local recurrence is the predominant site of recurrence in most patients with retroperitoneal sarcoma, with a median time to recurrence of greater than 5 years in patients presenting with primary disease.[9] Recurrent disease predicts further recurrence and eventual death. Hence, the ability of the surgeon to completely resect these tumors is the primary determinant of survival.

Prognosis Following Recurrence

The prognosis of patients who undergo resection of recurrent STS depends on the site of recurrence. Patients with extremity STS who develop a local recurrence fare better than those with local recurrence in the retroperitoneum or abdomen. Indeed, patients with local recurrence in the extremity have an excellent chance of long term survival following re-resection and radiation therapy. Local recurrence in a retroperitoneal or visceral site is much less likely to be resectable, and patient survival diminishes significantly when it occurs. Distant recurrence is a difficult problem for all patients with STS. The most common site of distant recurrence (comprising approximately 80% of all recurrences) for those with extremity and trunk STS is the lung, while patients with STS of the retroperitoneum or viscera are more likely to recur in the liver or abdominal cavity. Surgeons have traditionally taken a very aggressive approach to metastatic STS in the lung, and in patients who can be completely resected, long term survival of 20-40% can be achieved. When all patients who develop lung metastasis are considered, however, the percentage of patients who benefit from an operative approach is less than 10%.

Treatment Options

Operative resection with negative surgical margins is the most effective method of treatment for most adults with STS, and certain general principles apply. STSs tend to produce a pseudocapsule of compressed tissue with tumor growth. This tissue is almost invariably invaded by tumor cells and should not be used as a plane of dissection at operation. Ideally, sarcomas should be resected with a wide margin of normal tissue (at least 2 cm). The definitive resection must encompass any prior biopsy or resection scars. When the surgeon is unable to obtain an adequate margin in one area (for example the aorta), it is futile to resect an important adjacent organ or structure to obtain a negative margin in another. Multimodality therapy should be considered in all patients who present with STS, but the surgeon must carefully consider the potential benefits against the risks involved of the use of adjuvant EBRT or brachytherapy in patients with extremity STS depends on the site, size, and grade of the tumor. Radiation therapy is uncommonly used for patients with retroperitoneal or visceral sarcoma, and is unlikely to provide a significant benefit for patients with small (< 5 cm), superficial extremity STS that is completely resected. Adjuvant or neoadjuvant chemotherapy should be considered in select patients with STS, particularly those with large, high grade extremity tumors. Again, patients with small STS (< 5 cm) or those with low grade tumors are unlikely to derive significant benefit from chemotherapy. A management algorithm for patients with extremity and trunk STS is illustrated in Figure 4.4, and for patients with retroperitoneal sarcoma in Figure 4.5.

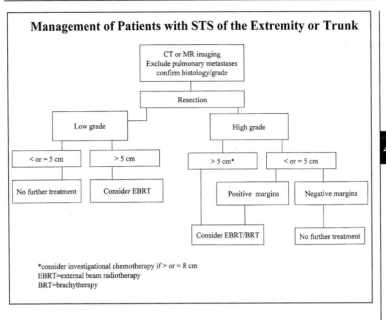

Fig. 4.4. Algorithm of management of patients with STS of the extremity or trunk.

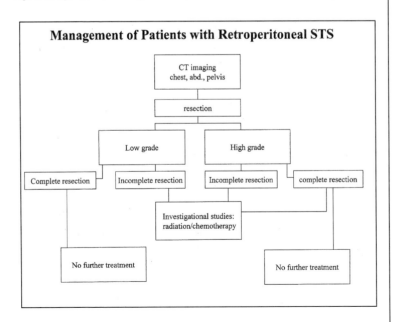

Fig. 4.5. Algorithm of management of patients with STS of the retroperitoneum.

Management of Patients with Extremity STS

Increased utilization of limb-sparing surgery in the treatment of most patients with extremity STS is the result of a concerted effort by surgeons and radiotherapists to decrease the morbidity of treatment without compromising local control or survival. Indeed, the vast majority of patients treated for STS of the extremity are able to retain a functional limb with a combination of surgery and radiation therapy. "Limb-sparing surgery" involves a wide resection of the tumor with several centimeters of surrounding normal tissue. A tumor confined to one anatomic muscle group should be resected using a compartmental en bloc resection (from muscle origin to insertion) when the tumor involves most of the musculature within the compartment. A sub-compartmental resection is preferable when this allows preservation of functional tissue without comprising surgical margins of resection.

Level I evidence (established by randomized clinical trials) supports the efficacy of adjuvant radiation therapy following operative resection in the management of patients with extremity STS.[14] This is true for patients with low as well as high grade STS. Patients with extremity or trunk tumors that are less than 5 cm in size may not require adjuvant radiotherapy if a wide margin of resection can be accomplished. Adjuvant radiation therapy should be used selectively in these patients.

Function-preserving resection, with preservation of vital structures, can be combined with brachytherapy to provide excellent local control in both primary and recurrent high grade STS of the extremity. This technique involves the placement of afterloading catheters (parallel hollow nylon catheters placed percutaneously) in the tumor bed following resection. A nomogram is used to determine the number of afterloading catheters needed for a given tumor bed and specific activity of the radioisotope sources. The surgeon delineates the tumor bed using small metal clips for radiologic identification and the radiation oncologist places the catheters in the operating room prior to closure. The treatment field generally extends several centimeters beyond the suspected confines of the tumor. On the fifth to sixth postoperative day the catheters are loaded with Iridium-192. A dose of 4200-4500 cGy is delivered in 4-6 days to the isodose contour incorporating the tumor volume determined at operation.

A major advantage of brachytherapy over EBRT is the shorter length of treatment—4-6 days as opposed to 5-6 weeks. A theoretical advantage lies in the absence of tissue hypoxia, known to diminish the effectiveness of radiotherapy, in the tumor bed. In addition, proximity of the catheters to the tumor bed allows more precise delivery of radiation with less exposure of normal tissue than external radiation therapy. The proximity of the radiotherapist to the surgeon at the time of operation also facilitates precise mapping of the tumor bed and delineation of areas at particular risk for recurrence. In a prospective randomized trial of patients with completed resected STS of the extremity conducted at our institution, the use of brachytherapy was shown to decrease the incidence of local recurrence in patients with completely resected high grade tumors [8]. There was no benefit in the group of patients with low grade tumors. This technique may be particularly useful in patients with tumors that arise between anatomically defined muscle compartments (antecubital space, popliteal space, groin, supraclavicular fossa, head and neck) in whom the margins of resection are compromised by proximity to major neurovascular structures.

Brachytherapy or EBRT should be administered following resection in these patients. Resection of major neurovascular structures should only be considered if these are the only compromised margins, and some function of the extremity will be maintained. Division of the sciatic nerve, for example, will result in a functional extremity if the patient is able to use a brace for foot drop and understands and accepts the high risk of damage to an insensate foot.

Patients with locally advanced STS of the extremity involving major neurovascular structures present a special challenge. Patients with locally recurrent STS of the extremity following resection and radiotherapy may still be candidates for limb-sparing surgery and brachytherapy, and should be referred to institutions with significant expertise with these patients. Figure 4.6 depicts an MRI of a patient with a recurrent liposarcoma of the upper right thigh following 2 years after resection and EBRT.

In the absence of effective treatment for systemic disease, amputation for local control should only be used in patients in whom limb-sparing surgery and radiotherapy would result in a functionless extremity. This is especially true in those at high risk for distant metastasis. Again, most patients with extremity STS who die of their disease do so because of distant metastasis, not local recurrence.

Management of Patients with Retroperitoneal STS

Retroperitoneal STSs are usually large at presentation and often displace (and less often, involve) adjacent structures such as the kidney, renal vein, pancreas, spleen, bowel, vena cava and aorta. Most patients who present with primary retroperitoneal sarcoma at our institution will undergo resection of at least one contiguous organ.[9] Delivery of adequate doses of radiation is hampered by toxicity to adjacent structures and chemotherapy remains ineffective for patients with this disease.

The completeness of resection and tumor grade are the most important determinants of survival in patients with retroperitoneal sarcoma. A complete bowel prep should be performed preoperatively. A transperitoneal approach is utilized, and the surgeon must be prepared to resect kidney, pancreas, colon, and major neurovascular structures when necessary in order to obtain grossly negative margins. Involvement of these structures is not an indication of an unresectable tumor. Negative operative margins can be difficult to obtain, but structures invaded by tumor should be resected when this will accomplish a complete gross resection. The kidney is the most commonly resected adjacent organ (followed by the colon and pancreas), but pathologic involvement of the kidney is uncommon.[15] Intra-operative assessment of microscopic margins is usually not helpful in retroperitoneal sarcoma, as the great majority will have limited margins based on their proximity to vital structures such as the vena cava and aorta, and the option of obtaining an additional surgical margin is not possible. A complete operative resection, however, can be obtained in most patients with primary retroperitoneal sarcoma and in about half of patients who present with recurrent disease.[9] Patients should only undergo a planned incomplete resection when palliation of significant symptoms is the goal of the operation.

EBRT has never been demonstrated to increase survival in patients with retroperitoneal soft tissue sarcoma. In addition, the use of radiotherapy in these patients is significantly restricted by potential toxicity to nearby structures, particularly the bowel. Delivery of tumoricidal doses is associated with the development of significant short and long term complications, including acute and chronic enteritis, obstruction and stricture. The liver and kidney are also quite susceptible to the toxicity of EBRT. For this reason, there has been an interest in developing alternative methods of delivering radiation to the tumor bed by using intraoperative radiation therapy (IORT). This is accomplished at the time of operation after the tumor has been removed. All uninvolved organs and areas are shielded with lead barriers, leaving the tumor bed exposed. Afterloading catheters are then placed and high doses of radio activity are delivered while the surgical, anesthesia, and nursing personnel monitor the patient remotely. Following delivery of the radiation, the barriers are removed and the operation is completed. Prospective randomized trials have demonstrated that using IORT for retroperitoneal sarcoma decreases local recurrence but does not improve survival. While complications are fewer compared to patients undergoing EBRT, it is still associated with significant morbidity, particularly peripheral neuropathy.

The use of adjuvant radiotherapy in retroperitoneal sarcoma should be limited to patients with high grade tumors at high risk for local recurrence in whom the therapy can be administered with acceptable morbidity. Tissue expanders and synthetic mesh slings to keep the small bowel out of the field should be employed. The use of IORT should be considered in select patients because it appears to provide better local control. In addition, it is less toxic when compared to a full course of EBRT in patients with retroperitoneal STS.

Management of Patients with Visceral Sarcoma

The goals of treatment of STS arising in a visceral organ are similar to those for retroperitoneal disease. Complete operative resection and tumor grade are the most important determinants of outcome. Radiation therapy is often hampered by the proximity of the tumor to adjacent, radiosensitive structures such as bowel, kidney, and liver. Current chemotherapeutic strategies are disappointing.

Intraoperative frozen section analysis is unreliable in the evaluation of the most common type of gastrointestinal sarcoma, leiomyosarcoma. The surgeon must make a decision regarding the extent of resection based on a clinical assessment of the nature of the tumor, its size and location. If there is a suspicion of malignancy, an adequate margin of normal tissue should be obtained. A small leiomyosarcoma of the stomach can be resected with a 2 cm margin of normal tissue, while a subtotal gastrectomy may be necessary for large tumors of the pylorus or cardia. A sarcoma of the duodenum may require pancreaticoduodenectomy, while small tumors of the rectum or anal canal can be resected via a transanal approach.

Genitourinary STSs behave like retroperitoneal sarcoma (kidney, ureter) or visceral sarcomas (prostate) and should be managed accordingly. Bladder sarcomas have a more favorable prognosis as they often present earlier due to hematuria and can be completely resected. Selected patients with small (< 2 cm) bladder sarcomas may be successfully managed with a transurethral resection alone.

Management of Patients with Uterine STS

Uterine sarcomas are characteristically locally aggressive with a propensity for early distant recurrence. The diagnosis is usually made with fractional dilatation and curettage. Cystoscopy and proctoscopy should be performed before operation to determine the extent of involvement of other pelvic structures. A total abdominal hysterectomy with bilateral salpingo-oophorectomy is standard therapy. Extra-uterine disease is often found at exploration and is a very poor prognostic sign. There is no evidence that the use of adjuvant radiotherapy improves survival in patients with uterine sarcoma, although it may reduce pelvic failure.

Management of Patients with Head and Neck STS

Because of anatomic constraints, it is often difficult to obtain a wide margin of normal tissue around a STS of the head and neck. In these patients, adjuvant radiation therapy in the form of brachytherapy or EBRT can be used to decrease the patient's risk of local recurrence. There may be some advantage to preoperative EBRT in patients with STS of the head and neck because the volume of treated tissue can be minimized, thereby sparing adjacent areas the functional and cosmetic morbidity of therapy.

Resection of Metastatic Disease

A significant prolongation of survival can be achieved in selected patients undergoing a complete resection of pulmonary metastases from STS. Patients should be considered for resection if local/regional control has been accomplished, metastases are limited to the lung, and they are a reasonable operative risk. When these criteria are met, 20-30% of these highly selected patients who undergo complete resection of their disease will survive long-term. Patient and tumor factors that are used to make a decision regarding pulmonary metastasectomy include the disease free interval, the number of metastases present, and whether the disease is unilateral or bilateral. It is important to remember that the vast majority of patients who develop lung metastases will not benefit from pulmonary resection, and that reports of surgical series are based on a highly selected subgroup of patients.

Adjuvant or Neoadjuvant Chemotherapy for Patients with STS

Single institution and national trials of adjuvant chemotherapy have failed to demonstrate an overall survival benefit for patients treated with adjuvant chemotherapy following resection of STS, although several prospective randomized studies suggest that it may prolong disease free survival. A recent meta-analysis suggests that adjuvant chemotherapy provides a minimal benefit in terms of disease free survival in patients with localized completed resected STS but no improvement in overall survival.[16] The use of adjuvant chemotherapy remains an area of significant controversy, and its use should be restricted, whenever possible, to patients participating in prospective clinical trials.

Patients at particularly high risk for death from their disease should be entered into investigational trials of neoadjuvant (preoperative) or adjuvant chemotherapy. These include patients with large (> 8 cm), high grade extremity, retroperitoneal or visceral tumors. As a rule, these patients have a probability of long-term survival of 20% or less. Likewise, patients with an excellent long-term prognosis should not be included in investigational trials. These include patients with completely resected low grade sarcomas and adults with small (< 5 cm) high grade extremity sarcomas.[17]

Regional Chemotherapy for Patients with Extremity STS

Regional chemotherapy is a therapeutic option for patients with advanced STS of the extremity, particularly when amputation is the only alternative. Regional chemotherapy is delivered by surgically isolating the vasculature at the root of the limb and applying a tourniquet proximally. A cardiopulmonary type pump oxygenator is used to maintain circulation and oxygenation of the limb through the cannulated vasculature at the root of the extremity. This procedure is called isolated limb perfusion (ILP). A hyperthermic, hyperoxic perfusion is then commenced which delivers high dose chemotherapy to the extremity only. The limb is perfused for approximately one hour. At completion, the extremity is flushed, the tourniquet is released, the catheters are removed, and the vessels are repaired.

The agent traditionally used for ILP procedures is melphalan. Response rates in patients with STS treated with hyperthermic melphalan, with or without other chemotherapeutic agents, are less than 30%, with very few complete responses. Recently, a combination of cytokines (TNFα ± IFNγ) and melphalan have been demonstrated to increase response rates, although response durability remains a significant problem. The technique of ILP requires that a patient be treated at a center with significant expertise in this therapy.

Chemotherapy for Patients with Advanced STS

Currently, a doxorubicin-based regimen should be considered standard therapy for patients with advanced, unresectable STS. Unfortunately, treatment is limited by cardiomyopathy, which occurs in about 10% of patients, and can result in death. Cardiac toxicity can be diminished by giving the drug as a continuous infusion as opposed to a bolus, without comprising therapeutic efficacy. Ifosfamide is the most promising recent addition to the chemotherapeutic armamentarium for patients with stage IV disease. Ifosfamide is a cyclophosphamide analogue that is activated by hepatic microsomes. When used in combination with doxorubicin, response of 30-35% can be achieved.[18] Hemorrhagic cystitis, the major toxicity of ifosfamide, can be prevented by use of the uroprotectant agent, mesna. Myelosuppression occurs with doxorubicin/ifosfamide in combination but can be diminished with the use of human granulocyte-macrophage colony stimulating factor.

Another regime with activity in patients with metastatic STS is MAID, a combination of mesna, doxorubicin, ifosfamide, and dacarbazine. Response rates in approximately half of all patients treated have been reported using this combination. Ongoing clinical trials comparing single agent and combination chemotherapies will further define the appropriate therapy for patients with stage IV disease.

Outcome of Treatment

Outcome of Treatment in Patients with Extremity STS

Factors that affect survival in patients with STS of the extremity include stage, size, grade, and depth. Histologic subtype is less important than tumor size and grade. In addition, patient age and sex do not independently affect disease-specific survival in adults with STS. Favorable clinical and pathologic factors for survival include localized vs. metastatic disease, low vs. high grade, small size (≤ 5 cm) vs. large size (> 5 cm), and superficial as opposed to deep anatomic location. The status of the pathologic margins of resection does not appear to impact on survival, although it does predict local recurrence. Patients with tumors which are < 5 cm in diameter have a particularly favorable survival, even if they are high grade.[17]

The most common site of metastases in patients with extremity STS is the lungs. Patients who can undergo resection of their pulmonary metastases have a more favorable survival than those who cannot. Predictors of survival in patients with lung metastases from STS include a long disease free interval, lower numbers of lesions, and unilateral as opposed to bilateral disease.

Outcome of Treatment in Patients with Retroperitoneal STS

The 5 year disease-free survival of all patients presenting to our institution with primary retroperitoneal sarcoma who underwent resection is 59%. Consistent predictors of survival in these patients are stage, tumor grade, and status of resection margins.[9] More than half of all patients will eventually recur, however, and subsequent resections of local recurrence are less likely to provide a favorable survival experience. STS of the retroperitoneum that recurs in discontinuous intra-abdominal sites or in the liver is not amenable to surgical cure.

Posttreatment Surveillance

Patients who have undergone a resection for STS should be followed closely for the development of local recurrence as well as metastatic disease. Appropriate follow-up is determined by the natural history of the particular sarcoma and the suitability of the patient to undergo resection of recurrent disease.

Patients with high grade sarcoma of the extremity and trunk who develop distant metastasis usually do so within the first 2 years. A baseline CT scan should be performed in these patients prior to operation and then postoperatively, a pulmonary imaging study is usually performed every 3-4 months for 3 years. It is reasonable and appropriate to alternate chest radiographs with CT scans, as there is no evidence that more sensitive imaging (i.e., CT scanning) results in a survival advantage to patients who develop metastases. After the first 3 years a chest radiograph or CT scan of the chest every 4-6 months is reasonable. After 5 years a yearly chest radiograph should be obtained. In patients with a lower risk of pulmonary metastases (i.e., those with low grade tumors or small, high grade tumors) a CT scan of the chest or chest radiograph is usually obtained every 3-6 months for the first 2-3 years, twice a year for the next 2-3 years, and yearly thereafter. Imaging of the involved extremity is of little value in most patients following treatment for extremity STS as physical examination will detect the majority of local recurrences. Patients with a difficult exam secondary to body habitus or radiation fibrosis may be exceptions to this general rule. Routine blood tests have no value in detecting recurrent disease in patients with extremity STS. It should be recognized that these recommendations are based on retrospective analysis of the efficacy of clinical follow up practice as well as knowledge of the natural history of the disease and have never been demonstrated to be cost effective in a prospective analysis.

Pulmonary metastases are less common than local recurrences in patients with retroperitoneal and visceral STS. For this reason, a CT scan of the abdomen and pelvis is a more appropriate imaging test postoperatively, as liver metastases are the most frequent distant site of recurrence in these patients. In general, patients with high grade tumors will experience a local recurrence sooner than those with low grade tumors. A reasonable follow-up schedule for patients with high grade retroperitoneal or visceral sarcoma is to perform CT scan imaging of the chest, abdomen and pelvis every 4 months for 3 years and then every 6 months for 2 more years. For patients with low grade tumors a CT scan of the abdomen and pelvis and a chest radiograph every 6 months for 5-6 years is reasonable and then yearly for an additional 2-3 years. It must be emphasized that these recommendations are general in nature. Indeed, it has never been demonstrated that patients who undergo intensive surveillance for recurrent disease fare better than those who do not. As the cost-effectiveness of posttreatment surveillance of cancer patients is studied in prospective trials a more evidence-based approach will undoubtedly guide practice patterns in the future.

Selected Readings

1.   Greenlee RT, Murray T, Bolden S et al. Cancer Statistics, 2000. CA. A Cancer Journal for Clinicians 2000; 50:7-33.
     Cancer statistics for the year 2000, published by the American Cancer Society.
2.   Pollock RE, Karnell LH, Menck HR et al. The National Cancer Data Base report on soft tissue sarcoma. Cancer 1996; 78(10):2247-57.
     Review of national statistics (USA) for soft tissue sarcoma.
3.   Brennan MF. The surgeon as a leader in cancer care: lessons learned from the study of soft tissue sarcoma. J Am Coll Surg 1996; 182(6):520-9.
     Important review of the role of the surgeon in the management of the patient with soft tissue sarcoma.
4.   Brady MS, Brennan MF. Soft tissue sarcoma. In: Allen-Mersh TG, ed. Surgical Oncology, ed. First Edition. London: Chapman & Hall, 1996; 401-420.
     General review: soft tissue sarcoma.
5.   Russo P, Brady MS, Conlon K et al Adult urological sarcoma. J Urol 1992; 147(4):1032-6; discussion 1036-7.
     Large, single institution experience with urologic sarcoma.
6.   Brady MS, Garfein CF, Petrek JA et al. Posttreatment sarcoma in breast cancer patients. Ann Surg Oncol 1994; 1(1):66-72.
     Paper describing the presentation, management and outcome of patients with breast cancer who develop treatment related soft tissue sarcoma.
7.   Brady MS, Gaynor JJ, Brennan MF. Radiation-associated sarcoma of bone and soft tissue. Arch Surg 1992; 127(12):1379-85.
     Large, single institution experience with radiation-associated sarcoma.
8.   Pisters PW, Harrison LB, Leung DH et al. Long-term results of a prospective randomized trial of adjuvant brachytherapy in soft tissue sarcoma. J Clin Oncol 1996; 14(3):859-68.
     Report of a large phase III trial evaluating brachytherapy in patients with resected soft tissue sarcoma.
9.   Lewis JJ, Leung D, Woodruff JM et al. Retroperitoneal soft-tissue sarcoma: Analysis of 500 patients treated and followed at a single institution. Ann Surg 1998; 228(3):355-65.
     Large, single institution review of the presentation, management and outcome of patients with retroperitoneal sarcoma.
10. Brennan MF, Casper ES, Harrison LB et al. The role of multimodality therapy in soft-tissue sarcoma [see comments]. Ann Surg 1991; 214(3):328-36; discussion 336-8.

Excellent review of the multidisciplinary management of patients with soft tissue sarcoma.

11. Gaynor JJ, Tan CC, Casper ES et al. Refinement of clinicopathologic staging for localized soft tissue sarcoma of the extremity: a study of 423 adults. J Clin Oncol 1992; 10(8):1317-29.
    Large, single institution experience with extremity soft tissue sarcoma.

12. Brennan MF. Management of extremity soft-tissue sarcoma. Am J Surg 1989; 158(1):71-8.
    Major review article on the multidisciplinary management of the patient with soft tissue sarcoma.

13. Donohue JH, Collin C, Friedrich C et al. Low-grade soft tissue sarcomas of the extremities. Analysis of risk factors for metastasis. Cancer 1988; 62(1):184-93.
    Key paper documenting incidence of pulmonary metastasis from low grade extremity soft tissue sarcoma.

14. Wylie JP, O'Sullivan B, Catton C et al. Contemporary radiotherapy for soft tissue sarcoma. Semin Surg Oncol 1999; 17(1):33-46.
    Review article on the role of radiation therapy in the management of the patient with soft tissue sarcoma.

15. Russo P, Kim Y, Ravindran S et al. Nephrectomy during operative management of retroperitoneal sarcoma. Ann Surg Oncol 1997; 4(5):421-4.
    Large, single institution experience with nephrectomy at the time of resection for retroperitoneal sarcoma.

16. Adjuvant chemotherapy for localised resectable soft-tissue sarcoma of adults: meta-analysis of individual data. Sarcoma Meta-analysis Collaboration. Lancet 1997; 350(9092):1647-54.
    Major meta-analysis of chemotherapy trials for patients with soft tissue sarcoma.

17. Geer RJ, Woodruff J, Casper ES et al. Management of small soft-tissue sarcoma of the extremity in adults. Arch Surg 1992; 127(11):1285-9.
    Key paper describing single institutional experience with patients with small (< 5 cm) soft tissue sarcomas.

18. Patel SR, Benjamin RS. New chemotherapeutic strategies for soft tissue sarcomas. Semin Surg Oncol 1999; 17(1):47-51.
    Major review article on the use of chemotherapy for patients with soft tissue sarcoma.

4

CHAPTER 5

Head and Neck Cancer

James C. Alex, Jason Klenoff

Introduction

Cancer of the head and neck region, which affects 53,650 people annually, is notable for its complex surgical anatomy and considerable potential for functional and aesthetic insult. Loss of the larynx, palate, tongue, or mandible can be particularly devastating to the patient. Fortunately, the last 20 years has produced several treatment and reconstructive advances. Myocutaneous flaps and neurovascular free tissue transfer have greatly reduced the loss of form and function for head and neck patients undergoing major ablative surgery. Induction chemotherapy and/or radiation therapy can be used in lieu of radical surgical resection for organ preservation. A consequence of these advances is that head and neck cancer management requires a multidisciplinary team including a head and neck surgeon, reconstructive surgeon, neurosurgeon, oromaxillofacial surgeon, chemotherapist, radiation therapist, prosthedontist, nutritionist, social worker and nurse manager.

Anatomy

The head and neck region is generally divided into the oral cavity, nasal cavity, pharynx, larynx, sinuses, and salivary glands (Fig. 5.1). The oral cavity begins at the vermilion border of the lips and extends posteriorly to the palatoglossal fold. It contains the lips, teeth, retromolar trigone, gingiva, buccal mucosa, hard palate and oral tongue. Behind the palatoglossal fold lies the palatine tonsils which are within the pharynx.

The pharynx is subdivided into three divisions: the oropharynx, nasopharynx, and hypopharynx. The anterior border of the oropharynx is the palatoglossal fold. Superiorly, it is bordered by the soft palate. Inferiorly, it descends to the level of the hyoid. The base of tongue, tonsils and vallecula are within the oropharynx. Above the oropharynx lies the nasopharynx. The region is bordered inferiorly by the soft palate and anteriorly by the choanae which are the limit of the nasal cavity. Both the adenoids and the opening of the Eustachian tube lie within the nasopharynx. From the level of the hyoid to the inferior border of the cricoid cartilage is the hypopharynx. The hypopharynx contains the piriform sinuses which lie on either side of the laryngeal inlet and are bordered medially by the aryepiglottic folds and laterally by the thyroid cartilage.

The larynx lies anterior to the hypopharynx and is subdivided into subglottic, glottic, and supraglottic regions. The supraglottic region extends from the tip of the epiglottis to the apices of both laryngeal ventricles and includes the false cords, the

Surgical Oncology, edited by David N. Krag. ©2000 Landes Bioscience.

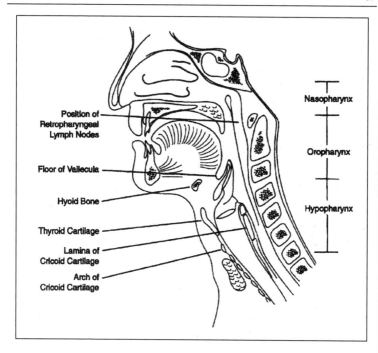

Fig. 5.1. Anatomic terminology for the head and neck region.

epiglottis, the arytenoids and aryepiglottic folds. The glottic region extends from the inferior border of the supraglottic region to 5 mm below the free edge of the true vocal cords. The subglottic region extends from the inferior margin of the glottic region to the inferior edge of the cricoid cartilage.

The lymphatic anatomy and drainage patterns are important in understanding and managing the regional spread of head and neck cancer. By convention, the neck is divided into five main lymph node regions called "levels" (Fig. 5.2). Level I refers to the submental and submandibular triangles which are bounded by the posterior belly of the digastric, the hyoid bone, and the body of the mandible. Level II denotes the upper jugular lymph nodes which are contained within the area over the cephalad one third of the sternocleidomastoid muscle. The superior border is the skull base and the inferior border the hyoid bone. Level III contains the middle jugular nodes and corresponds to the area over the middle third of the sternocleidomastoid. This region is marked by the hyoid bone superiorly and the omohyoid muscle and cricoid cartilage inferiorly. Level IV contains the lower jugular nodes, which are over the caudad third portion of the sternocleidomastoid muscle. This region is bordered superiorly by the omohyoid muscle (and cricoid cartilage) and inferiorly by the clavicle. The region bordered by the posterior border of the sternocleidomastoid muscle, the anterior border of the trapezius, and the clavicle inferiorly defines level

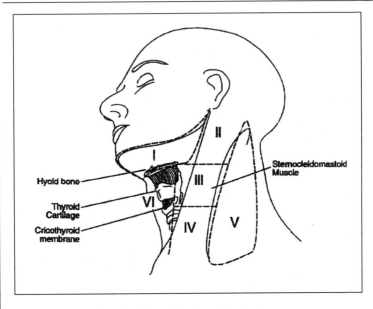

Fig. 5.2. Descriptive levels of the neck regional lymphatics.

V and is also called the posterior triangle. Although not commonly used, level VI refers to the tracheoesophageal lymph nodes. It is bordered superiorly by the hyoid, inferiorly by the suprasternal notch and laterally by the strap muscles.

In addition to facilitating communication about nodal disease between physicians, this staging system also allows the comparison of lymph node draining patterns from various head and neck primaries. For instance, Shah (1990) demonstrated that squamous cell carcinomas of the oral cavity tend to metastasize to the levels I, II and III. Carcinomas of the oropharynx, hypopharynx and larynx metastasize to levels II, III, and IV. Level V nodal disease was only found in the presence of other nodal disease and is considered a poor prognostic sign. By knowing the primary tumor site and understanding the pathways of lymphatic tumor spread, one can predict which region of neck is at highest risk for metastatic disease. Conversely, the location of clinically positive neck lymph nodes often suggests the site of the primary cancer.

Scope of the Problem: Epidemiology

In North America, 6.6% of all cancers arise in the head and neck region. Approximately one third of head and neck patients die from their disease. Although squamous cell carcinoma constitutes 90% of aerodigestive cancers, many other epithelial and nonepithelial tumors can arise in this region (Table 5.1). Head and neck malignancies tend to occur in men (male to female ratio of 5:1) with the largest percentage of cases occurring in 60 to 69 year old age range.

Probably the most important etiologic factors in head and neck cancer are tobacco and alcohol. Greater than 80% of head and neck cancers (with the exception of

Table 5.1. Common malignant tumors of the head and neck region

Epithelial	Nonepithelial
Squamous cell carcinoma and variants	Soft tissue sarcoma
Transitional cell carcinoma	Rhabdomyosarcoma
Adenocarcinoma	Leiomyosarcoma
Minor salivary gland malignancies	Fibrosarcoma
Adenoid cystic carcinoma	Liposarcoma
Mucoepidermoid carcinoma	Angiosarcoma
Acinic cell carcinoma	Myxosarcoma
Carcinoma ex-pleomorphic adenoma	Hemangiopericytoma
Melanoma	Connective tissue sarcoma
Olfactory neuroblastoma	Chondrosarcoma
Undifferentiated carcinoma	Osteosarcoma
	Lymphoreticular tumors
	Lymphoma
	Plasmacytoma
	Giant cell tumor

5

salivary gland cancers, nasopharyngeal cancers, and some lip cancers) may be attributed to tobacco and alcohol consumption. There is about a 4-fold increase in head and neck cancers among smokers. Alcohol increases the relative risk of head and neck cancer by a similar rate. As expected, the more one smokes or drinks, the greater the risk. For those who chose to smoke and drink, the risk is multiplicative. It takes 16 years of tobacco free existence before a former smoker will have the same probability of developing a head and neck cancer as a nonsmoker.

There are also occupational risk factors for head and neck cancer. Painters, construction workers, metal or plastic workers, and machine operators have all been found to have higher rates of head and neck cancer in probable association with the heavy metals and solvents to which they are exposed.

Another risk factor is income. Low-income individuals represent a disproportionately high number of head and neck cancer cases and advanced stage cancer cases.

Worldwide, head and neck cancer is more predominant in developing countries where 12% of male and 7% of female cancers arise in the head and neck. In developed countries, 7% of male and 2% of female cancers arise in the head and neck. This difference reflects in part the greater user of tobacco (both chewing and smoking) and alcohol in developing countries. For example, the Indian custom of chewing "paan" has been implicated as the cause of that country's high incidence of buccal and pharyngeal carcinoma. This method of chewing tobacco involves rolling sliced betel-nut, tobacco dust, slaked lime, and other ingredients in a green leaf. Chewers have about an eight times higher risk of developing carcinoma of the buccal mucosa. In Europe, the increase in the incidence of head and neck cancer has paralleled the increase in alcohol and cigarette consumption. France demonstrates an opposite trend. In this country, national alcohol consumption has declined and so too has the incidence of oropharyngeal and laryngeal cancers.

The predominance of various types of the head and neck cancers differs around the world. In South India, the tongue and mouth are the major sites for head and neck cancer, while in Bombay and Western India, hypopharyngeal and laryngeal cancers predominate. In France and Eastern and Central Europe, almost half of head and neck cancers are oropharyngeal and hypopharyngeal. In contrast, almost half of Spain's head and neck cancers are laryngeal. Nasopharyngeal cancer predominates in Hong Kong and constitutes two thirds of male and three fourths of female head and neck cancer cases. Nasopharyngeal cancer is also highly prevalent in Southeast Asia. Environmental factors appear to influence the occurrence of this disease. When Chinese immigrants from areas with a high rate of nasopharyngeal carcinoma, such as the Kwantung province, relocate to the United States, the incidence of developing nasopharyngeal carcinoma in the same group of people decreases.

Diagnosis

Diagnosis begins with a thorough history and physical (Table 5.2). Primary tumors, depending on their location in the head the neck region, can elicit different complaints. Often the nature of the signs and symptoms will directly lead the physician to the location of the affected site. Oral cavity carcinomas often present with a painful and/or bleeding lesion. Speech quality may be affected especially in the case of a tongue lesion. Carcinomas involving areas such as the mandible, alveolar ridge, or retromolar trigone can elicit dental complaints. Sinus and nasal cavity tumors have a tendency to cause epistaxis and nasal obstruction. They can also invade upward into the orbit and cause periorbital edema and diplopia. Nasopharyngeal tumors, through local growth, often occlude the ipsilateral Eustachian tube blocking drainage from the middle ear and causing a unilateral otitis media with ear pain. Oropharyngeal and hypopharyngeal carcinomas may cause swallowing difficulties such as dysphagia, or odynophagia. Patients may also present with coughing resulting from tumor induced aspiration. As oropharyngeal and hypopharyngeal lesions grow and encroach upon the larynx, hoarseness, shortness of breath and increased work of breathing result. A patient with early stage laryngeal carcinoma can also present with these symptoms. Another sign of oropharyngeal and laryngeal carcinomas is otalgia, which is referred pain via the glossopharyngeal and vagus nerves.

Important diagnostic questions include whether the patient smokes or abuses alcohol, has diabetes or cardiopulmonary disease, or has had a previous cancer or recent weight loss.

After a complete history is taken, a comprehensive ear, nose and throat exam should be performed. The routine ear exam should particularly note the presence of a serous otitis media which is suggestive of a nasopharyngeal mass. The nasal exam using a nasal speculum or an endoscope should assess the septum, turbinates, sinus meati and nasopharynx for possible pathology. When examining the oral cavity it is important to palpate as well as visualize. Bimanual palpation of the floor of mouth and base of tongue is critical in determining submucosal and subepithelial masses. A mirror exam is utilized for visualization of the nasopharynx as well as the hypopharynx and larynx. Both vocal cords should be mobile and fully adduct to midline. Increasingly, fiberoptic endoscopy is used as an adjunct to or in lieu of the mirror exam. The neck should be visually examined and palpated for irregularities in normal structures such as the thyroid, larynx and lymph nodes. Neck masses are evalu-

Table 5.2. Important elements of the head and neck cancer work up

History	Risk factors, duration of symptoms, associated medical conditions, symptoms of metastatic disease
Physical examination	Airway status, extent of primary tumors, appearance of tumor (e.g., fungating, ulcerating, nodular), vocal cord fixation, presence of nodal neck disease
Diagnostic endoscopy and biopsy	Assess primary and determine skip areas, thoroughly palpate neck under general anesthesia, biopsy within and beyond the borders of the tumor
Laboratory tests	
Albumin, transferrin	Nutritional status indicators
Calcium	Hypercalcemia may signify metastatic disease
Alkaline phosphatase	Elevates with bony metastasis, liver disease, and other conditions
Electrolytes, creatinine, Blood count, liver enzymes	Routine preoperative screening
Imaging studies	
CT scan neck with contrast	Useful in delineating subglottic extension, cartilaginous involvement, extralaryngeal spread and nodal disease
CT scan of chest/chest x-ray	Helps detect metastatic disease as a separate primary in the chest
MRI	Most useful in detecting occult nodal disease
Bone scan	Detects bony metastatic disease
Barium swallow	Evaluates hypopharynx and esophagus for extension of tumor or separate primary
Other studies	
Pulmonary function test	Preoperative evaluation of pulmonary reserve

5

ated for location, size, shape, relationship to other structures, consistency and mobility. Finally, a thorough cranial nerve exam should be performed.

Two other critically important elements in the head and neck work-up are radiologic studies and pathologic evaluation of biopsies and fine needle aspirates.

CT and MRI are the primary imaging modalities used to determine tumor size, invasion, and relationship to important anatomical structures. CT and MRI also reveal the extent of regional nodal disease and at times identify clinically nonpalpable nodal metastases. In patients suspected of having head and neck disease, a chest x-ray should be obtained to rule out pulmonary metastases and lung secondary primaries—the latter which occurs in 5% of the head and neck population. When indicated, other useful head and neck radiologic studies include a panorex film for the mandible, a barium swallow for the esophagus and facial plain films for the sinuses and head and neck soft tissues.

Pathologic diagnosis of mucosal lesions may be facilitated by means of a punch biopsy performed under local anesthesia in the office. For a small oral lesion, a full excisional biopsy can be performed. When the tumor is beyond the field of direct vision, endoscopic visualization and biopsy should be undertaken in the operating room. For hypopharyngeal and laryngeal lesions, triple endoscopy consisting of bronchoscopy, laryngoscopy, and esophagoscopy is used. Nasal masses are visualized and biopsied using a rigid nasal endoscope. Biopsy allows histologic evaluation of the lesion and comparison to adjacent biopsy sites, and if a recurrence, to previously excised tissue.

If a patient presents with a palpable neck mass and no primary has been found after a thorough head and neck exam, diagnostic work up proceeds with a fine needle biopsy. This quick, simple office procedure allows for precise cytologic evaluation of a mass and has significantly decreased the amount of open neck mass biopsies performed for tissue diagnosis. The accuracy of this technique approaches 100% for cervical lymphadenopathy caused by squamous cell carcinoma. However, diagnosis by fine needle aspiration is difficult with some inflammatory diseases and lymphoma and open biopsy may be necessary.

Once the tumor type and the extent of local regional and systemic disease is determined, the patient's cancer can be staged and management determined.

Staging

The TNM system is utilized for staging of head and neck cancers and differs for each type of cancer (Table 5.3). The T stage is based on several factors including tumor size, sub-site location and local invasion. The nodal classification applies to all head and neck cancers. N0 denotes no evidence of nodal metastases. N1 denotes a single ipsilateral node measuring less than 3 cm in diameter. N2 is generally subdivided into a, b, and c. Having one ipsilateral node between 3 and 6 cm in greatest dimension is equivalent to N2a. N2b is the classification for having more than one ipsilateral node, none greater than 6 cm. N2c denotes the presence of bilateral or contralateral nodes, none greater than 6 cm in diameter. Lastly, N3 indicates nodal disease greater than 6 cm. The M stage denotes the presence (M1) or absence (M0) of distant metastases.

The stage of disease is designated as I-IV depending on their TNM classification. Generally, T1 and T2 lesions without nodal disease are stage I and II respectively. A T1 or T2 primary with nodal disease or T3N0 primary is stage III. Stage IV disease is assigned to T4 tumors, smaller cancers with extensive nodal disease and any malignancy with distant metastases.

Treatment

Treatment options for head and neck cancer include surgery, radiation, and chemotherapy. In general, smaller lesions (Stage I) can be treated with surgery or radiation with equal efficacy. More extensive disease (Stage II–IV) is treated with combined modalities. A notable exception is nasopharyngeal carcinomas in which all stages are initially treated with radiotherapy.

The advantage to surgery is that treatment occurs in one procedure. Should the cancer return, repeat resection is possible. However, whether a primary or a recurrence, the resectibility of the tumor, the functional morbidity of the resection and

Table 5.3. TNM staging, by site, of head and neck cancer

Lip, Oral Cavity, and Oropharynx
T1   Tumor is less than 2 cm in greatest diameter
T2   Tumor is between 2 cm and 4 cm
T3   Tumor is greater than 4 cm
T4   Tumor involves adjacent structures such as bone, or skin

Hypopharynx
T1   Tumor is less than 2 cm in greatest diameter, and only involves one
       hypopharyngeal subsite
T2   Tumor involves more than one subsite, or is 2 cm to 4 cm in greatest
       diameter
T3   Tumor is greater than 4 cm, or fixation of the hemilarynx
T4   Tumor invades adjacent structures such as thyroid or cricoid cartilage

Supraglottis
T1   Tumor limited to one subsite, and vocal cords have normal mobility
T2   Tumor involves more than one subsite, and cords remain unfixed
T3   Tumor limited to the larynx with vocal cord fixation, and/or involves the
       postcricoid area, or pre-epiglottic tissues
T4   Tumor invades through the thyroid cartilage, and/or extends into soft
       tissues of the neck, thyroid, and/or esophagus

Glottis
T1   Tumor limited to T1a—one vocal cord, or T1b—both vocal cords
T2   Tumor extends to the supraglottis, or subglottis, and/or impaired vocal cord
       mobility
T3   Tumor limited to the larynx with vocal cord fixation
T4   Tumor involves adjacent structures outside the larynx, or invades through
       the thyroid cartilage

Subglottis
T1   Tumor is limited to the subglottis
T2   Tumor extends to the vocal cords with normal or impaired mobility
T3   Tumor limited to the larynx with vocal cord fixation
T4   Tumor involves adjacent structures outside the larynx, or invades through
       cricoid or thyroid cartilage

Regional Lymph Nodes
N0   No regional lymph node disease
N1   One involved ipsilateral lymph node less than 3 cm in greatest diameter
N2   N2a–One involved ipsilateral lymph node between 3 cm and 6 cm in
       greatest diameter
       N2b–Multiple involved ipsilateral lymph nodes, none greater than 6 cm in
       greatest diameter
       N2c–Bilateral, or contralateral lymph nodes, none greater than 6 cm in
       greatest diameter
N3   An involved lymph node greater than 6 cm in greatest diameter

Distant Metastasis
M0   No distant metastasis
M1   Distant metastasis

Adapted from Flemming ID, et al. AJCC cancer staging handbook. Lippincott-
Raven; New York. 1998.

5

the operative and perioperative risks to the patient must be considered. The prospects for functional reconstruction using regional and microvascular free flaps must also be weighed. These techniques have significantly increased post surgical rehabilitation and function of such structures as the tongue and mandible and in doing so have decreased the threshold for surgical resection.

A major advantage of radiation therapy is that it maximizes the preservation of tissue and tissue function. In addition, patients do not require general anesthetic. However, radiation therapy requires multiple treatments which occur over several weeks. The patient must be committed to completing all treatments to insure efficacy. Because radiation effects the rapidly dividing cells of both the primary cancer and the mucous membranes, patients often complain of dysphagia, odynophagia, or xerostomia and/or a decreased sense of taste. Other serious complications include soft tissue ulcer, dental caries, spinal cord injury, orocutaneous fistulas, osteoradionecrosis and chondronecrosis.

A third treatment modality, chemotherapy, has been traditionally utilized for the palliation of recurrent and metastatic disease. The standard single agent of choice is methotrexate. However, several other drugs have been used as single agents or as part of combination therapies, including 5-fluorouracil, cisplatin, and bleomycin. More recently, chemotherapy has been combined with radiation therapy for use in organ sparing protocols. The Veterans Affairs study in 1991 (Department of Veteran Affairs, Laryngeal Cancer Study Group, 1991) used two cycles of combined cisplatin and fluorouracil on patients with advanced laryngeal squamous cell carcinoma (stage III and IV). Responders were given one more cycle of chemotherapy and then definitive radiation therapy. Nonresponders underwent surgical salvage. Of those assigned to the chemotherapy arm the rate of laryngeal preservation after two years was 66%. Two year survival rates between a surgery and chemoradiation were not significantly different. Chemoradiation has, therefore, become a viable option for organ sparing in certain patients.

In treating head and neck cancer, the regional lymphatics must also be addressed. Treatments include a neck lymph node dissection and radiation therapy. Lymphadenectomy allows for excision of any current nodal metastases while also removing the lymphatic system utilized by head and neck cancers for future spread of disease.

The classical form of neck dissection is the radical neck dissection. This procedure removes the lymphatic tissue from levels I-V, the accessory nerve, the sternocleidomastoid, and internal jugular vein. Loss of the spinal accessory nerve results in denervation of the trapezius muscle which causes shoulder drop, shoulder pain, and difficulty in abducting the arm more than 90°. Transection of the internal jugular vein predisposes to increased facial edema, although the contralateral vein eventually compensates. Sacrifice of both internal jugular veins greatly increases the risk of facial and cerebral swelling and can cause intracranial complications.

In the mid 1960s, the modified radical neck dissection became the popular form of lymphadenectomy in cases where lymph node disease had not spread to nonlymphatic structures. Proven to be oncologically sound, this dissection spares the spinal accessory nerve, the internal jugular vein, and/ or the sternocleidomastoid muscle.

The next refinement in head and neck lymph node resection was the selective neck dissection. Unlike the modified radical neck dissection in which all five lymphatic levels are removed, the selective neck dissection focuses on specific levels. For instance, since smaller N0 oral carcinomas tend to spread to levels I, II, and III, the surgeon can choose to excise only these three levels. This type of neck dissection is termed a supraomohyoid neck dissection. An extended supraomohyoid neck dissection also includes level IV. A lateral neck dissection, which includes levels II-IV, might be more useful for a oropharyngeal, hypopharyngeal, or laryngeal carcinoma. Excision of levels II-V is termed a posterolateral neck dissection, and may be utilized for posterior cutaneous lesions. An anterior compartment neck dissection excises level VI and is used for thyroid, hypopharyngeal, tracheal, and laryngeal carcinomas.

Any of these neck dissection types may be extended to include tumor involved structures not normally excised in a routine neck dissection. For example, neurovascular structures such as the marginal mandibular nerve, the hypoglossal nerve, and the carotid artery may need to be sacrificed to insure a complete oncologic resection.

While neck dissection is a mainstay in treatment of the N+ neck, its role in the N0 neck is less clear. It is generally accepted that the N0 neck should be treated in cases where the risk of occult metastases is greater than 20%. The risk of regional micrometastases is usually determined by the site of the primary lesion and its T stage. When management of the regional lymphatics is indicated, either surgery (a modified or selective neck dissection) or radiation therapy are equally effective treatments for clinically nonpalpable (N0) nodal disease. Although N1 nodal disease can be treated with radiation therapy, most clinically apparent neck disease is treated with neck dissection. If surgery is chosen to treat the primary tumor and the neck, postoperative radiation therapy is not required if the surgical margins are absolutely clear around the primary, and if there are no more than two positive lymph nodes both measuring less than 3 cm and both without extracapsular spread. More extensive disease requires combined surgery and radiation therapy.

Nasal Cavity and Paranasal Sinuses

Sinonasal cancers are rare and represent only 3% head and neck malignancies. Squamous cell carcinoma is the most common sinonasal tumor (90%) followed by adenocarcinoma (6%). The maxillary sinus is the most common site of origin. Often a tumor grows asymptomatically in a sinus only to manifest itself late in the disease process after significant growth and invasion has occurred. On average, symptoms such as nasal obstruction and rhinorhea are present for six months before medical attention is sought. Other presenting symptoms include localized pain, epistaxis, and swelling. In advanced stages, visual changes may occur indicating orbital invasion and disease extension.

Surgical resection is the mainstay of treatment. Radiation therapy is used for lymphoreticular and unresectable tumors or for poor surgical candidates. It is also used adjunctively for positive margins and perineural and perivascular invasion. Because the lymphatic drainage of the nasal cavity and paranasal sinuses is sparse, less than 10% of patients have node metastases at initial presentation. Distant metastases on initial presentation occurs in less than 7% of patients. The 5 year survival rates for sinonasal tumors varying according to site and are as follows: nasal cavity–60%, maxillary cavity–50%, ethmoid sinus–50%, sphenoid sinus–15% infratemporal fossa–50% and pterygopalatine fossa–50%.

Nasopharyngeal

Nasopharyngeal carcinoma, like sinonasal tumors, grows asymptomatically in the early stages. Approximately 75% of patients with nasopharyngeal carcinoma present with a neck mass. A serous otitis media causing aural fullness and discomfort may also be present initially and results from tumor obstruction of the Eustachian tube. Hence the maxim, "a unilateral serous otitis media is cancer until proven otherwise." With advanced disease, patients may develop a 6th nerve palsy and complain of visual symptoms.

The highest incidence of nasopharyngeal carcinoma occurs in the southern provinces of China. Environmental factors such as a diet high in nitrosamines (found in salted dry fish) and exposure to the Epstein-Barr virus are felt to play an etiologic role. Anti-EBV antibody levels are used as prognostic indicator in the follow-up of these patients.

The initial treatment for nasopharyngeal carcinoma is radiation therapy. Due to the location of the primary and relative inaccessibility of the first draining retroparotidean lymph nodes, this carcinoma is not amenable to surgical excision. The nasopharyngeal area has a very rich lymphatic supply and 70-90% of patients will develop metastatic nodes. Therefore, radiation therapy is used to treat both the local and regional disease. Only after radiation failure is surgery considered as a treatment option.

The World Health Organization (WHO) has divided nasopharyngeal carcinomas into three distinct histologic types: squamous cell carcinomas (type I), nonkeratinizing carcinomas (type II), and lymphoepithelioma (type III). Type III, which is the most sensitive to radiation therapy has a 65% 5 year survival rate. The 5 year survival rate for types I and II are 15% and 45%, respectively.

Oral Cavity

Squamous cell carcinoma accounts for 95% of oral cavity malignancies. Presenting symptoms include bleeding, otalgia, odynophagia and dysphagia. In later stages, the patient may develop a decreased ability to open the mouth (trismus). Of critical concern is the accurate determination of mandibular involvement. When tumor involvement the of mandible is suggested by physical examination, both high-resolution CT scan and Gd-DPTA enhanced MRI are highly valuable in confirming the diagnosis. Depending on the degree of invasion, part or all of the involved mandibular segment will require resection.

T1 and T2 lesions of the oral cavity may generally be treated with either surgery or radiation therapy. The choice of modality must take into account tumor accessibility, functional morbidity, and patient tolerance of the treatment modality selected. One must also consider the decreased effectiveness of radiation when bony involvement is present, as well as the risk of osteoradionecrosis. Larger lesions, such as T3 and T4 carcinomas, usually require surgery and radiation. The decision on whether or not to treat the N0 neck depends on the to risk of micrometastases. When the risk is greater than 20% treatment of the neck by surgery or radiation therapy is indicated. When nodal disease is present clinically (N+), a neck dissection is indicated.

Survival rates for oral cavity carcinomas are site specific. The 5 year survival rates for buccal mucosa carcinoma are approximately 75%, 65%, 30%, and 20% for stages I-IV, respectively. For the floor of the mouth, the corresponding stage specific rates are approximately 90%, 80%, 65%, and 30%. The 5 year survival rate for alveolar ridge or retromolar trigone carcinomas is about 65% for all stages. Oral tongue carcinomas, on the other hand, have 5 year survival rates of approximately 75% for stage I and II, and 40% for stage III, and IV.

Any discussion regarding oral cavity carcinoma should include leukoplakia which is a common finding on clinical exam. Leukoplakia is a white lesion of the oral cavity which can not be rubbed off and is not another clinical entity, such as lichen planus, candidiasis, white sponge nevus, or lupus erythematosus. In a study of leukoplastic lesions, Waldron and Shafer (1975) found 80.1% had no evidence of epithelial dysplasia. However, 12.2% showed mild to moderate epithelial dysplasia, 4.5% showed severe dysplasia or carcinoma in situ, and 3.1% showed infiltrating squamous cell carcinoma. Thus, the leukoplastic lesion runs a small but significant risk of containing or developing carcinoma. For this reason, these lesions should be biopsied and followed closely.

Oropharynx

Although squamous cell carcinoma is the most common histologic type of oropharyngeal malignancy, approximately 16% of oropharyngeal tumors are lymphomas. Common presenting symptoms include: sore throat, referred otalgia, dysphagia, foreign body sensation, trismus, hemoptysis and voice change. The rate of lymph node metastases various according to site and is approximately: 70% for the tongue base, 60% for the palatine tonsil and 40% for the soft palate.

T1 or T2 lesions can generally be treated by surgery or radiation. However, because even small tumor excisions in this region (i.e., base of the tongue or soft palate) severely impact on function, radiation therapy is often the preferred method of treatment. Larger lesions often require combined surgery and radiation therapy. Because the oropharynx has a rich lymphatic network and high rate of micrometastases, treatment of the N0 neck—either by neck dissection or radiation therapy—is recommended.

Tonsillar carcinoma's survival rate ranges from the 90% for Stage I to 30% for stage IV. The survival rate for base of tongue ranges from 65% for Stage I to 10% for Stage IV.

Hypopharynx

The three major areas of the hypopharynx where cancer presents are the pyriform sinus (60-70%), the lower posterior pharyngeal wall (25-30%) and the postcricoid region (5%). Greater than 95% of the malignancies in this region are squamous cell carcinoma. Most of the remaining 5% are adenocarcinomas arising from glandular structures in the mucosa or from islands of ectopic gastric mucosa. The most common complaint on presentation is throat pain followed by dysphagia and otalgia. With further growth, hoarseness may occur.

As with other head and neck cancers, radiation or surgery can be used to treat smaller lesions of the hypopharynx. Surgery consists of a partial laryngopharyngectomy which preserves the glottis and creates no more functional morbidity than radiation therapy. In larger T3 and T4 lesions, treatment consists of surgery and postoperative radiation. The incidence of occult metastases in hypopharyngeal carcinoma is between 50-80% and necessitates treatment of the regional lymphatics.

The overall survival rate for hypopharyngeal carcinoma is approximately 40%, but site, stage, and the presence or absence of positive lymph nodes has a significant influence on the eventual outcome.

Larynx

Laryngeal carcinomas are categorized as either glottic, supraglottic, or subglottic. Glottic cancer accounts for 75% of laryngeal malignancies. Supraglottic tumors are found 22% of the time while subglottic cancer occurs in less than 3% of laryngeal tumors.

Carcinomas of the glottis are generally favorable for two reasons. First, because they involve the vocal chords, patients tend to present early in the disease process with complaints of hoarseness and foreign body sensation. Second, the glottis has minimal lymphatic drainage and consequently many early glottic lesions present without regional neck disease. T1 and T2 cancers can be treated by radiation therapy, chordectomy, vertical partial laryngectomy or hemilaryngectomy. A newer method of endoscopic laser excision has also shown good results and is gaining acceptance. Stages III and IV laryngeal carcinomas have been traditionally treated with a combination of laryngectomy and radiation. More recently, organ sparing protocols consisting of chemotherapy and radiation therapy has been advocated for advanced laryngeal cancer. In a landmark VA study (1991), a regimen of cisplatin and 5-flurouracil followed by radiation therapy allowed 64% of patients to retain laryngeal function and did not compromise overall survival. Due to the low incidence of occult metastases the N0 neck is not generally treated. The 5 year survival for glottic carcinomas stages I to IV are 85%, 66%, and 50% respectively.

Unfortunately, the signs and symptoms of supraglottic carcinomas present later than glottic carcinomas. Patients may initially complain of throat discomfort and/or otalgia. Later, with advanced disease, dysphagia, and voice change may become evident. Surgical treatment is preferred over radiation therapy with surgical salvage because, although the five-year survival is the same, the chance of eventual voice preservation is greater with surgery. The traditional surgical procedure performed is a supraglottic laryngectomy which removes false vocal cords, aryepiglottic folds, epiglottis, pre-epiglottic space, superior half of the thyroid cartilage, and the hyoid bone. Breathing, phonation and swallowing are preserved. The resection does increase the risk of aspiration, and therefore adequate pulmonary function must be accessed preoperatively. If there is a vocal cord fixation, involvement of a significant part of the base of tongue, or involvement of the piriform sinus apex, supraglottic laryngectomy is not an appropriate oncologic procedure. For larger tumors (T3 and T4), a near total or total laryngectomy along with radiation therapy are recommended. For all T stages, one must treat the regional lymph nodes due to the high rate of occult metastases. The 5 year survival rates for supraglottic carcinoma are 90%, 81%, 64%, and 50% for stages I through IV respectively.

Subglottic carcinomas are relatively rare, compromising only 1-2% of laryngeal carcinomas. In the early stages, subglottic carcinoma does not cause overt symptoms. Treatment is similar to other head and neck cancers in that early lesions may be treated with a single modality, and advanced lesions typically required combined regimens. The survival rate for subglottic carcinomas treated with radiation and surgical salvage is approximately 36%.

Parotid Tumors

The major salivary glands are the parotid, submandibular, and sublingual glands. There are also a multitude of minor salivary glands spread throughout the upper respiratory tract. Approximately 75% of salivary tumors occur in the parotid, and of these nearly 80% are benign. Submandibular gland tumors (20% of salivary tumors) are benign 50% of the time. Minor salivary gland tumors are malignant 75% of the time. FNA is useful in the diagnosis of a salivary mass and for preoperative planning. An experienced cytologist can distinguish benign from malignant salivary gland disease 95% of the time. Because the treatment for a mass in the tail of a parotid is surgery, CT and MRI scans, though helpful in delineating a tumors location and extent of growth, are not mandatory. When examining a patient, facial nerve weakness, fixation to the skin, trismus and lymph node metastasis are considered indicative of malignancy.

Pleomorphic adenoma is the most common salivary gland tumor, accounting for 60-70% of all benign salivary gland tumors. When presenting in the parotid gland, 90% will be in the superficial lobe. Macroscopically, they are encapsulated. However, upon closer inspection they exhibit transcapsular pseudopodial growth. Consequently, a wide excisional margin is recommended to prevent recurrence. Postoperative radiotherapy may be utilized to decrease the incidence of recurrence. However, the morbidity of this treatment does not offset the benefits and is not recommended. The major risk of these tumors is that of malignant degeneration. Within the first five years of diagnosis there is a 1.5% risk of degeneration which increases to 10% if the tumor is observed for greater than 15 years.

The second most commonly encountered benign tumor is Warthin's tumor, otherwise known as papillary cystadenoma lymphomatosis. It usually presents in men who are in their sixties. There is a 10% chance of bilaterality. Treatment is excision, and recurrence is rare.

In children, hemangiomas are the most commonly encountered salivary gland tumors and the tumor's unique soft consistency facilitates the diagnosis. As with hemangiomas which occur elsewhere in the body, these lesions have a tendency to involute over time and do not necessitate intervention.

The most common malignant salivary gland tumor is mucoepidermoid carcinoma. These tumors may be classified as high, intermediate or low grade depending on the ratio of mucus cells to epidermoid cells within the tumor. The more epidermoid cells, the higher the grade, and the more aggressive the tumor. The treatment modality of choice is wide local excision sparing the facial nerve whenever possible. For all malignancies greater than 4 cm, postsurgical radiation therapy improves survival and locoregional control.

Adenoid cystic carcinoma is the second most common malignant tumor of the salivary glands. Its defining characteristic is its tendency for perineural invasion. This carcinoma tends to present more rapidly with facial nerve symptoms and commonly recurs. Wide local excision of the primary along with any involved nerves and postoperative radiation therapy are mandatory for locoregional control.

Table 5.4. The incidence of parotid and submandibular gland tumor types

Tumor Type	Parotid Incidence (%)	Submandibular Incidence (%)
Benign		
Pleomorphic adenoma	59	36
Warthin's tumor	7	< 1
Malignant		
Mucoepidermoid	8	12
Malignant Mixed	4	10
Acinous Cell	4	< 1
Adenoid Cystic	3	25
Squamous Cell	2	7
Adenocarcinoma	< 1	7

Adapted from Bailey et al. Head and Neck Surgery—Otolaryngology. Philadelphia: J.B Lippincott 1998.

When tumor involvement necessitates facial nerve excision, as with adenoid cystic carcinoma, nerve grafting is an option. Typically the sural nerve or the greater auricular nerve are harvested for this purpose.

The incidence of various tumor types of the parotid and submandibular glands are listed in Table 5.4. In the parotid, acinous cell and low-grade mucoepidermoid carcinoma have a five-year survival of approximately 90%. All other malignant parotid tumors have a five-year survival in the range of 42-70%.

Suggested Reading

1.   Cummings CW. Otolaryngology Head & Neck Surgery. Mosby; New York. 1998.
     One of the most popular reference books for otolaryngologists. It is lengthy but quite comprehensive.
2.   Department of Veterans Affairs, Laryngeal Cancer Study Group. Induction chemotherapy plus radiation compared with surgery plus radiation in patients advanced laryngeal cancer. N Engl J Med 1991; 324: 1685.
     Landmark study which assesses the effectiveness of a treatment regime which allows for laryngeal preservation.
3.   Jussawalla DJ, Deshpande VA. Evaluation of cancer risk in tobacco chewers and smokers. Cancer 1971; 28: 244-52.
     An excellent article on the risks of cancer in patients who use tobacco products.
4.   Lee KJ. Essential Otolaryngology. Appleton & Lange; Stamford, CT. 1999.
     A wonderfully concise book of otolaryngology facts and figures written in outline form.
5.   Sankaranarayanan R, et al. Head and neck cancer. Anticancer Res 1998; 18: 4779-86.
     An excellent review of epidemiologic trends in head and neck surgery.
6.   Shah JP. Patterns of cervical lymph node metastasis from squamous carcinomas of the upper aerodigestive tract. Am J Surg 1990; 160: 405-9.
     A classic paper on the lymphatic spread of head and neck cancer.
7.   Waldron CA, Shafer WG. Leukoplakia revisited. Cancer 1975; 36: 1386-92.
     An informative paper on a common clinical finding on routine physical exam-leukoplakia.
8.   Wang CC. Radiation therapy for head and neck neoplasms. Wiley-Liss; New York. 1997.
     An excellent text book written by one of the pioneering leaders in this field.

CHAPTER 6

Esophageal Cancer

Philip D. Schneider

Scope of the Problem

In this Chapter, we will discuss squamous cell carcinoma and adenocarcinoma of the esophagus from the level of the cricopharyngeus muscle to the esophagogastric junction. This includes adenocarcinomas arising at the esophagogastric junction. Pharyngeal cancers and cancers of the gastric cardia are excluded. Esophageal cancer before the 1980s referred not only to an anatomic location but also usually only to squamous cell carcinoma. In fact, adenocarcinomas of the esophagogastric junction were considered gastric cancers and were excluded from analysis or were treated as esophageal cancer without specific regard to the histology. Thus, comparison of modern data with historical data is difficult, especially when attempting to evaluate the outcome of these histologic types of cancer.

The increased prevalence of esophageal cancer in certain areas of the world such as Gonbad in Iran; Linxian in Hunan, China; and the coastal southeastern United States has led to a high index of suspicion in these areas about swallowing complaints and to aggressive screening programs. In these areas, squamous cell cancers predominate. The United States has an incidence of squamous esophageal cancer of 3-8 per 100,000 population, and, not surprisingly, the mortality rates are essentially the same for this lethal disease. Yet, were it not for an observation made by Blot et al (1991), esophageal cancer as a public health concern would prompt a low level of concern in most Western countries. Blot et al noted a 10% increase in the number of lower esophageal adenocarcinomas during the previous decade for men and women of all races. The National Cancer Institute-sponsored SEER (Surveillance, Epidemiology, and End Results program) combined tumor registry data to form the basis for estimates of cancer risk in the general population. For the decade of the 1980s, SEER reported an incidence of adenocarcinoma of the esophagus of 0.6-5.1 per 100,000 population—the 10% increase, while noting that the number of squamous cell carcinomas of the esophagus remained stable. Lower esophageal adenocarcinoma, thereby, was recognized as the most rapidly increasing visceral malignancy in the United States.

Risk Factors for Development of the Esophageal Cancer

Two general clinical presentations for esophageal cancer appear to reinforce the epidemiological and biochemical evidence that the etiologies for squamous cell carcinoma and adenocarcinoma are different. The presentations are noted in Table 6.1.

There is little doubt that potent environmental factors increase the risk of squamous cell carcinoma. Nitroso-compounds from food and tobacco products, as well

---

Surgical Oncology, edited by David N. Krag. ©2000 Landes Bioscience.

Table 6.1. Comparative signs and symptoms for patients with squamous cell and adenocarcinoma of the esophagus

Squamous cell carcinoma	Adenocarcinoma
Smoking, snuff-chewing	Occasional tobacco use
High consumption of alcoholic spirits	Variable alcohol consumption
Marked weight loss at presentation	Modest weight loss, if any
Thin body habitus	Robust body habitus
Marked dysphagia, odynophagia	Mild dysphagia, rare odynophagia
No symptoms of esophageal reflux	Long-standing history of esophageal reflux and, often, recent improvement in reflux symptoms
Frequent cardiovascular comorbidity	Infrequent cardiovascular comorbidity

as ethanolic spirits play a role. Less common associations with squamous cell carcinoma include iron deficiency (Plummer-Vinson syndrome) and scleroderma.

Effective screening programs have been developed in regions where definite environmental contribution to the etiology has been noted epidemiologically. Individuals with any odynophagia or dysphagia can be slated for endoscopy for diagnosis. Where endemic, cytology from gauze covered balloons drawn back from the stomach can lead to secondary screening and diminish primary screening costs. In Western populations, short of identifying individuals at greater risk and subjecting them to screening endoscopies, careful attention to symptoms is essential. This must always be coupled with the concern that any upper gastrointestinal symptoms such as dysphagia or odynophagia may be the forerunner of one of these uncommon but nonetheless treatable and curable malignancies.

Adenocarcinoma, due to its rapidly increasing incidence, is the subject of intense investigation. Its apparent association with a history of reflux esophagitis and, especially, with Barrett's changes of the esophageal mucosa has directed studies to the biochemistry and molecular biology of mucosal injury with reflux. This vague association of adenocarcinoma with tobacco use is in marked contrast to smoking and squamous cell carcinoma. More intriguing is the speculation that widespread use of H-2 receptor blockers and proton pump inhibitors used in the treatment of esophageal reflux disease is a contributing factor. These agents, which prevent reflux of acid into the esophagus, do not correct the physiology of reflux. It is speculated that this leads to a bilious injury of the mucosa that may be carcinogenic. Without more information, one still must presume that any patient with reflux esophagitis, and, particularly, refluxing patients with the changes of mucosal metaplasia or Barrett's esophagus must be followed carefully and subjected to periodical routine endoscopy.

Barrett's Esophagus

The British surgeon, Norman Barrett, first described the phenomenon of columnar epithelial metaplasia of the normally squamous esophageal mucosa in 1950. He first surmised that congenital rests of columnar epithelium accounted for these findings. It is now universally acknowledged that the process is related to injury and healing in the lower esophagus and, thus, it is an adaptation rather than a congenital anatomic aberration. Although Barrett's changes are frequent in patients with reflux symptoms, 40% of patients with Barrett's changes have minimal or no reflux symptoms.

The association of Barrett's mucosa with adenocarcinoma was soon made with estimated rates of 40%. In recent years, more careful studies place cancer risk in Barrett's in the range of 2-4%. This risk is higher with the type of Barrett's changes termed special columnar epithelium, as opposed to the cardiac and fundic histologic variety of Barrett's, although the former's substantial risk appears to be related to the higher risk of dysplasia with specialized columnar Barrett changes. Of great concern is the finding that correction of reflux by medical or surgical measures is frequently unsuccessful in reversing Barrett's changes and reducing the cancer risk. Blot et al's observations about increasing adenocarcinoma of the lower esophagus and observations that Barrett's mucosal changes are found in as many as 65% of patients with adenocarcinoma have, appropriately, focused a great deal of investigation into the causes and modification of these changes.

Reversal of advanced Barrett's changes seldom occurs, whether the condition is managed medically using inhibitors of gastric acid production and prokinetics to stimulate gastric emptying or whether it is managed surgically with antireflux surgery. Some surgeons have expressed an interest in earlier antireflux surgery in patients followed for reflux who display early Barrett's changes. The availability of laparoscopic antireflux procedures, which have reduced morbidity, is giving impetus to earlier intervention in anticipation of higher rates of reversal of the changes of Barrett's esophagus. For the present, however, the major focus is placed on careful follow-up of patients found to have Barrett's metaplasia. Subsequent endoscopic biopsies are performed to determine if any dysplasia has developed. Severe dysplasia merits consideration of prophylactic intervention, as discussed later in the section "Treatment Options."

Methods of Diagnosis

Without question, the diagnostic test of choice for esophageal cancer is upper gastrointestinal endoscopy. Occasionally, a barium cine esophagram and upper GI series are called for. These latter two tests serve to:

1. rule out or further define functional disorders of the esophagus that may be in the differential diagnosis of dysphagia, and
2. delineate additional pathology before endoscopy. For instance, the reflex spasm of the cricopharyngeus in response to severe reflux may lead to a Zenker's pulsion diverticulum that presents a hazard to the patient having an endoscopy and is a potential complicating factor in the correction of esophageal problems.

The endoscopist has several tasks of importance when doing the initial diagnostic screen. First, do no harm. The management of esophageal cancer is immeasurably compromised by perforation of the esophagus. Second, diagnose the array of problems and the extent of the problems within the esophagus. This requires careful biopsies with precise cataloguing of location. Frequently, endoscopists who are not surgeons fail to remind themselves of the technical imperatives faced by the surgeon who must make decisions based upon the knowledge of distance from the incisors and relationship to the esophagogastric junction as well as details of the pathology. If possible, the study should completely evaluate the stomach and duodenum to rule out additional pathology and provide information about potential technical problems if surgery is planned. Additional endoscopies may be required for complete staging, as discussed later in the Chapter. The endoscopist must not presume to place feeding tube enterostomies until staging is complete. Esophageal dilation, however, is a valuable intervention that can greatly facilitate further evaluation while providing immediate palliative benefit to the patient with esophageal obstruction.

Staging

In most programs, the plan for further evaluation is influenced by the local and institutional treatment philosophies and treatment protocols. Esophageal cancer is clearly a disease whose treatment is in transition. In some institutions, the biology of metastasis is interpreted to absolutely prohibit surgery for cervical and midesophageal cancers. In these instances, once the diagnosis of carcinoma is made, further evaluation is designed to stage the patient and discover the risks of planned radiation therapy or combined radiation therapy and chemotherapy and to prepare patients for the rigors of that treatment program. For lower esophageal and esophagogastric junction cancers, not only is pretreatment staging required, but the evaluation also will change if surgery is not planned. For example, determining the presence of stage III disease, at some institutions, excludes the patient from eligibility for surgery, whether a favorable result occurs with radiation and chemotherapy or not.

Suffice it to say that preoperative staging is imperative. Yet, there is also controversy about the best means to stage. Table 6.2 lists staging options. Some of the procedures may be linked to placement of enteral feeding tubes that are useful to assist with nutrition during possible radiation and/or chemotherapy and surgery.

The extent of the diagnostic laparoscopy is also a matter of debate: determining whether the absence of hepatic or peritoneal metastatic disease is sufficient for staging of the abdomen before instituting therapy or whether more rigorous evaluation of celiac nodes and the lesser sac should be undertaken. Laparoscopy provides an opportunity to place a feeding tube into the small bowel. If the patient is not going to undergo surgery, a gastrostomy may be placed for ease of use. In patients previously operated upon, an open feeding enterostomy may be desirable. Table 6.3 provides a summary of the current staging schema for esophageal cancer.

Further Preoperative Evaluation

Despite numerous editorials and consensus discussions at national meetings indicating that surgery alone remains the standard of care for esophageal malignancies, there is no major oncology program that is not employing a tripartite, multimodal approach to the treatment of stage II and occasional stage III patients. Stage I patients receive surgery unless, as mentioned, cervical or midesophageal cancers are treated with radiation with or without chemotherapy by institutional fiat. Thus, for patients where surgery is part of the plan, a thorough evaluation of comorbid cardiovascular and pulmonary disease and overall health must be undertaken. The combination of abdominal and, occasionally, thoracic surgery is a stout physiologic test for any patient. For patients with squamous cell carcinoma, heavy smoking-related diseases frequently compromise recovery. Complications of all of the surgical approaches challenge the most robust of adenocarcinoma patients as well.

Table 6.2. Staging interventions for esophageal cancer

Upper gastrointestinal endoscopy and biopsy
Esophageal, endoscopic ultrasound with/without biopsy of suspicious para-esophageal/celiac lymph nodes by endoscopic needle aspiration
CT scans of neck/chest/abdomen
Diagnostic laparoscopy
Bronchoscopy if the CT shows tracheo/bronchial flattening or invasion
CEA, CA 19-9

Table 6.3. The staging of esophageal cancer

T	T0:	No tumor evident in the esophagus
	Tis:	Carcinoma in situ
	T1:	Tumor invades the lamina propria or submucosa
	T2:	Tumor invades the muscularis mucosa
	T3:	Tumor invades the adventitia
	T4:	Tumor extends into adjacent structures
N	N0:	No regional node metastasis
	N1:	Regional node metastases
M	M0:	No known distant metastasis
	M1:	Distant metastasis that also includes nonregional lymph node metastasis as follows:

Lower esophagus	M1a:	celiac lymph node metastasis
	M1b:	other distant metastasis
Mid esophagus	M1a:	not used
	M1b:	nonregional lymph node metastasis and/or other distant metastasis
Upper esophagus	M1a:	cervical lymph node metastasis
	M1b:	other distant metastasis

Stage 1	TIS		
	T1N0M0		
Stage II	IIA	T2N0M0	
	IIB	T1N1M0 T3N0M0 T2N1M0	
Stage III	T3N1M0		
	T4, any N, M0		
Stage IV	Any T, any N, M1		
	IVA	Any T, any N, M1a	
	IVB	Any T, any N, M1	

6

Treatment Options

Patients with Barrett's esophagus demonstrating severe dysplasia should undergo total esophagectomy. Transhiatal esophagectomy is generally accepted as the ideal treatment. Several investigators are studying photodynamic mucosal ablation with laser light and ultrasonic destruction as alternatives to surgery in patients with Barrett's esophagus. However, when carcinoma in situ is discovered, virtually all experts recommend esophagectomy.

For early stage patients—usually stage I— surgery will be the sole or principal intervention for the management of this disease. Again, there is no evidence that any combination of radiation and chemotherapy is as good as surgery, let alone better than surgery, at providing the best chance for cure.

The best operation for esophageal cancer is still contested. Individual surgeons with unique expertise report excellent results using their specific techniques and usually claim them to be superior to alternative techniques. Although the virtues of transhiatal versus wide-field esophagectomy often become the focus of these discussions, a steady evolution of great importance has occurred. There can be no doubt that the ideal operation for esophageal cancer, based on anatomy, pathology, and outcomes analysis, includes a total esophagectomy. There is simply too much information to argue that any operation with an anastomosis of esophagus to stomach, colon, or jejunum in the chest is satisfactory in terms of operative morbidity or in terms of oncology. Thus, for esophageal and esophageal junction cancers, three general operations are employed:

1. Three-incision esophagogastrectomy with a cervical esophagogastrostomy (or esophagocolostomy) (Fig. 6.1.A). This is an extension of the standard Lewis approach, which uses a laparotomy and right thoracotomy with thoracic esophagogastrostomy. This operation is inadequate for dealing with proximal extension of lower esophageal cancers.

2. Transhiatal esophagogastrectomy with cervical esophagogastrostomy (Fig. 6.1.B).

3. Three-incision, wide-field esophagogastrectomy with colon interposition and cervical esophagocolostomy (Fig. 6.1.A).

Tables 6.4 and 6.5 summarize the major points in preparation and performance of the above procedures.

In general, each operation employs a pyloromyotomy or pyloroplasty to assist with gastric remnant emptying because the vagus nerve is disrupted by the dissection and emptying problems due to pylorospasm occur in as many as 20% of patients. An enteral feeding tube is also advisable since this may provide important enteral access during a complicated postoperative course. In the section of this Chapter, "Special Considerations," a new operation made possible by advances in video-assisted laparo/thoracoscopic surgery permits an excellent mediastinal and celiac node dissection and has a lower operative morbidity than any of the above three operations. As with transhiatal esophagectomy, a cervical esophagogastrostomy completes the reconstruction.

For stage II and, occasionally, stage III disease, preoperative chemo/radiation has produced an astonishing rate of complete and partial responses to treatment. Although only a single study has shown an overall increase in survival with the combined approach versus surgery alone, the prospect of new agents and reduced morbidity is very encouraging. A complete pathological response, which occurs in 2.5-40% of patients receiving neoadjuvant therapy, has become a biological/therapeutic indicator of improved chances for better survival after surgery, compared to patients undergoing surgery after a poor or no response to neoadjuvant therapy. Statistically, however, chemotherapy and radiation used with currently understood selection criteria does not improve overall survival. Yet, as noted, virtually no large university or community oncology program fails to use the treatment combination for patients deemed suitable.

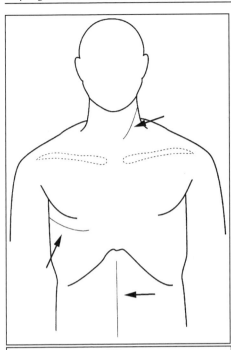

Fig. 6.1A. The commonly used incisions for the Lewis esophagectomy or wide-field esophagectomy. The classic Lewis operation places the esophagogastrostomy in the chest and no neck incision is performed.

6

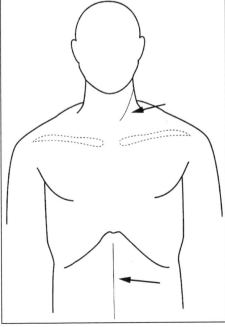

Fig. 6.1B. The incisions for a transhiatal esophagectomy with a cervical esophagogastrostomy.

Table 6.4. Preoperative preparation

Day before surgery
    Standard bowel preparation including cathartics and oral antibiotics
    Overnight hydration, volume monitoring as required
Day of surgery
    Prophylactic, intravenous antibiotics
    Arterial pressure monitoring catheter
    Large bore intravenous catheters
    Swan-Ganz catheter
    Double lumen endotracheal tube if thoracic approach will be used
    Positioning as appropriate
    DVT prophylaxis

Table 6.5. Technical considerations during the performance of esophagectomy

General
    Thorough abdominal exploration
    Biopsies of suspicious peritoneal and hepatic lesions
    Biopsy of suspicious nodes beyond the region of the resection
    Pyloromyotomy or pyloroplasty
    Conclude operation with feeding tube enterostomy
Transhiatal esophagectomy
    Mobilize right colon to expose the duodenum and head of the pancreas
    Kocherize the duodenum completely
    Divide the greater and lesser omentum, preserving the right gastroepiploic
        artery on the greater curve of the stomach and the right gastric artery
    Divide the left gastroepiploic artery and the short gastric vessels.
    Preserve the spleen (see Fig. 6.2A).
    Open the hiatus and place traction on the esophagus and clear the esophagus
        of attached lymphoareolar tissue and vagus nerve fibers. Stay tightly on the
        esophagus.
    Mobilize and divide the esophagus via a cervical incision
    Pull the esophagus and stomach into the abdominal field, divide the stomach
    with suitable distal margins and preserve as much greater curve as possible
    Pull the cardia portion of the greater curve into the cervical incision and
        perform the cervical esophagogastrostomy by hand or stapler (see Fig. 6.2B).
Three incision esophagectomy
    Prepare stomach as for transhiatal esophagectomy
    Perform thoracotomy with mediastinal dissection of lymphoareolar tissue and
        esophagus
    Mobilize and divide the esophagus via a left, or right, cervical incision
    Divide the stomach as for a transhiatal esophagectomy
    Bring the stomach to the neck in the posterior mediastinum or retrosternal
        position. Resect the medial third of the clavicle for retrosternal conduits to
        provide room to enter the neck
    Perform a cervical esophagogastrostomy (see Fig. 6.2B).
Wide field esophagectomy
    Prepare the right or left colon for interposition by taking care to preserve the
        venous drainage without compromise
    Perform abdominal and thoracic incisions to permit en bloc resection of
        proximal stomach, celiac lymph nodes, spleen, mediastinal lymph nodes,
        and thoracic esophagus
    Place colon conduit in retrosternal position
    Perform cologastrostomy and cervical esophagogastrostomy (see Fig. 6.3).

6

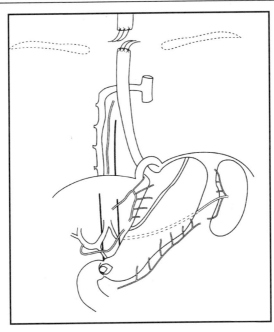

Fig. 6.2A. The general approach to tumor resection and preparation of the stomach for positioning in the neck after esophagectomy.

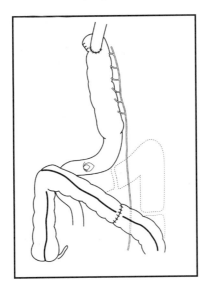

Fig. 6.2B. The appearance of the stomach after transhiatal esophagogastrectomy and cervical esophagogastrostomy.

6

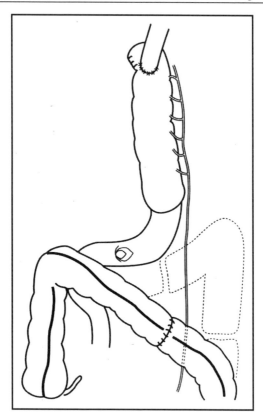

Fig. 6.2C. The appearance of the colon after esophagogastrectomy by either the modified Lewis procedure, or after wide-field esophagectomy showing the esophagocolostomy and cologastrostomy.

Neoadjuvant Chemotherapy

A number of regimens have evolved in an effort to reduce treatment-related morbidity. Most current regimens are variations of combination chemotherapy regimens consisting of cis-platinum, 5-fluorouracil, and other agents used in conjunction with radiation therapy designed to take advantage of the radio-sensitizing properties of cis-DDP and 5-FU. A typical regimen involves 3 cycles of cis-DDP and 5-FU, with radiation therapy timed to begin at the initiation of chemotherapy and continued to a total dose of 4000-4500 cGy. The surgery is timed to allow complete recovery from the neutropenia of chemotherapy, which is usually 3-6 weeks after the last chemotherapy cycle. Although it is assumed that delays beyond 8 weeks will lead to technical difficulties with the surgery, the onset of fibrosis is variable and longer delays due to morbidity of the neoadjuvant treatment occasionally require that surgery be postponed beyond the 6-week ideal. The effectiveness of treatment

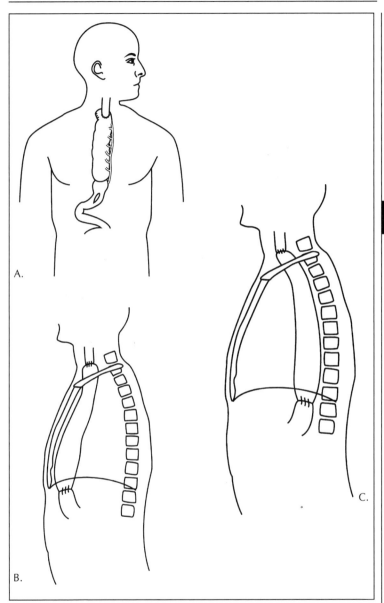

Fig. 6.3. The position of the gastric or colonic conduit after esophagogastrectomy (A). The conduit is either in the retrosternal position (B), or the posterior mediastinal position in the bed of the esophagus (C). The posterior mediastinal position may provide 2 cm or more of additional length to reach the neck compared with the substernal position.

6

can be assessed in the operative specimen and chemotherapy continued for two additional postoperative cycles.

Treatment Options for Unresectable Disease

For patients proceeding to surgery with the expectation of resection but who are found to be unresectable due to metastasis or locally advanced disease, a number of options are available. Some authors advocate palliative resection with transhiatal esophogectomy whenever possible. Others argue that whatever the morbidity, resection is hardly justified for a disease with such poor life expectancy at its advanced stages. The potential for endoscopic dilation, laser or radiofrequency ablation with dilation and stenting, and radiation with or without chemotherapy in the patient found to be unresectable are recognized options. The choice is usually determined by institutional capability, as there is no evidence favoring one treatment over another in terms of overall improved survival.

For patients presenting as stage IV, radiation and chemotherapy offer the best hope for palliation. Photodynamic therapy and various permutations of stenting via endoscopy are attractive for patients without the option of combined chemoradiation therapy.

Palliative bypass, previously often alluded to, has all but disappeared. The only surgical therapy with appeal in this circumstance is transhiatal esophagectomy for patients with near complete esophageal obstruction, good performance status and a resectable primary with limited metastatic disease.

Outcome of Therapy

The median survival for untreated esophageal cancer is less than 8 months; there have been no 5-year survivors.

For optimally treated disease the survival for both squamous cell carcinoma and adenocarcinoma is similar. Survival by stage is noted in Table 6.6.

Complications are an important aspect of the treatment of esophageal cancer. The initial reports of the combined approach to therapy reported treatment-related mortality near 25%. Modern iterations are much less toxic, and several studies report an absence of treatment-related deaths when nutrition is maintained and complications of neutropenia and sepsis are avoided. Nonetheless, the therapies are morbid and require close scrutiny of recovering patients through all phases of their therapy. In table form, the types of complications are as follows in Table 6.7.

A few details uniquely relevant to esophageal cancer surgery, as opposed to problems associated with chemotherapy and radiation therapy, are worth addressing at this point. These include:

1. preoperative contracted intravascular volume,
2. choice of procedure,
3. a positive resection margin,
4. anastomotic leaks, and
5. vocal cord paralysis.

Patients with esophageal cancer are not candidates for chemotherapy and radiation if they are completely obstructed. More often than not, the patient can be dilated to a luminal diameter that allows modest oral intake and the swallowing of oral secretions. Nonetheless, support with enteral tube feeding is important to

Table 6.6. Five-year survival after completing treatment for squamous cell carcinoma and adenocarcinoma by stage.

Stage I:	80%
Stage IIA:	40%
Stage IIB:	25%
Stage III:	15%
Stage IV:	0%

Table 6.7. Treatment related complications

Chemotherapy/radiation
    Nutrition
    Anemia
    Renal dysfunction/electrolyte abnormalities
    Hypovolemia
    Postoperative pulmonary lymphedema
Surgery
    Hemorrhage
    Sepsis, early and late if the spleen is removed
    Anastomotic leak
    Wound infection
    Pleural effusion
    Cardiac arrhythmias
    Vocal cord paresis/paralysis
        Aspiration/aspiration pneumonia
        Hoarseness/stridor
    Chylothorax
    Tears of tracheobronchial tree
    Swallowing difficulty
        Compromised deglutition
        Poorly emptying gastric or colonic conduits
    Delayed gastric emptying/pyloric stances
    Failed colon interposition with ischemic colon
    Chronic aspiration pneumonia
    Poor nutrition
    Depression

6

consider. In spite of having received apparently adequate nutrition preoperatively, cancer patients often are volume depleted. Patients have a type of isotonic dehydration and may not be anemic due to vascular volume contraction. Once anesthetized or in the midst of vasodilation for any reason, these patients can quickly become hypotensive. Vigorously address this potential complication preoperatively and consider preoperative optimization of preload and cardiac output using pulmonary capillary wedge pressure and cardiac output determinations.

Important technical nuances of each surgical procedure are numerous. Some can easily be avoided by selecting the proper procedure. In general, no single operative approach will work in all instances. The esophageal surgeon must have an array of operations to choose among and an awareness of technical modifications to employ when required. The patient must not be fitted to the Procrustean bed of a specific operation for which he or she may be ill suited.

A frequent problem in the past was the problem of the positive proximal esophageal margin after the resection. By determining to do total esophagectomy on virtually all patients with esophageal cancer, this is less often a concern. More pressing is the solution to the need for a larger gastric margin in some esophagogastric junction cancers and the loss of the stomach as a conduit that will reach the neck. For such a patient, a compromise anastomosis in the chest may be the best solution. Alternatively, colonic interposition may be best. Intermediate, temporizing options exist including "parking" the gastric remnant either substernally or subcutaneously or within the abdominal cavity and simultaneously performing a cervical esophagostomy. Later, free jejunal interposition or colonic interposition may be performed to the subcutaneous, retrosternal, or intraabdominal gastric remnant.

Postoperatively, the greatest fear is sepsis due to anastomotic disruption. As with all surgery, the greatest humility is required in evaluating the patient who is doing less well than was hoped. One must constantly assume the worst with regard to the evaluation and management of technical failures and vigorously use CT and other imaging studies along with interventional or intraluminal invasive maneuvers to salvage such situations. The nature of the reconstruction is such that no surgeon should ever permit an intervention based only on a radiologist's interpretation of a diagnostic study. With full knowledge of how the patient's surgery was performed, only the surgeon, having seen the imaging in question, can permit an intervention. For example, the esophagogastrostomy is most often constructed in an end-to-side fashion. An air-fluid level slightly above and away from the anastomosis is often noted on postoperative water soluble contrast/barium swallow. In the septic patient, this must not immediately be construed to be a leak that requires surgery or drainage. The clinical picture and knowledge of the procedure should dictate the surgeon's choice for management. Nevertheless, the availability of talented, aggressive, interventional radiologists is as much a means to minimize the morbidity of complications in these procedures as it is in other areas of surgery.

Vocal cord paresis or paralysis is relatively common. Aside from diligent careful attention to avoiding recurrent nerve injury, the general guideline is to aggressively deal with these two problems. The first clue is often the first postoperative swallow test. One must allow the radiologist performing the initial study to follow a small amount of water-soluble contrast with barium so that an assessment of aspiration can be accomplished. This has the parenthetic advantage of permitting a more accurate assessment of the anastomosis. Even after a normal study, and, especially if the study is indicative of aspiration, the patient should be observed with aspiration precautions. When in doubt, the cord in question can be brought to the midline with Gelfoam. Later, persistence of the problems can be dealt with by Teflon injection or surgery. If the problem is more profound, a tracheostomy may be required. Chronic, untreated aspiration is a ticking bomb. In the near term and midterm, it may be an unrecognized source of pulmonary sepsis and late postoperative death.

Posttreatment Surveillance

A realistic approach to following patients postoperatively must take into account that recurrences other than local, anastomotic recurrences will ultimately prove fatal.

First and foremost, patients are followed postoperatively and thereafter for their symptom status. Although protocols may dictate periodic blood work (including CEA or CA 19-9, if elevated prior to treatment), chest x-rays, CT scans, and, occasionally, PET imaging and bone scans, one must constantly ask how the information obtained will actually benefit the patient. The altered biology attending neoadjuvant therapy has increased the number of patients with central nervous system recurrence compared with the relatively low incidence of such metastases in the past when only radiation or surgery were the mainstays of therapy. Does that mean we should routinely obtain MRI brain studies? Of course we should not. It does suggest that central nervous symptoms should be promptly investigated.

As chemotherapy improves, early detection of recurrence may have benefit. Until and unless that happens, a careful interim history and physical exam provide the best means of following these patients. For squamous cell carcinoma, a continued surveillance for metachronous upper aerodigestive cancers is prudent.

Special Considerations

A few talented individuals at a very few centers have begun to perform video-assisted laparoscopic/thoracoscopic esophagectomy with gastric pull-up for esophageal cancer. The decrease in morbidity in the few reported series is spectacular. The thoroughness of the celiac and mediastinal dissection is encouraging in terms of numbers of nodes obtained. Perhaps the greatest contribution of these groups is their uniform use of laparoscopic staging. Exclusion of stage IV cases from neoadjuvant regimens is a positive contribution to future therapy.

Technology no doubt holds more surprises for both the curative and palliative management of such patients. Yet the neoadjuvant approach has yet to be proven superior to surgery alone and palliative maneuvers are seldom as good as resection at restoring quality of life for the esophageal cancer patient. There is definitely no role, as yet, for the use of radiation and chemotherapy alone for the curative management of the lower esophageal cancer patient outside of a clinical trial. Skepticism about the use of radiation and/or radiation/chemotherapy for cervical and midesophageal cancers as the sole treatment will be supported when improved agents and radiation schedules yield a true survival advantage for lower esophageal cancer.

The technical advances of minimally invasive surgery have an important investigative role in the prevention of esophageal cancer. Eventually, widespread familiarity with video-assisted laparoscopic/thoracoscopic esophagectomy will alter the management of Barrett's esophagus, Barrett's with severe dysplasia, and established esophageal malignancies. Laparoscopic antireflux surgery is generally available, and a better understanding of how and when to intervene when Barrett's is detected might have important and immediate applicability as a cancer prevention measure.

Selected Readings

1. Blot WJ, Devesa SS, Kneller RW et al. Rising incidence of adenocarcinoma of the esophagus and gastric cardia. JAMA 1991; 265:1287-1289.
   The paper based on SEER registry data, which noted the astonishing, increasing incidence of lower esophageal adenocarcinoma during the 1970s. Discusses the overall incidence of squamous cell carcinoma and discusses possible etiologies for the findings.

2. Orringer MB, Marshall , Iannettoni MD. Transhiatal esophagectomy: Clinical experience and refinements. Ann Surg 1999; 230:392-400.
    A recent update on technical aspects and outcomes of Dr. Orringer's singular experience. Access to the original work and to the experience of others using this technique is provided.

3. DeMeester TR, Peters JH, Bremner CG et al. Biology of gastroesophageal reflux disease: pathophysiology relating to medical and surgical treatment. Annu Rev Med 1999; 50:469-506.
    Summary of the medical and surgical failings with regard to esophagogastric junction reflux and discussion of the biological implications with regard to the etiology of Barrett's esophagus and adenocarcinoma in Barrett's.

4. DeMeester TR. Esophageal carcinoma: Current controversies. Semin Surg Oncol 1997; 13:217-233.
    This paper summarizes the points of difference regarding the various approaches to esophageal cancer. It provides a detailed discussion of technical and theoretical issues relating to the therapy of this disease.

5. Skinner DB, Belsey RHR. Resection for potentially curable esophageal cancer. In: Skinner DB, Belsey RHR, eds. Management of Esophageal Disease. Philadelphia: WB Saunders,1988:736-763.
    This Chapter remains the most complete discussion of en bloc esophagectomy in the literature. Results of the technique may be accessed via Dr. DeMeester's review noted above (Ref. 4).

6. Nguyen NT, Schauer PR, Luketich JD. Combined laparoscopic and thoracoscopic approach to esophagectomy. J Am Coll Surg 1999; 188:328-332.
    Clearly the wave of the future, this article summarizes the available data from the few centers currently employing this approach.

CHAPTER 7

Gastric Cancer

Stephen A. Shiver, Brian W. Loggie

Scope and Epidemiology

The incidence of gastric cancer in the United States has decreased approximately fourfold over the last 60 years.[1] Though the exact cause of this decline is unknown, it represents a dramatic trend in cancer epidemiology. Whereas it was once the country's leading cause of cancer related mortality, it is now ranked seventh (Table 7.1).[2] Despite the declining incidence, the disease continues to cause significant morbidity and mortality. In 1998, there were 22,600 newly diagnosed cases of gastric cancer in the United States and 13,700 deaths attributed to the disease.[2] An exception to the declining overall rate of gastric malignancy is adenocarcinoma arising in the gastric cardia. The incidence of cancer arising at this site has rapidly increased, paralleling the rising incidence of distal esophageal and gastroesophageal junction tumors.[3]

Gastric cancer takes on a much more prominent role when considered on a world-wide scale. It continues to be among the leading causes of cancer related deaths, surpassed only recently by lung cancer. Marked geographic variation is the rule. Countries of the Far East have the highest incidence. In Japan, the disease accounts for 40-50% of cancer related deaths. The lowest incidence occurs in the Middle East, United States, and Canada. The European continent appears to have an intermediate risk.[3] Pathological differences also exist based on geography. The intestinal type (as compared to the diffuse type) of gastric cancer appears to predominate in those regions with high incidence whereas the diffuse type predominates in areas where the incidence is low, such as the United States. This trend has practical importance because the diffuse type is generally associated with a more aggressive course and subsequently a poorer prognosis.[3]

Gastric cancer continues to be a disease of primarily older individuals. It is extremely rare prior to the age of 30 and the incidence increases with age. Historically, there has been a 2-1 male to female ratio of gastric cancer.[3] Recent United States statistics support the male predominance: 14,300 of the 22,600 newly diagnosed cases in 1998 involved men.[2] According to ethnicity, African-Americans have a higher incidence than the white population.

Risk Factors

Several risk factors for gastric carcinoma have been proposed, but none has received more attention than diet. Dietary differences have been postulated to be partially responsible for the marked geographical differences in gastric cancer rates. The majority of such data comes from case control studies in which the dietary habits of

Surgical Oncology, edited by David N. Krag. ©2000 Landes Bioscience.

Table 7.1. Leading causes of cancer related deaths among men living in the United States, 1998

Lung Cancer	32%
Prostate	13%
Colon and Rectum	9%
Pancreas	5%
Leukemia	4%
Non-Hodgkins Lymphoma	4%
Stomach	3%

Cancer Statistics, 1998

patients with gastric cancer are compared to individuals without the disease. Nitroso compounds are believed to be involved in gastric carcinogenesis. Precursors of nitroso compounds may be found in smoked, cured, or preserved foods. Other agents, such as ascorbic acid and beta carotene, may reduce the risk of developing gastric cancer. Presumably, this occurs because of their ability to act as free radical scavengers. A diet high in fruits and vegetables is also thought to reduce the risk.[3]

A substantial research effort has focused on the role of *Helicobacter pylori* as a potential gastric carcinogen. At present, it is felt that *Helicobacter pylori* infection increases the risk but is not in and of itself sufficient to produce the disease. In addition, it appears to be more closely linked to the intestinal type histologic pattern than it is to the diffuse type disease. Evidence suggesting *Helicobacter pylori's* role in gastric cancer includes:

1. Increased incidence of childhood *Helicobacter pylori* infections in geographic areas with a high incidence of gastric cancer
2. Prospective serologic studies which showed a 3-6 fold increased risk for developing gastric cancer in patients with *Helicobacter pylori* infection.[3] Since *Helicobacter pylori's* putative involvement in gastric carcinogenesis has been suggested, the prospect of chemoprevention of gastric cancer by eradicating the organism has been entertained. Eradication of the bacterium from the stomach does result in regression of gastritis and intestinal metaplasia.[4] However, the etiology is multifactorial and it is extremely unlikely that eradicating the organism will prevent all gastric cancer. The precise impact of chemoprevention is not known and must await the results of ongoing clinical trials.

Atrophic gastritis is an important risk factor and may occur with *Helicobacter pylori* infection or in other clinical settings such as pernicious anemia. Previous gastrectomy confers a cumulative risk over time such that after 15 years the relative risk of developing gastric cancer is 3-5 times that of controls. Additional risk factors that have been proposed include the presence of adenomatous polyps and blood group A.[3]

Screening

There is no organized screening program for gastric cancer in the United States or other Western countries. However, screening is performed in Japan because of the much higher incidence of the disease. The most common method of mass screening in that country involves gastric x-ray studies using standard contrast techniques. Since instituting the program, the death rates from gastric cancer have declined in Japan. However, some argue that an actual biologic difference exists between the tumors in Japan and Western countries and that it is this biological difference that accounts for differences in survival. The Japanese maintain that the disease itself is no different and the improved survival is due to detecting the malignancy at an earlier, curable stage. The proportion of early gastric cancers in Japan has been as high as 53% in some series. This issue continues to be controversial. Given the low incidence in the United States, it is unlikely that mass screening programs will be initiated in the near future.

A new approach to screening involves serologic testing. In Japanese patients with and without gastric cancer, studies have noted differences in pepsinogen I levels and in the pepsinogen I/pepsinogen II ratio. Pepsinogen I levels are often decreased while pepsinogen II levels increase in response to gastric atrophy. This type of testing may allow for more selective screening with imaging techniques such as endoscopy.[5]

Diagnosis

Patients presenting with early gastric malignancy have no pathognomonic signs or symptoms. Late or delayed diagnosis is common. Nonspecific symptoms such as epigastric pain, weight loss, early satiety, nausea, and vomiting may be present. No physical findings are present for early stage disease. Late stage disease may present with left supraclavicular adenopathy, hepatomegaly, ascites, or cachexia. The presence of ascites or palpable masses indicates advanced disease. A palpable umbilical nodule (Sister Mary-Joseph Node) is uncommon but when present indicates dissemination. Heme positive stool may be noted.

The persistence of nonspecific symptoms eventually precipitates a diagnostic workup. Fiberoptic endoscopy is the diagnostic method of choice. It allows direct visualization and inspection of the entire gastric mucosal surface. Multiple biopsies of all suspicious lesions should be obtained to reduce the likelihood of sampling error. All gastric ulcers should be considered to be malignant until proven to be benign by biopsy and subsequent follow up. Features associated with, but not specific for, malignant gastric ulcerations include an irregular ulcer base, irregular border, associated mass, and irregular mucosal folds at the periphery of the ulcer.

The upper gastrointestinal series continues to play a role in the diagnosis of gastric neoplasms. Though it is significantly less expensive than endoscopy, it has the obvious disadvantage of not being able to directly visualize the lesion and perform concomitant biopsy. However, it can be quite sensitive for detecting gastric cancer when a double contrast technique consisting of both air and barium is utilized (Fig. 7.1). In a Japanese study involving 695 patients with known early gastric cancer, the upper gastrointestinal series detected 89.6% of the lesions whereas endoscopy detected 97.5%. The remaining 2.5% were missed by both modalities.[6]

There are no laboratory abnormalities specific for gastric cancer. A complete blood count might show a hypochromic, microcytic anemia secondary to chronic blood loss from the gastrointestinal tract. Liver function studies may show an elevated alkaline phosphatase in the setting of hepatic metastases.[7]

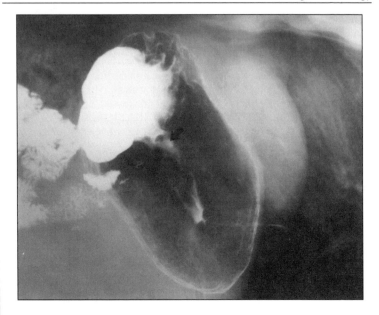

Fig. 7.1. Subtle appearance of a gastric carcinoma on an upper gastrointestinal series. Note the thickened mucosal folds suggesting the presence of a mass lesion (Courtesy Mike Chin, MD).

Preoperative Evaluation

Since gastric cancer is for the most part a disease of the elderly, significant medical comorbidities may increase the risk associated with performing a major abdominal operation. The patient's overall medical condition should be optimized before proceeding with surgery.

The initial step is a thorough history and physical exam. In order to determine whether the patient can tolerate major abdominal surgery, particular attention should be given to the pulmonary and cardiovascular systems . A chest x-ray should be performed to look for pulmonary pathology. A chest film might also show lesions suspicious for metastatic disease which would then prompt further evaluation with a chest CT scan. If a patient smokes, every effort should be undertaken to encourage smoking cessation preoperatively. In the setting of severe pulmonary disease, an extensive evaluation including pulmonary function tests and arterial blood gas analysis may be needed to help determine if the patient is a reasonable surgical candidate. Any history of angina or known coronary artery disease should be sought. At a minimum, a preoperative EKG should be obtained. This will serve as a baseline and might possibly uncover unknown cardiac problems. In certain circumstances, further noninvasive testing may be warranted. This may include a traditional treadmill stress test or a pharmacologic stress test such as a dobutamine echocardiogram.

Screening laboratory studies are usually obtained including a complete blood count, electrolyte panel, and liver function tests. The complete blood count may indicate anemia from blood loss or vitamin deficiency. The electrolyte panel serves as a baseline and may reveal medical problems such as renal insufficiency. As previously mentioned, elevation of liver function tests may suggest the presence of hepatic metastases.

Staging

Accurate staging in gastric cancer is critical. It is of great importance both in treatment planning and determining the ultimate prognosis. The most widely accepted staging system at present is based on the American Joint Commission on Cancer (T)umor, (N)odes, and (M)etastasis system (Table 7.2).

Modern radiological techniques have continued to revolutionize the staging process and abdominal CT imaging is currently the most widely used staging modality.[6] To be effective, the scan must be of high quality and include both oral and intravenous contrast. Fine cuts (5 mm) of the upper abdomen including the stomach and liver should be obtained.[6] A properly performed and interpreted CT scan can detect lymphadenopathy, invasion of adjacent structures, and hepatic metastases. Figure 7.2 shows the typical appearance of a gastric tumor on CT scan.

CT findings may be used to predict resectability using the TNM system. CT identification of a T4 lesion, N3 lesion, or evidence of distant metastases implies unresectability. There are limitations to CT based staging strategies and disease can be overstaged or understaged. CT scans cannot differentiate between the histologic layers of the gastric wall. Thus, it can only determine whether or not the tumor extends beyond the wall itself. Detection of extragastric spread depends on the ability to detect obliteration of the normal perigastric fat planes. These planes may be disturbed by inflammation or severely diminished in the setting of cachexia. With regard to nodal disease, CT scanning detects nodal metastases based on increased nodal size. However, normal-sized nodes can harbor malignant cells and nodes can be enlarged secondary to reactive hyperplasia.[6] Detecting distant disease can also be problematic. CT scans are excellent at identifying large visceral metastases but may miss small (< 1 cm) hepatic metastases and peritoneal implants.[7]

A promising new modality is endoscopic ultrasonography. CT scanning is superior in detecting distant disease but ultrasound is superior in locoregional staging. Thus, ultrasound is complimentary to CT evaluation. Unlike CT, it has the ability to detect five layers of the gastric wall that correspond to histologic layers. Accuracy in determining tumor depth averaged 82-92% in several studies.[8]

Laparoscopy is being used with increased frequency to stage numerous abdominal tumors including gastric cancer. It is particularly useful in detecting small hepatic metastases and peritoneal implants that are often missed by noninvasive modalities. Combining laparoscopic technology with ultrasound offers the intriguing possibility of enhancing presurgical Tumor and Nodal staging capabilities.[9] At present, the role of laparoscopy continues to evolve.

Additional modalities are being investigated and may be of increased use in the future. Positron emission tomography (PET) is based on the tendency of tumor cells to take up glucose more readily than normal cells. Though it is being used extensively in some tumors such as small cell lung cancer, its use in gastric cancer is still experimental. Magnetic resonance imaging at present is not felt to be superior to CT scanning though work in this area is ongoing.

Table 7.2. Staging of gastric cancer

Primary Tumor

TX	Primary tumor cannot be assessed
TO	No evidence of primary tumor
Tis	Carcinoma in situ
T1	Tumor invades lamina propria or submucosa
T2	Tumor invades muscularis propria
T3	Tumor invades adventitia
T4	Tumor invades adjacent structures

Regional Lymph Nodes

NX	Regional lymph nodes cannot be assessed
NO	No regional lymph node metastases
N1	Metastasis in perigastric lymph nodes within 3 cm of edge of primary tumor
N2	Metastases in perigastric lymph nodes more than 3 cm from edge of primary tumor, or in lymph nodes along left gastric, common hepatic, splenic, or celiac arteries

Distant Metastasis

MX	Presence of distant metastases cannot be assessed
MO	No distant metastases
M1	Distant metastasis

American Joint Commission on Cancer Staging of Gastric Cancer, 1988

Stage Grouping

Stage O	Tis, NO, MO
Stage 1A	T1, NO, MO
Stage 1B	T1, N1, MO
	T2, N0, MO
Stage II	T1, N2, MO
	T2, N1, MO
	T3, N0, M0
Stage IIIA	T2, N2, MO
	T3, N1, MO
	T4, NO, MO
Stage IIIB	T3, N2, MO
	T4, N1, MO
Stage IV	T4, N2, MO
	Any T, any M, M1

American Joint Commission on Cancer Staging of Gastric cancer, 1988

Treatment Options

The first documented gastrectomy for gastric malignancy was performed in 1881 by Billroth. Gastric cancer continues to be primarily a surgically managed disease. The only hope of curing the disease centers on surgical resection.[1]

Simply viewed, surgery for gastric cancer involves resecting the tumor and restoring gastrointestinal continuity. The goal should be to resect all gross and microscopic disease. Depending on the location of the tumor, this may involve distal gastrectomy, subtotal gastrectomy, or total gastrectomy. Gastrointestinal continuity is usually restored via a gastroduodenostomy or gastrojejunostomy. If a total gastrectomy is needed, a roux-en-Y esophagojejunostomy may be performed. Our preference is

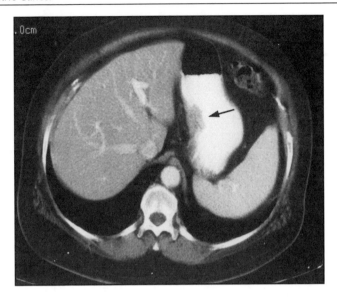

Fig. 7.2. CT appearance of a gastric carcinoma. Note the thickened gastric wall along the lesser curvature (Courtesy Mike Chin, MD).

7

to reestablish gastrointestinal continuity in a Billroth II or roux-en-Y fashion (Fig. 7.3). If there is an anastomotic recurrence, it is more likely to be treatable in this setting than in the setting of a duodenal anastomosis. Bile reflux gastritis can complicate all reconstruction methods. Some have advocated the use of Hunt-Lawrence pouch to replace the gastric reservoir function. Studies suggest that there is no long term benefit to this approach and we have no experience with it at our institution.

Surgical controversies continue to exist with regard to the extent of gastric resection and lymph node dissection. Much of this controversy arises from the markedly different survival rates among patients treated surgically in Japan and Western countries such as the United States. As previously mentioned, some authorities believe that these differences are due to the biologic behavior of the cancer itself, not differences in surgical technique. In general, the patients in Western countries tend to be older, have the diffuse form of the disease, and have a greater proportion of proximal tumors. All of these factors predict a worse prognosis. Also, early gastric cancer comprises approximately 40% of cases in Japan and only 6-10% in the United States.[1] Despite these observations, many Japanese investigators conclude that their more aggressive surgical treatment results in improved survival even when the stage of the disease is taken into account.[10]

Recent Western data suggest that less than a total gastrectomy is possible in selected cases. This is desirable since total gastrectomy is associated with higher postoperative morbidity than a partial gastrectomy. A prospective, randomized trial involving 169 patients with gastric cancer showed that the 5 year survival was the

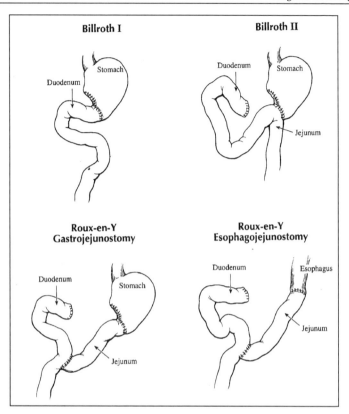

Fig. 7.3. Various reconstruction methods following gastric resection.

same (48%) when subtotal or total gastrectomy was performed.[7] Thus, the general Western view is that total gastrectomy is not required in all cases.

Views on the extent of lymph node dissection continue to differ among surgeons. In general, the Japanese tend to be more aggressive with regard to lymphadenectomy. There is conflicting available data and this issue is yet to be resolved. Lymphadenectomy is currently classified according to the level of lymph node dissection D1-D4. D1 involves resection of perigastric lymph nodes. D2 dissection removes nodes along the left gastric, celiac, and common hepatic vessels as well as at the splenic hilus.[9] D3/D4 resections include even further lymph node removal. It does appear that an extended lymph node dissection improves staging accuracy but its effect on long term survival is still unclear.[1] Some retrospective studies have suggested an improved survival for D2 resection compared to D1 resection alone.

However, numerous prospective randomized trials showed no survival difference among those undergoing D1 and D2 resections.[7] Another confounding factor involves the morbidity and mortality associated with extended resections. The Western experience often cites higher morbidity and mortality rates for more aggressive lymphadenectomy than do the Japanese. Since there appears to be increased morbidity and no definite increased survival, many Western surgeons are reluctant to routinely perform extended lymphadenectomy.

Because gastric cancer is often associated with peritoneal metastases, additional surgical approaches are needed if one is going to attempt a complete resection. A novel experimental approach under development is cytoreduction and intraperitoneal perfusion chemotherapy. In the absence of any known visceral metastases, this may be performed at the time of the original resection. The method is designed to expose small volume peritoneal disease to a high concentration of chemotherapeutic drug. It involves meticulously searching for and removing as much peritoneal disease as possible. The abdomen is then perfused with a heated chemotherapeutic agent such as mitomycin C.

Unfortunately, resection with curative intent for gastric cancer is only possible in 50-60% of newly diagnosed cases. Obviously, there is great interest in developing other potential therapeutic options. There is no proven effective systemic adjuvant chemotherapy. The impact of neoadjuvant chemotherapy on downstaging, resectability, treatment morbidity, and long term survival is currently undergoing evaluation. Radiation therapy has no standard role in treating gastric cancer at the present time. Neoadjuvant irradiation for gastroesophageal tumors and some proximal gastric tumors may be of benefit in selected cases. Surgery alone remains the standard of care. Further research needs to be performed to determine which if any additional modalities will favorably impact survival.

Outcome of Treatment

Since surgery is the only effective treatment modality, long term survival is critically dependent on the ability to resect all gross and microscopic disease. An R0 resection, defined as complete gross resection with negative microscopic margins, carries the best prognosis. Unfortunately, most patients with gastric cancer present with advanced disease and an R0 resection is not possible.

Predictably, prognosis is adversely affected by deeper gastric wall invasion and nodal involvement. The worst prognosis is obviously seen in the setting of metastatic disease. It is ultimately the stage at the time of diagnosis that determines long term survival. If the disease is confined to the mucosa (Stage 0), an R0 resection is the rule and the 5 year survival approaches 100%. However, the 5 year survival rates decline rapidly with increasing stage. Stage I 90%, Stage II 50%, Stage III 15%, and Stage IV less than 1%.[11]

Follow Up and Posttreatment Surveillance

We choose to follow patients rather frequently in the immediate postoperative setting to ensure that the patient is having an uneventful recovery. Restoration of adequate oral caloric intake must be established. Thereafter, we see patients in clinic approximately every 4 months for the first two years postoperatively. A directed history and physical exam is performed at each clinic visit. If the patient continues to do well, the frequency of follow up may then be cautiously decreased. Depending on the extent of gastric resection, patients may require monthly B12 injections because of the loss of intrinsic factor. This may be coordinated with the patient's primary care physician.

Despite the risk of locoregional and metastatic disease, no accepted standards exist with regard to postoperative surveillance. Routine laboratory studies are rarely useful and there are no tumor markers sensitive or specific for disease recurrence. A yearly chest x-ray is often obtained and some advocate periodic endoscopic examination. If recurrent disease is strongly suspected, additional studies may be indicated. CT scanning or PET evaluation may detect locoregional or distant disease. Other than anastomotic recurrences, treatment options and outcomes are very poor for recurrent disease.

Selected Readings

1. Brennan MF, Karpeh MS Jr. Surgery for gastric cancer: The American view. Seminars in Oncology 1996; 23(3):352-359.
   Summary of current American approach to management of gastric cancer.
2. Lardis SH, Murray T et al. Cancer Statistics, 1998. CA A Cancer Journal for Clinicians 1998; 48(1):6-30.
   Summary of incidence and survival statistics.
3. Neugut AI, Hayek M et al. Epidemiology of gastric cancer. Seminars in Oncology 1996; 23(3):281-291.
   Comprehensive manuscript on epidemiology including current views on Helicobacter pylori *and diet.*
4. Hill MJ. Chemoprevention of gastric cancer by Helicobacter pylori eradication? Projects in Gastric Cancer Research 1997: 9-16.
   Discusses the etiologic role of Helicobacter pylori *in gastric carcinogenesis and the prospects for future prevention of the disease.*
5. Yoshida S, Daizosaito. Gastric premalignancy and cancer screening in high risk patients. Amer J Gastroenterol 1996; 91(5):839-843.
   Discusses the utility and results of screening for gastric cancer in Japan.
6. Halvorsen RA, Yee J et al. Diagnosis and staging of gastric cancer. Seminars in Oncology 1996; 23(3):325-335.
   Covers current diagnostic modalities for gastric cancer and their role in the modern staging process.
7. Karpeh MS, Brennan MF. Surgical oncology forum gastric carcinoma. Ann Surg Oncol 1998; 5(7):650-656.
   Excellent article which addresses and discusses multiple controversial areas with regard to surgical management including extent of gastric resection and lymphadenectomy.
8. Pollack BJ, Chak A et al. Endoscopic ultrasonography. Seminars in Oncology 1996; 23(3):336-346.
   Discusses present and potential future role of endoscopic ultrasound in the staging process.
9. Conlon KC, Karpeh MS. Laparoscopy and laparoscopic ultrasound in the staging of gastric cancer. Sem Oncol 1996; 23(3):347-351.
   Discusses present and potential future role of endoscopic ultrasound in the staging process.
10. Maruyama K, Sasako M et al. Surgical treatment for gastric cancer: The Japanese approach. Sem Oncol 1996; 23(3):360-368.
    Summary of current Japanese approach to the surgical management of gastric cancer.
11. Kawai H, Watanabe Y. The impact of mass screening on cancer mortality in Japan. Gastrointes Endosc 47(3):320-323.
    Discusses the impact of mass screening on the Japanese population.

CHAPTER 8

Colon and Rectal Cancer

Neil Hyman

Scope of the Problem

Approximately 130,000 Americans develop colon or rectal cancer each year. This accounts for 10% of the cancers diagnosed each year, as well as 10% of the cancer-specific deaths. Colorectal cancer is the third most common site of malignancy in both men and women, and the third most lethal cancer in either sex. Unlike the situation with many other cancers, the incidence and mortality rates are relatively similar between males and females. When men and women are combined, colon and rectal cancer is second only to lung cancer in the incidence of annual deaths.

Of the estimated 130,000 patients who develop colon and rectal cancer each year, approximately 56,000 will die of their disease. Fortunately, both the incidence and overall mortality associated with colon and rectal cancer have recently begun to decline. Although the reasons for this are unclear, it seems likely related to more aggressive screening modalities and earlier diagnosis. Over the past twenty years, there has been a statistically significant improvement in five year colorectal cancer survival rates from approximately 50-63%. The improved survival rate combined with the recent trend towards a lower incidence is indeed a hopeful trend. However, the overall lifetime risk of an American developing a colon or rectal cancer remains approximately 5-6%.

Risk Factors for Development of Disease

The incidence of colon and rectal cancer is clearly higher in developed countries than in developing countries. The highest incidences worldwide are seen in Australia, New Zealand, North America and northern and western Europe. Incidence rates are low in Africa and Asia (except Japan).

These large geographic differences suggest an effect of various environmental exposures, probably dietary in nature. This concept is reinforced by the fact that people who emigrate from countries of low incidence to high incidence, tend to ultimately assume the higher risk rate of their adopted country.

It has usually been reported that diets high in meat and animal fats, but low in fiber, tend to promote the development of colorectal cancer. Intake of various fruits and vegetables has been felt to be beneficial. This may be related to the more rapid transit associated with high fiber diets, with a less prolonged exposure of putative carcinogens to the colonic mucosa. Alternatively, the protective effects of fruits and vegetables may be related to various vitamins or other micronutrients that are found in these foods. In fact, various micronutrients have been implicated both experimentally and in vivo as affording protection against colorectal cancer. These

Surgical Oncology, edited by David N. Krag. ©2000 Landes Bioscience.

include calcium, selenium, and various other antioxidants. There has been considerable interest in the chemoprotective role of aspirin and various nonsteroidal antiinflammatory agents (NSAIDS).

For years, the development of colon and rectal cancer has been linked to the adenoma-to-carcinoma sequence. It was noted that many colorectal cancers still had residual benign adenomatous tissue and that the relative distribution of adenomatous polyps approximated the distribution of colon and rectal cancer. Further, individuals with familial adenomatous polyposis (FAP), in which hundreds or thousands of adenomas develop, inevitably would develop cancer. Most compelling was the evidence from the National Polyp Study (NPS) that documented the marked reduction in the expected incidence of colorectal cancer in patients undergoing colonoscopy when all adenomatous polyps were endoscopically resected.

However, dramatic advances in molecular biology have markedly modified our understanding of colorectal carcinogenesis. A multistep mutation pathway has emerged involving sequential activation of oncogenes and loss or inactivation of tumor-suppressor genes. In essence, as a neoplasm progresses from adenoma to carcinoma, an increasing number of genetic alterations is noted. Through a series of genetic losses and mutations, normal mucosa becomes hyperproliferative, develops into an adenoma and ultimately progresses to carcinoma and then ultimately to metastasis. The genetic alterations associated with colorectal cancer is an intense area of research that hopefully will serve as a basis for a better understanding of the biology of colorectal neoplasia and new clinical approaches to diagnosis and treatment.

Approximately 5% of colorectal cancers derive from a definable, inherited colorectal cancer syndrome. These include FAP (familial adenomatous polyposis), in which hundreds or thousands of adenomas carpet the colonic and rectal mucosa. The onset of polyps is usually in the teenage years and there is an almost inevitable progression to colorectal cancer without colectomy. HNPCC (hereditary nonpolyposis colorectal cancer) may be distinguished from FAP in that affected members develop very few polyps. However, these appear to progress rapidly into cancers, characteristically situated in the right colon. Two thirds of cancers associated with HNPCC are located proximal to the splenic flexure compared to a far more distal predominance of sporadic cancers. Other hereditary syndromes predisposing to colorectal cancer include Gardner's syndrome, Turcot's syndrome, Muir-Torre syndrome and juvenile polyposis syndromes.

Outside of these defined inherited syndromes, it is apparent that genetic predisposition plays a more subtle role in the development of sporadic colorectal cancer. Patients with a first degree relative with colorectal cancer or adenomatous polyps probably have a risk of developing colon and rectal cancer that is in the range of twice that of the rest of the population. Of particular importance is a family history of adenomas or carcinomas occurring at a young age, across multiple generations, or involving multiple family members.

In addition to genetic considerations, it has long been known that inflammatory bowel disease, specifically ulcerative colitis is associated with a progressive increase over time in the probability of developing colorectal cancer. More recently it has become apparent that Crohn's colitis also increases the risk of malignancy, although probably not to the extent of ulcerative colitis.

Screening for the Problem

Screening for colorectal cancer is cost effective and just makes good sense.

First, the prognosis of colorectal cancer is very much dependent on stage. As such, early detection will lead to a much greater chance of cure. However, the real goal of screening programs should be colorectal cancer prevention. It must be remembered that the vast majority of colorectal cancers are derived from the adenoma-to-carcinoma sequence. Evidence suggests that in most cases, an adenoma has probably existed for 8-10 years prior to degeneration into a malignancy. As such, this provides for a wide window of opportunity to detect and remove adenomas prior to the development of cancer.

For many years, it was presumed that screening for adenomas with subsequent removal would prevent colorectal cancer. However, it was really the National Polyp Study (NPS) that provided powerful evidence for the utility of endoscopic polypectomy. Patients undergoing colonoscopy with removal of polyps when found, enjoyed a 76-90% risk reduction in the probability of developing a colorectal cancer over the subsequent ten years. Several other studies confirm the powerful protective effect of screening endoscopy and polypectomy on cancer prevention.

Appropriate screening modalities for colorectal cancer that have proven efficacy include fecal occult blood testing, flexible sigmoidoscopy and colonoscopy. Fecal occult blood testing has the advantage of being the least expensive and invasive modality, but has a substantially smaller effect on reducing the incidence and mortality of colorectal cancer than colonoscopy. Flexible sigmoidoscopy is probably intermediate in its impact.

Screening recommendations for colorectal cancer are constantly in evolution as further studies continue to reinforce the value of routine lower endoscopic screening of even asymptomatic patients. In essence, the issue is one of defining the most cost effective strategy for any given patient, balancing the cost, complication rate and invasiveness of any given procedure against its protective effect against cancer related morbidity and mortality.

Typical screening recommendations for "average risk" patients are listed in Table 8.1. Patients at higher risk for colorectal cancer are generally advised to undergo earlier and more frequent screening, usually with colonoscopy. Risk factors include:

1. A personal history of colorectal cancer or adenomatous polyps,
2. a strong family history of colorectal cancer or polyps (cancer or polyps in a first degree relative younger than 60 or in two first degree relatives of any age),
3. a history of inflammatory bowel disease,
4. a family history of one of the hereditary colorectal cancer syndromes (e.g., FAP or HNPCC).

Further, first degree relatives of individuals with colon polyps are also probably appropriate for screening colonoscopy. With the mounting evidence documenting the efficacy of screening lower endoscopic examinations, particularly colonoscopy, endoscopic screening is becoming more vigorously advocated for ever more subgroups of patients, perhaps even those without any risk factors.

Of note, the National Polyp Study also showed that a three year interval for follow-up colonoscopy was as safe as a one year follow-up. If it can be shown that

Table 8.1. American Cancer Society guidelines for colorectal screening ("average risk" patients).

Beginning at age 50, digital rectal examination plus:
- • Annual fecal occult blood testing and flexible sigmoidoscopy every 5 years
- or -
- • Colonoscopy every 10 years
- or -
- • Double contrast barium enema every 5-10 years

even longer intervals between examinations are equally safe, then colonoscopic screening will become even more cost effective.

In summary, screening modalities for colorectal cancer, particularly sigmoidoscopy and colonoscopy, are of proven benefit in reducing mortality from colon and rectal cancer. However, the cost effectiveness of the various strategies (e.g., sigmoidoscopy vs. colonoscopy) for various subgroups of patients continues to be debated. Over the past several decades, there has been a clear shift of colorectal cancer more proximally, away from the rectum and towards the right colon. As such, it seems likely that there will be a continuing shift towards screening modalities that afford total colonic evaluation.

Methods of Diagnosis

As with other malignancies, appropriate diagnostic workup begins with a history and physical examination. Cancers of the right vs. left colon present in a distinct and different fashion. Tumors on the left side will often present with a change in bowel habits and/or blood per rectum noticeable to the patient. Stool in the sigmoid colon and rectum is generally bulky and well formed. As such, when the already narrow caliber colon is compromised by a tumor, the patient will often experience obstructive symptoms such as cramping or note narrowing of the stool caliber. Since the tumor is relatively close to the anal verge, blood is often recognizable.

Conversely, the wide diameter right colon contains relatively liquid stool. As such, one is far less likely to develop a lesion of sufficient diameter to create obstructive symptoms. Mixing and digesting of any blood that is created by the lesion is far less likely to be noticed by the patient. As such, right colon malignancies classically present with anemia and guaiac positive stool. The anemia may be picked up on routine blood work or as part of the assessment of an older patient presenting with fatigue or angina. Strong consideration should be given to a right sided colon cancer in any patient presenting with iron deficiency anemia, particularly with occult blood in the stool. Many such patients have an easily palpable mass in the right lower quadrant.

Digital rectal examination is an important part of any thorough physical examination. As the adage goes, there are really only two reasons why a physician should not perform a digital rectal examination:

1. the patient does not have a rectum,
2. the doctor does not have a finger!

Based on a thoughtful history and physical examination, the appropriate diagnostic tests can be ordered. Rigid proctosigmoidoscopy, flexible sigmoidoscopy, barium enema or colonoscopy may be utilized depending on the expected site of the malignancy.

Preoperative Evaluation

A thoughtful assessment of comorbidities, including cardiovascular and pulmonary risk factors is critical in allowing for the appropriate choice of procedures and adjunctive modalities. For example, a patient with advanced COPD may do better with an expeditious procedure performed under regional rather than general anesthesia. Alternatively, a patient with significant coronary artery disease and a short life expectancy may be more appropriately selected for local excision of a rectal cancer rather than radical resection.

The cost effectiveness of staging workups prior to surgical resection is controversial. Patients should undergo a total colonic evaluation, either with colonoscopy or barium enema, since approximately 5% of patients will have more than one cancer in the colon. A chest x-ray is also obtained to rule out pulmonary metastases.

However, the relative merits of obtaining a preoperative CT scan of the abdomen and pelvis, as well as obtaining the serum marker CEA (carcinoembryonal antigen) is debatable. The rationale for obtaining a preoperative CT scan is to assess the local effects of the tumor, for example identify any adjacent organ invasion, and to look for metastatic disease elsewhere in the peritoneal cavity, such as liver metastases deep in the parenchyma that might not be otherwise noted at laparotomy. It can also serve as a "baseline" for follow-up imaging studies. Others would point out that the CT rarely changes management and only adds to the expense of the workup. Even if there is metastatic disease, a palliative resection would still be warranted. As such, any local extent or metastatic disease can be defined at laparotomy. With respect to lesions deep in the parenchyma of the liver, this can be excluded by intraoperative ultrasound.

A much more compelling argument for a preoperative CT scan can be made in patients with rectal cancer. Patients with extrarectal extension are often best treated with preoperative irradiation. Further, patients with metastatic disease may be well palliated with radiation therapy alone, completely avoiding surgery altogether.

As for preoperative CEA assessment, approximately two thirds of colorectal cancers are associated with elevated CEA. If CEA is elevated preoperatively, this test can then be repeated postoperatively when it will presumably come down to normal if there was no metastatic disease. This then affords a noninvasive method to assess the patient longitudinally for recurrence.

Staging

Historically, colorectal cancer had been staged by the Dukes classification system. As with other malignancies, the TNM method of staging is now used to provide for standardization of staging classifications. The TNM system very closely parallels the Dukes staging system (Table 8.2). A stage I lesion involves only partial thickness of the bowel wall. Stage II denotes a transmural lesion without lymph node or distant metastases. Stage III disease indicates that there are lymph node metastases, but no distant disease. Stage IV signifies metastatic disease.

Table 8.2. TNM and Dukes staging systems for colorectal cancer

TNM Stage		Dukes Stage
I $T_1N_0M_0$, $T_2N_0M_0$	A	
II $T_3N_0M_0$, $T_4N_0M_0$	B	
III any T, $N_{1,2}M_0$	C	
IV any T, any N, M_1	D	
T_1: tumor invades submucosa	N_0: node negative	M_0: no metastases
T_2: tumor invades muscularis propria	N_1: 1-3 + nodes	M_1: metastasis present
T_3: tumor invades serosa/pericolic fat	N_2: ≥ 4 + nodes	
T_4: tumor invades adjacent peritoneal surface (e.g., adjacent organ involvement)		

Treatment Options

Colorectal cancer is a surgical disease. Specifically, surgery remains the only treatment modality capable of producing a cure. Further, even in cases of metastatic disease, surgery will usually be required to effectively palliate symptoms. Otherwise most of these patients will suffer the often miserable local effects of the tumor, such as obstruction, bleeding, perforation or pain prior to succumbing to disseminated disease. However, chemotherapy and/or radiation therapy frequently play an important adjunctive role in maximizing cure rates in conjunction with surgery, or providing for improved duration and quality of life for those patients with metastatic disease.

Surgery

Approximately 80% of patients diagnosed with colorectal cancer have no overt evidence of metastatic disease and are therefore amenable to a potentially curative resection. In general, the following principles must be kept in mind: Resections are based on the presumed lymphatic drainage of the tumor. Lymph flow from the cancer generally parallels the venous drainage back towards the major mesenteric arteries and veins. With the exception of rectal cancer, adequate margins are generally quite easy to obtain and not the major factor in determining the resection.

The major lymphovascular pedicles are ligated at their origin thereby allowing for a wide mesenteric lymphadenectomy. A defined segment of the bowel is thereby rendered ischemic and must be resected. This will generally define the extent of bowel resection rather than the precise location of the tumor in the specific segment, as obtaining tumor free margins are generally quite straightforward.

An example would be useful. A carcinoma of the ascending colon would be predicted to have lymphatic drainage extending towards the origin of the ileocolic vessels and the right colic vessels (if present). As such, surgery would include ligation of the ileocolic lymphovascular pedicle close to its' origin as well as the right colic pedicle. The terminal ileum and right colon are thereby rendered ischemic and are resected. The apron of lymph nodes along the vascular pedicles are included with the specimen. The ileum is then anastomosed to the proximal transverse colon completing the right hemicolectomy. Gastrointestinal function is usually well preserved despite even extensive colectomy. Therefore, generous resections of the colon are well tolerated with relatively minor effects on subsequent bowel function.

On the other hand, rectal cancer presents a somewhat different situation. First, the rectum is confined within the bony pelvis. As such, obtaining wide lateral margins around the tumor can be more problematic. Further, exposure in the pelvis is substantially more difficult than with the abdominal colon. Obtaining an appropriate distal and radial margin around the tumor can be more challenging. For these reasons, local recurrence is much more common after rectal resection than colonic resection. From the functional standpoint, the rectum plays an important reservoir function and works closely in concert with this sphincter musculature. Resection of even a small portion of the rectum can lead to substantial problems with evacuation and/or continence.

Despite these potential restrictions, there has been a clear emphasis towards sphincter preservation in the treatment of rectal cancer based on several parallel developments. For many years, it has been appreciated that the lymphatic flow of a rectal cancer is generally cephalad, towards the origin of the inferior mesenteric vessels. More recently, it has become clear that extensive distal margins are unnecessary. In general, it is uncommon for a rectal cancer to have microscopic extension for more than 1 cm below the visualized or palpable inferior edge of the tumor. As such, the historic admonitions calling for 5 cm distal margins are no longer heeded and distal margins of 2 cm (or perhaps 1 cm) are quite acceptable, if the dissection has been conducted appropriately. More emphasis is now placed on complete excision of the mesorectal tissue in the appropriate plane with attention paid to obtaining tumor free lateral soft tissue margins. As opposed to the relative similarity of results obtained amongst surgeons with colon cancer, there appears to be marked variability in recurrence rates after surgical resection of rectal cancer. This has created an increased focus on surgical technique and appropriate mesorectal clearance.

Historically, virtually all rectal cancers were treated by a combined abdominal and perineal approach with complete removal of the rectum and anus and creation of a colostomy. This is known as an abdominoperineal resection (APR). It is now clear that carcinomas of the upper and middle thirds of the rectum can be treated by anterior resection with preservation of the low rectum and anal canal. This implies that the upper portion of the rectum and sigmoid colon are removed along with the lymph nodes associated with the inferior mesenteric vascular pedicle. The descending colon is then anastomosed to the rectum at the appropriate level. The term low anterior resection usually denotes that the cancer was in the mid rectum and a low pelvic anastomosis was required.

APR is still often required for tumors of the low rectum, although multiple treatment options have emerged to extend the frontiers of sphincter salvage to even these low lying malignancies. Tumors less than one quarter to one third of the circumference of the rectum are often suitable for local excision if they are superficial. Rectal ultrasound has emerged as a very accurate method for determining depth of penetration preoperatively. If ultrasound reveals that such a tumor is confined to the submucosa (T1), then a local excision may be performed with results that appear equivalent to more radical surgery. Local excision may also be offered to selected patients with T2 tumors, but adjunctive chemoradiation appears to be necessary to achieve similar results to more ablative surgery. For circumferential tumors or those with transmural extension into the perirectal fat (T3), then radical surgery is still required. When such a tumor is located more than 1-2 cm above the pelvic floor, reconstruction may still be possible via coloanal anastomosis.

Special Circumstances

Obstruction

Colorectal cancer is the most common cause of large bowel obstruction. With large bowel obstruction, the cecum is at risk for perforation in accordance with the law of Laplace. It must be remembered that the majority of patients with obstructing colorectal cancer present with potentially curable disease (Stage II or III). As such, this by no means necessarily represents a biologically "advanced" or "incurable" situation.

The most immediate compelling issue pertains to prevention of bowel perforation. The second goal is providing for an oncologically sound resection. The third goal is maintenance or restoration of function.

Management of an obstructing sigmoid colon cancer demonstrates these principles quite nicely. Historically, a cancer of the sigmoid colon was treated in three stages. The first stage was a transverse colostomy that would decompress the colon and prevent cecal perforation. At a second stage, the sigmoid colon cancer was resected and an anastomosis created. At yet a third procedure, function was restored by closure of the transverse colostomy.

Most often, this sequence of surgeries is not necessary. At an initial stage, the sigmoid colon can be resected with the proximal end brought out as a colostomy and the distal end sutured or stapled closed (Hartmann procedure). This allows for both decompression of the colon as well as removal of the tumor in one stage. When the colon is distended and obstructed with proximal fecal loading, it has generally not been considered safe to perform a primary anastomosis. Colonic continuity is established at a second stage.

Most recently, several innovative options have been presented to allow for single stage surgery. If the tumor is in the proximal sigmoid colon, a subtotal colectomy may be performed with anastomosis from the ileum to the distal sigmoid colon, allowing for accomplishment of all three goals at one time. Another alternative is sigmoid resection with on-table lavage. In this procedure, the colectomy is performed and a catheter is placed through the appendiceal lumen or terminal ileum. Warm saline is then infused until the colon is clean. Anastomosis may then be performed. Still another option is to place an intraluminal self expanding metallic stent through the malignant stricture to allow for re-establishment of a lumen. This will then facilitate mechanical bowel preparation and definitive surgery at a later date.

Perforation

A colorectal cancer may erode completely through the bowel wall and cause perforation. These perforations may be contained (localized perforation) or allow contamination of the entire peritoneal cavity by fecal content (free perforation). In addition to bacterial soiling, tumor cells thereby gain access to adjacent organs or other peritoneal surfaces. For this reason, the prognosis for a perforated colorectal cancer is clearly worse than is the case with nonperforated colorectal malignancies.

Metastatic Disease

Unfortunately, the vast majority of patients with metastatic colorectal cancer will succumb to their disease. Historically, 5-FU based chemotherapy has produced a measurable response in approximately 30-50% of treated patients. However, even when successful, chemotherapy simply prolongs survival. Cures with chemotherapy are rarely, if ever noted.

The role of surgery in patients with distant disease is to provide for palliation so the patient may live out the rest of their life without the potentially miserable local symptoms of bleeding or obstruction. A small subset of patients with metastatic disease have surgically resectable deposits confined to one organ such as the liver or lung. Some of these patients may be eligible for potentially curative resection.

Outcome of Treatment

Survival in colorectal cancer is directly related to stage. Approximately 90% or more of patients with Stage I disease are cured. Five year survival for Stages II, III and IV disease are approximately 75%, 50% and 5% or less respectively. There is considerable variability in the reported outcomes, which is further confused by the usage of adjuvant therapy. Nonetheless, the previously mentioned numbers should serve as a rough guideline.

With respect to colon cancer, there has been no improvement in survival associated with the use of adjuvant therapy for Stage I or II disease. However, there is evidence for a modest improvement in disease free and overall survival for patients with Stage III disease who receive 5-FU based chemotherapy. A wide variety of dosing schedules and methods of administration have been explored. Typically, 5-FU is used in combination with Levamisole or with Leucovorin. The most commonly utilized adjunctive regimen at present is probably 5-FU intravenously with Leucovorin for an approximately six months' course.

As opposed to colon cancer where surgical treatment failure is usually related to distant disease, there are multiple areas of concern with rectal cancer. Subsequent to surgical treatment of the primary cancer, approximately one third of recurrences are local, one third are systemic and one third are a combination of the two. As such, adjuvant therapy for rectal cancer often includes external beam radiation therapy, a local modality. The usual recommendation has been for postoperative chemotherapy in conjunction with radiation therapy for Stage II or Stage III rectal cancer. However, there is a clear trend towards treating many, if not all, of these patients with preoperative radiation rather than postoperative radiation. This emerging controversy probably bears some discussion.

A survival benefit has been shown with combined modality chemotherapy and radiation therapy administered postoperatively to patients with Stage II or III rectal cancer. This allowed for complete pathologic staging of the tumor, ensuring that patients with Stage I or IV disease would not be submitted to "unnecessary" irradiation. However, a number of issues have limited this approach. Approximately 50% of patients develop a surgical complication or are not sufficiently recovered to allow for initiation of postoperative radiation therapy in a timely fashion. Further, the local tissues are rendered somewhat ischemic by surgical dissection and therefore would be expected to respond less well to irradiation. After surgery, the bowel is fixed by adhesions and the same loops of bowel, more or less fixed in the radiated field, are vulnerable to radiation damage.

On the other hand, prior to surgery, the tumor is well oxygenated and responds better to radiation. There are no adhesions to tether the small bowel in the pelvis and by use of various positioning techniques, the small bowel may be kept safely out of harm's way. Since the radiation is given before surgery, the descending colon brought down into the pelvis for anastomosis at the time of surgery will not be made fibrotic by treatment and may serve as a better reservoir. The irradiated rectum is resected and replaced with descending colon that was out of the radiation field.

The introduction of rectal ultrasonography has allowed for reliable preoperative staging, largely negating the concerns about not knowing the pathologic stage prior to treatment. Further, concerns about excessive surgical complications in patients who were treated with preoperative radiation therapy have generally not been realized. However, there does seem to be an increased risk of perineal wound infection after abdominoperineal resections in patients who receive preoperative radiation therapy.

The addition of sensitizing chemotherapy to the radiation seems to maximize the efficacy of treatment. For these reasons, an increasing number of major centers are utilizing preoperative regimens involving both chemotherapy and radiation therapy for Stage II and Stage III disease. Approximately 15% of patients treated with preoperative chemoradiation therapy will have no residual tumor demonstrable at the time of surgical resection.

Side Effects of Treatment

Surgery

Resection of a colorectal cancer carries with it significant potential for short and long term adverse side effects. Colon resection requires a major laparotomy with the usual risks of bleeding, wound infection or intra-abdominal abscess. Among the most feared complication of colorectal resection is that of anastomotic leak. Reported leak rates vary, but approximately 5% of patients will develop a clinically significant leak. Many of these will require a temporary colostomy for management with further subsequent surgery to reverse the colostomy.

In the long term, there is an approximately 10-15% risk of developing an adhesive small bowel obstruction. Adhesions are an inevitable consequence of abdominal surgery and there are no specific techniques available to completely obviate the possibility of this complication. Although many obstructions will resolve with nasogastric decompression, many others will require surgical lysis of adhesions. Surprisingly, most patients undergoing resection of a colon cancer will enjoy bowel function similar to their preoperative state. The remaining colon is usually able to compensate quite well for loss of even a considerable length of large intestine.

However, resection of even small portions of the rectum can create significant adverse functional consequences. The rectum acts as a storage reservoir for stool and works in concert with the anal canal to allow for socially acceptable bowel function. With loss of the rectal reservoir, bowel function can be substantially disturbed. Patients will often experience frequent small bowel movements and incomplete evacuation. There is also a risk of imperfections in continence that may be expected to increase as the anastomosis is created in a more distal location. In other words, the more rectal reservoir that is resected and the lower the anastomosis, the more bowel function is likely to be disturbed.

In summary, colon resection is usually well tolerated, even when extensive. On the other hand, rectal resection will usually create more disturbance in bowel function. Fortunately, many of the functional disturbances will improve with time. Recently, there has been an increasing emphasis on techniques that preserve function. For example, many surgeons add a surgically created rectal reservoir (colonic J pouch) to procedures that require low or ultra low anastomoses in an attempt to make the colon act more like a rectal reservoir.

Chemotherapy

When the issue of chemotherapy is raised with patients, most patients tend to have dramatic images of the expected toxicity. Specifically, nausea, vomiting and hair loss are felt by many to be uniformly associated with all kinds of chemotherapy. As mentioned earlier, traditional adjuvant chemotherapy for colorectal cancer has been 5-FU based. These regimens are usually reasonably well tolerated. The most common adverse effects are diarrhea, stomatitis, neutropenia and leukopenia. However, these side effects are actually seen in a minority of the patients, generally in the range of 20%. Older patients seem particularly vulnerable to these toxicities. As such, an individualized risk/benefit ratio is particularly appropriate when considering adjuvant therapy for patients with advanced age.

Radiation Therapy

8

Although radiation therapy is effective in reducing the rate of local recurrence after resection for selected patients with rectal cancer, there is a very real functional price to pay for this benefit. Not only is the tumor bed irradiated, but normal surrounding structures such as the urinary bladder, adjacent loops of bowel, bone and soft tissue are also affected. The degree of damage and the risk of complications is directly proportional to the amount of radiation given, the method of fractionation administered and the tolerance of the tissue in the field to radiation. Quality of life studies of patients undergoing adjuvant radiation have shown substantial functional detriment. Usually after surgery, the bowel that is brought down to be anastomosed to the distal rectum will dilate over time and improve its ability to serve as a reservoir. However, radiation induces scarring and pelvic fibrosis thereby impairing the ability of the colon to serve as an appropriate reservoir replacement for the rectum. Patients experience an increased clustering of bowel movements, more stool frequency and have a higher risk of incontinence. Urinary disturbances, sacroileitis and perhaps an increased incidence of adhesive obstructions may result. As mentioned earlier, many of these adverse effects may be less when radiation therapy is given preoperatively.

However, the devastating effects of local recurrence must also be kept in mind. Recurrent rectal cancer can be extremely painful and cause horrific symptoms. As such, the dangers of local recurrence must be weighed against the potential functional compromise associated with radiation therapy. This highlights the critical importance of defining exactly which subgroups of patients truly benefit from external beam radiation. With an increasing emphasis on surgical techniques, such as total mesorectal excision, the appropriate role of radiation therapy, particularly for Stage II rectal cancer, will need to be carefully evaluated.

Posttreatment Surveillance

There is no consensus on the value of posttreatment surveillance for recurrence of colorectal cancer. Most patients who develop recurrent cancer will develop multiple foci of metastatic disease. As such, the majority will not be candidates for curative re-resection. Further, chemotherapy plays only a palliative role in terms of extending length and quality of survival. It rarely, if ever, cures metastatic disease. Since neither surgery nor chemotherapy is likely to provide for a second opportunity for cure, it may be argued that surveillance programs simply identify those patients who will die of disease earlier.

However, there are several important arguments to the contrary. First, a small but significant percentage of patients do develop resectable disease localized to one organ, such as the liver or lung. These patients may be cured by hepatic resection or pulmonary resection. The occasional local recurrence may also be re-resectable for cure. Overall, it appears that careful postoperative surveillance may improve overall survival by somewhere in the range of 2%.

It is important to understand the likely patterns of recurrence when assessing the utility of follow-up modalities. Approximately 80% of recurrent disease is demonstrable within two years of resection and 95% within three years. As such, follow-up is most intensive during the first two to three years. Standard follow-up includes a thorough history and physical examination seeking clinical features suspicious of recurrence. These would include change in bowel habits, blood in the stool, fatigue, weight loss or localized pain in the region of the previous resection. It should be remembered that when colon cancer recurs, it will tend to do so systemically. On the other hand, rectal cancer recurs locally and systemically with roughly equal frequency.

In conjunction with history and physical, most physicians monitor CEA. Carcinoembryonal antigen was first described in the 1960s. The hope was that it would serve as a screening test, much like PSA does for prostate cancer. Unfortunately, approximately one third of colorectal cancers are not associated with an elevated CEA. Further, there are many reasons to have an elevated CEA other than colon cancer. The relatively wide range of "normal" values further hinders the utility of CEA as a screening test.

However, when one is diagnosed with colorectal cancer, CEA can be very useful in follow-up. When the diagnosis is made, most patients will have some degree of elevation of CEA. After resection, the CEA will return to normal, assuming all tumor has been resected. This then serves as a tumor marker that can be followed as an "early warning device". The CEA will typically elevate six weeks to six months prior to any clinical evidence of recurrent disease. This provides the opportunity to allow for early diagnosis of recurrent disease. There are no other blood tests that are as sensitive as CEA and it is not clear that any other serum chemistries or tumor markers add much to simply following the CEA.

Because of the propensity for local recurrence, patients undergoing rectal resection undergo periodic proctoscopic examination of the anastomotic site. Full colonoscopy is typically performed one year after resection to assess for local recurrence as well as search for new adenomas that may later develop into malignancies. Further follow-up colonoscopies are typically performed at 3-5 year intervals thereafter.

Many would add yearly chest x-rays to search for metastatic disease as this is a relatively inexpensive modality. The use of adjunctive follow-up imaging studies such as liver sonography or CT scanning is advocated by some.

From the pragmatic standpoint, the efficacy of intensive follow-up of patients with Stage I disease or older patients with multiple comorbidities can be questioned. In essence, if the risk of recurrence is exceedingly low or the patient would not potentially be a candidate for resection of recurrent disease, such as hepatic resection, then there seems little utility for intensive follow-up. Conversely, young healthy patients with node positive disease present a much stronger case for very careful follow-up.

One should not overlook the human element. Most patients with cancer of any type are frightened and tend to view most any symptom in light of their cancer history. Seeing their physician at scheduled intervals is usually of enormous value in reducing anxiety and improving the patient's sense of well being.

Suggested Readings

1. Winawer SJ, Zauber AG, Ho MN et al. Prevention of colorectal cancer by colonoscopic polypectomy: The National Polyp Study Workgroup. N Eng J Med 1993; 329:1977-1981.
   Proved the powerful protective effect of polypectomy on the development of colorectal cancer.

2. Moertel CG, Fleming TR, Macdonald JS et al. Levamisole and Fluorouracil for adjuvant therapy of resected colon carcinoma. N Eng J Med 1990; 322:352-358.
   Provided convincing evidence of the value of adjuvant chemotherapy for Stage III colon cancer.

3. Gastrointestinal Tumor Study Group: Prolongation of the disease-free interval in surgically treated rectal carcinoma. N Eng J Med 1985; 312:1465-1472.
   Established the value of combined modality chemoradiation in patients with resectable rectal cancer.

4. Vogelstein B, Fearon ER, Hamilton SR et al. Genetic alterations during colorectal-tumor development. N Eng J Med 1988; 319:525-532.
   Classic paper describing the multistep model of colorectal carcinogenesis, ushering in the era of molecular biology.

5. Bodmer WF, Bailey CJ, Bodmer J et al. Localization of the gene for familial adenomatous polyposis on chromosome 5. Nature 1987; 328:614-616.
   Defined the genetic basis for FAP, opening the door to genetic testing for hereditary colorectal cancer syndromes.

8

CHAPTER 9

Carcinoma of the Anal Region

Peter Cataldo

Scope of the Problem

Tumors of the anal canal and anal region are relatively rare, accounting for only 5% of all anorectal malignancies. The incidence of anal canal cancers are only 0.9 per 100,000.However, these rare malignancies do have a significant impact on quality of life. They are often diagnosed late as patients are hesitant to consult their physician regarding problems in the anal and perianal area. In the past, these lesions and their treatment often resulted in loss of sphincter function and necessitated a permanent colostomy. Recent advances in diagnosis, workup and especially treatment has fortunately minimized the number of patients who require a permanent colostomy.

Risk Factors

Age is a risk factor as anal canal cancer occurs most commonly in the sixth and seventh decades (mean age 62). For anal canal lesions, there is significant female predominance with a female to male ratio of 5:1. For anal margin lesions, the sexual predominance is just the opposite with a 4:1 male to female ratio. This may be due to the association between human papilloma virus virus, homosexual, anal receptive intercourse, and malignant neoplasms of the anal margin.

Anal carcinoma has been increasing in the United States over the past 30 years. It is more common in women, more common in blacks than whites, and significantly more common in metropolitan than rural areas. The recent proliferation of the AIDS virus is most likely, partially responsible for the increased incidence of anal neoplasia. For example, the incidence of anal neoplasia in men in the San Francisco Bay area is three times greater than that for the remaining United States. The relative risk of anal carcinoma in a homosexual male is currently 10.3 versus the normal population. Approximately 1% of gay males with the HIV virus develop carcinoma of the anal region. However, fully 15% of these individuals may harbor anal dysplasia. Currently most individuals with the AIDS virus die of systemic disease before anal carcinoma becomes clinically manifest.

Like cervical cancer, human papilloma virus has been implicated in the pathogenesis of squamous cell carcinoma of the anal margin and the anal canal. Epidemiologic and demographic data support the claim that individuals at risk for human papilloma virus transmission are also at risk for anal carcinoma.

Both anal and cervical cancer are associated with certain genotypes of the human papilloma virus. There have been approximately 60 genotypes identified, 20 of which affect the anal and genital regions. Types 16 and 18 have been traditionally associated

Surgical Oncology, edited by David N. Krag. ©2000 Landes Bioscience.

with cervical and anal cancer. In addition, types 31, 33, 34, 35 are associated with high grade dysplastic lesions. DNA from HPV 16 and 18 has been shown to be incorporated directly into the host genome. This may account for its ability to induce carcinogenesis.

Other groups at risk for anal carcinoma are immunocompromised patients. Patients receiving kidney, pancreas, liver, and heart transplants all have a high risk of squamous neoplasia including squamous cell carcinoma of the anal canal. This is presumably related to immunosuppressive therapy. In addition, individuals undergoing chemotherapy for other malignancies are also at risk for squamous cell carcinomas. In this group, carcinoma develops at an earlier age and is often multifocal. Often these present as in situ lesions. Despite their benign microscopic appearance, they are often clinically aggressive secondary to the immunocompromised nature of their host.

Human papilloma virus is a risk factor for the development of epidermoid carcinoma. Almost one half of individuals who develop squamous cell carcinoma of the anus have a prior history of genital warts. Other etiologic factors include venereal disease such as gonorrhea, herpes simplex type II and Chlamydia. Smoking also has a positive correlation with the disease. The incidence of anal and perianal carcinomas is also increased in individuals with Crohn's disease particularly with significant perianal Crohn's disease. Carcinoma may also arise in chronic anal fistulae.

Screening for Anal Carcinoma

Although screening programs are in their infancy, benefits may be shown for early diagnosis and treatment of epidermoid carcinoma of the anus. There have been no studies to identify survival benefits in individuals screened for squamous cell carcinoma of the anus but current programs do exist and recommendations are being developed.

Premalignant lesions have been identified on cytologic samples obtained from the anal canal and have been described as anal intraepithelial neoplasia (AIN) or anal squamous intraepithelial lesions (ASIL).

The technique for screening is as follows. A cytology brush is moistened and passed into the anal canal. The swab is rotated while mild pressure is exerted on the walls of the anal canal. The swab is then withdrawn and the material smeared onto a glass slide which is immediately placed in 95% alcohol. Smears are then stained with Papanicolaou stain as for cervical smears. If a positive sample is obtained, then the anal canal should be inspected carefully with the adjunct of topical 3% acetic acid. This increases the visibility of small squamous lesions. Visible lesions should then be excised or biopsied as appropriate.

Controversy exists as to what action should be taken when biopsies are positive. Current recommendations are extrapolated from the cervical data. It is recommended that low grade AIN be followed with no treatment. However, high grade AIN such as Grade II or Grade III should be treated with excisional therapy. These lesions are either locally excised or destroyed with electrocautery.

It is clear that the population at highest risk will benefit most from screening programs. Epidemiologic data indicates that homosexuals practicing anal receptive intercourse and individuals with advanced HIV infection are at highest risk. Specific recommendations by one group are as follows. Annual cytologic screening should be considered in the following groups:

1. HIV negative men with a history of anal receptive intercourse.
2. HIV positive men with CD 4 counts less than 500/mm^3.
3. Women with high grade cervical intraepithelial neoplasia.
4. HIV positive woman with CD 4 counts less than 500/mm^3.

Clinical Presentation and Diagnosis

Prior to any discussion on neoplasms of the anal canal, it is essential to define the anal and perianal anatomy. There has been much confusion in the past regarding the anatomy of this region. Based on this, the World Health Organization described standard nomenclature for the anatomy of the anal canal, the anal verge and the anal margin. The anal canal extends from the anorectal ring to the anal verge. This is approximately 2 cm above and 2 cm below the dentate line respectively. The tissue distal to the anal verge is described as the anal margin (see Fig. 9.1).

The anal margin contains modified squamous epithelium such that it contains no skin appendages or hair follicles. The lining of the anal canal is divided into three distinct histologic types. The upper anal canal is lined by columnar or rectal type epithelium. The mid portion of the anal canal is lined by cloacogenic or transitional epithelium. Finally, the distal portion of the anal canal is lined by nonkeratinizing squamous epithelium. Electron micrographic studies have shown that squamous epithelium extends into the upper canal and that columnar epithelium can extend to below the dentate line. The distinction between the anal canal and the anal margin is an important one. Often, lesions of the anal margin can be treated with local excision without impairing sphincter function. However, lesions of the anal canal are often more invasive and require combined chemoradiation for cure.

Lymphatic drainage also differs between the anal margin and the anal canal. Above the dentate line in the upper portion of the anal canal, lymphatic drainage generally follows the superior rectal vessels into the retroperitoneum. In the mid anal canal, drainage migrates along the internal pudendal nodes to the obturator nodes. Below the dentate line, the primary lymphatic drainage is via the inguinal nodes.

Individuals with neoplastic lesions of the anus often present late. More than 50% of the individuals have had symptoms for greater than two years prior to presentation. In addition, 10-20% of individuals will have palpable inguinal metastases at the time that they are diagnosed.

Delay in diagnosis is due to a combination of patient embarrassment and fear, and missed diagnosis. Between one quarter and one third of individuals with malignancy of the anal region have been misdiagnosed with benign pathology such as hemorrhoids, fissure, fistula, eczema, or abscess. A careful history, a thorough examination, and a biopsy of any suspicious lesions should make incorrect diagnosis very unlikely.

The common presenting signs are bleeding, pain, drainage and the presence of an anal mass (Fig. 9.2). It is not surprising that anal cancers are commonly misdiagnosed as benign lesions as the symptoms for both overlap significantly.

In any person with significant anal complaints, a complete anal and regional pelvic examination should be performed. These patients are often in pain and very

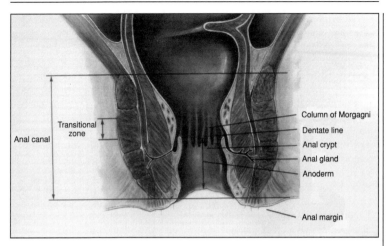

Fig. 9.1. Anatomy of the normal anal canal. Reprinted with permission from Gordon PH, Nivatvongs S. Principles and Practice of the Colon, Rectum and Anus, 2nd edition. St. Louis: Quality Medical Publishing, 1999.

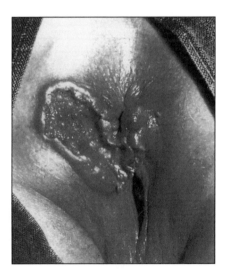

Fig. 9.2. Perianal squamous cell cancer. Reprinted with permission from Gordon PH, Nivatvongs S. Principles and Practice of the Colon, Rectum and Anus, 2nd edition. St. Louis: Quality Medical Publishing, 1999.

anxious regarding their examination. It is imperative to put patients at ease so that a thorough exam can be performed. The patient should be examined in the left lateral decubitus position with the buttocks slightly off the examining table. The buttocks should be spread gently and the anal and perianal tissues carefully inspected. Areas of redness, induration and/or ulceration should be noted. Following this, a careful digital rectal examination should be performed remembering to not only examine the rectum but to carefully examine the anal canal. Bimanual palpation with a second hand placed on the perineum greatly aids in identifying perirectal masses. Once this has been completed, the function of the anal sphincter should be assessed by checking anal tone and by asking the patient to squeeze while feeling for contraction of the puborectalis and the external anal sphincter. Once the rectal exam has been completed, a vaginal exam should be performed in females and a penile and scrotal exam in males. This is particularly important considering the coincident identification of cervical neoplasia and/or genital warts. Following this, the patient should be placed in the supine position and the inguinal area examined thoroughly. The inguinal region is a prime nodal bearing area particularly for lesions of the distal anal canal and the anal margin.

For externally visible lesions, a biopsy can be performed in the office. A punch biopsy or a small incisional biopsy is often adequate. One should obtain a biopsy which includes both the abnormal mass and a small amount of normal surrounding tissue. The location of lesions should be carefully marked prior to sending the samples to a pathologist.

For lesions within the anal canal, endoanal ultrasound may be helpful. In small series, treatment regimens have depended upon the presence or absence of invasion of the anal sphincters. Sphincter invasion can be identified by ultrasonography. For less accessible lesions within the anal canal, examination under anesthesia with biopsy is necessary. At that time, the extent of the lesion can be fully determined and, if appropriate, ultrasonography performed.

Once the diagnosis of cancer has been made, a staging workup is appropriate. Patients should undergo colonoscopy to rule out proximal lesions. CT scan of the abdomen and pelvis should be performed to look for regional adenopathy as well as for hepatic metastases. A chest x-ray will identify pulmonary lesions. Additional workup may be necessary based on any abnormalities identified on history and physical examination.

Staging

The American Joint Committee on Cancer along with the World Health Organization has developed a staging system for anal cancer. As with other cancers, staging has important implications as the five year survival is 80% or greater for early lesions while less than 20% of individuals with advanced lesions survive 5 years. Tumor stage is based on size of the lesion and invasion of adjacent organs. Nodal status is based on the location of metastatic nodes generally identified by CT scan or MRI. Finally, the presence or absence of distant metastases are noted. Slightly different staging systems have evolved for anal canal vs. perianal cancer (see Table 9.1 for complete staging systems).

Premalignant Conditions

Bowen's Disease

Bowen's disease was first described by John T. Bowen in 1912. It is a slow growing intraepithelial squamous cell neoplasia of the skin. It occurs most commonly

Table 9.1. TNM staging system for anal and perianal cancer

Anal carcinoma	Perianal Skin

Primary Carcinoma (T)

Anal carcinoma

Tis Carcinoma in situ
T0 No evidence of primary carcinoma

T1 Carcinoma 2 cm or less in greatest dimension
T2 Carcinoma more than 2 cm but not more than 5 cm in greatest dimension
T3 Carcinoma more than 5 cm in greatest dimension
T4 Carcinoma of any size invading adjacent organ(s) (e.g., vagina, urethra, bladder); involvement of the sphincter muscle(s) alone is not classified as T4
TX Primary carcinoma cannot be assessed

Regional Lymph Node (s) (N)
N0 No regional lymph node metastasis
N1 Metastasis in perirectal lymph node(s)
N2 Metastasis in unilateral internal iliac and/or inguinal lymph node(s)
N3 Metastasis in perirectal and inguinal lymph nodes and/or bilateral internal iliac and/or inguinal lymph nodes
NX Regional lymph nodes cannot be assessed

Distant Metastasis (M)
M0 No distant metastasis
M1 Distant metastasis
MX Presence of distant metastasis cannot be assessed

Stage Grouping

Stage 0	Tis	N0	M0
Stage I	T1	N0	M0
Stage II	T2	N0	M0
	T3	N0	M0
Stage IIIA	T1-3	N1	M0
	T4	N0	M0
Stage IIIB	T4	N1	M0
	Tany	N2	M0
	Tany	N3	M0
Stage IV	Tany	Nany	M1

Histopathologic Grade (G)
G1 Well differentiated
G2 Moderately differentiated
G3 Poorly differentiated
G4 Undifferentiated
GX Grade cannot be assessed

Primary Carcinoma (T)

Perianal Skin

Tis Carcinoma in situ
T1 Carcinoma 2 cm or less in greatest dimension
T2 Carcinoma more than 2 cm but not more than 5 cm in greatest dimension
T3 Carcinoma more than 5 cm in greatest dimension
T4 Carcinoma invades deep extra-dermal structure (e.g., cartilage, skeletal muscle, or bone)

Regional Lymph Nodes
N0 No lymph node metastasis
N1 Regional lymph node metastasis
NX Regional lymph nodes cannot be assessed

Distant Metastasis (M)
M0 No distant metastasis
M1 Distant metastasis
MX Presence of distant metastasis cannot be assessed

Stage Grouping

Stage 0	Tis	N0	M0
Stage 1	T1	N0	M0
Stage II	T2-4	N0	M0
Stage III	T4	N0	M0
	Tany	N1	M0
Stage IV	Tany	Nany	M1

Histopathologic Grade (G)
G1 Well differentiated
G2 Moderately differentiated
G3 Poorly differentiated
G4 Undifferentiated
GX Grade cannot be assessed

9

Modified from Beahrs OH, Henson DE, Hutter RVP et al. Manual for Staging of Cancer. American Joint Committee on Cancer, 4th ed. Philadelphia: JB Lippincott, 1992:83-87, 137-139; Hermanek P, Sobin L. TNM Classification of Malignant Tumors, 4th ed. New York: Springer 1987:50-52.

between the ages of 60 and 70 and generally presents as a pruritic, scaly, eczematoid lesion of the perianal tissues. Often the symptoms have been present for months to years prior to diagnosis. Biopsy reveals intraepithelial neoplasia and characteristic "Bowenoid" cells. These cells are large, multinucleated giant cells with vacuolization giving a halo effect.

In the past, it was thought that greater than 50% of individuals with Bowen's disease harbored internal malignancies. More recent review has indicated that this is a rare occurrence. More recent data suggests that 40% of individuals will develop other cutaneous premalignant or malignant lesions. However, internal malignancies occur in less than 5% of individuals. Therefore, further internal investigation is not currently recommended.

Treatment

Wide local excision is the treatment of choice. Bowen's disease often extends well beyond visible borders and therefore regional "mapping" with biopsies is recommended. Beck recommends serial biopsies 1 cm peripheral to the margins of the lesions. If these biopsies are positive, an additional set of biopsies should be performed. Once the true extent of the lesion is identified this area is excised. The anal and perianal tissues are then primarily closed or reconstructed with local flaps or skin grafts. Insuring negative margins with this mapping technique significantly reduces the rates of local recurrence.

Paget's Disease

Paget's disease was first described by Sir James Paget in 1874 and first identified in the anal region by Darier and Couillaud. Unlike Paget's disease of the breast, extramammary Paget's disease is not invariably associated with underlying malignancy. It is thought to represent a benign neoplasm of the apocrine glands. It may be found in the axilla and in the anogenital region. Coexisting internal malignancies have been found in 50% of individuals and therefore a thorough investigation has been recommended.

Paget's disease presents similar to Bowen's, with an often neglected pruritic, scaly, eczematous lesion of the perianal skin. Diagnosis is made by biopsy and identification of the characteristic Paget's cells. These cells contain a mucoprotein called sialomucin which stains positively with the PAS stain.

Treatment

Treatment is similar to that of Bowen's disease. Wide local excision is recommended in order to prevent local recurrence. If intervening carcinoma has occurred, then abdominoperineal resection is the treatment of choice. Unfortunately in this circumstance, distant metastases are common and the prognosis is poor. The most common site of metastases include the inguinal and pelvic lymph nodes, the liver, bone, lungs and brain.

Verrucous Carcinoma

A spectrum of disease exists beginning with benign anogenital warts and ending with invasive epidermoid carcinoma. Somewhere in the transition from benign to malignant disease fits the giant condyloma acuminatum or Buschke-Lowenstein tumor. These terms generally refer to a low grade verrucous carcinoma which has developed in a giant genital wart. Microscopically, the lesion appears benign but clinically it is locally aggressive. These lesions are often greater than 10 cm and may reach enormous size invading and destroying local tissues to include the anal sphincters, and into the pelvic cavity. Metastases are generally not reported.

Treatment

Treatment is local excision, often in stages. Local excision can be difficult and postoperative hemorrhage is common. In individuals where anal sphincters have been involved or destroyed, abdominoperineal resection may be necessary.

Malignant Neoplasms of the Anal Canal

Malignant lesions can be subdivided into squamous cell and transitional cell (this includes cloacogenic and basaloid). Squamous cell carcinomas are nonkeratinizing and arise from squamous epithelium in the distal anal canal. Cloacogenic cancers arise from transitional cells within the mid to upper anal canal. Distinction between these two lesions is mainly academic as both behave similarly and are treated with the same regimens. Almost all series put these two lesions under the term epidermoid carcinoma of the anal canal.

Lymph node metastases are present in 10-20% of the individuals at the time of diagnosis and eventually develop in an additional 10-25%. Lymph node metastases are directly related to the size of the primary lesion, the depth of invasion, and the histology. Thirty percent of individuals with muscular involvement have nodal disease while this occurs in 58% of individuals when the primary lesion is beyond the sphincters. Nodal metastases are rare in lesions less than 2 cm. High grade carcinomas have been associated with a 35-50% incidence of positive metastatic lymph nodes. Regional lymphatic spread is a marker of systemic disease. Most individuals who succumb to epidermoid anal cancer die as a result of systemic rather than local disease.

9

Treatment

Local excision has been rarely employed for anal cancer. With the advent of endoanal ultrasound, it is recommended by some for well differentiated lesions contained within the submucosa. A small series from the Mayo Clinic reports a 100% five year survival (one patient required abdominoperineal resection for local recurrence) in 13 individuals undergoing local excision. Similar results are reported from a small series at St. Marks in London.

Abdominoperineal resection had been the treatment of choice for many years. However, cure rates were average, morbidity was significant and all patients required a permanent colostomy. In the 1970s, Dr. Norman Nigro pioneered chemoradiation as a primary treatment for this disease. Currently the "Nigro protocol" is the standard treatment for epidermoid carcinoma of the anal canal.

After biopsy and staging workup, patients are scheduled for primary chemoradiation. The original Nigro protocol describes 5-FU 1,000 mcg/m^2 per 24 hours as a continuous infusion over four days. This was combined with mitomycin C 15 mg/m^2 in an intravenous bolus on day one. Patients then underwent external beam radiation for a total of 3,000 rads to the perineal, pelvic and inguinal regions over 28 days. A repeated four day infusion of 5-FU was started on the final day of radiation. Current regimens are similar but often include up to 4,500 rads to the pelvic region. Several studies have questioned the need for concomitant chemotherapy. The data is summarized as follows. Chemotherapy is not associated with an increased morbidity or mortality. It has been associated with decreased local recurrence and increased survival rates.

Three months following chemoradiation, the anal region is carefully assessed for residual tumor. In up to 80% of individuals, no residual tumor is seen. If residual tumor is present following chemoradiation and confined to the pelvis, then abdominoperineal resection is indicated. Cisplatin based chemotherapy has been recommended for extra pelvic metastases.

Tumors of the Anal Margin or Perianal Region

Natural history of squamous cell carcinoma of the anal margin is similar to that of squamous cell carcinoma elsewhere in the body. The majority of lesions can be treated with local excision if negative margins can be obtained. If local excision will not result in negative margins, then the Nigro protocol should be utilized.

Posttreatment Surveillance

There has been no established protocol for follow-up of individuals treated for epidermoid carcinoma of the anus. Follow-up similar to that for colorectal cancer is generally employed. Serial examinations are performed at three month intervals. Examination includes a general physical examination as well as careful examination of the perianal and inguinal areas. Serial chest x-rays and CT scans of the pelvis and abdomen are also included in most follow-up regimens.

Prognosis

As previously mentioned, early lesions treated with local excision have been associated with an excellent overall prognosis. If local recurrence should occur, it can often be treated with either chemoradiation or an abdominoperineal resection.

Abdominoperineal resection, the standard treatment in the past, is associated with a relatively high failure rate. Local recurrence rates from 27-50% have been reported and the five year survival range is from 24-62%.

Nigro protocol is much more successful in eradicating anal cancer. Local control rate ranges between 66 and 81%. Five year survival rates range from 52-88%. T1-3 N0 lesions after chemoradiation have an 80% five year survival rate while node positive lesions have a 52% five year survival rate. Significant negative prognostic factors included positive nodes at times of presentation as well as T4 lesions.

Anal Melanoma

The anorectum is the third most common site for melanoma following the skin and eye. Female to male ratio is 2:1 and the disease affects people most commonly in their sixties. Rectal bleeding, anal pain and the presence of a mass in the anal canal are the most common symptoms. The lesion is often advanced at the time of presentation and systemic symptoms are common.

On physical examination, a pigmented lesion can occasionally be seen in the anal canal. However, a significant percentage of these lesions are amelanotic and therefore may present without any pigmentation.

Treatment

Treatment consists of either local excision of abdominoperineal resection. Prognosis is equally poor with both treatments. Five year survival ranges from 5-22%. Most authors advocate local excision if possible but if the primary lesion is not amenable to local excision then abdominoperineal resection is appropriate. Local treatment should be directed towards relieving or preventing local symptoms as cure following surgical excision is rare. Unfortunately, most melanomas are resistant to chemotherapy and radiation therapy as well.

Adenocarcinoma

Adenocarcinoma is very rare and most actually represent distal extensions of distal rectal tumors. True adenocarcinoma of the anal canal is thought to arise from the anal glands deep within the crypts. These lesions generally develop extramucosally. They present as an extraluminal mass within the anal canal or as lymph node metastases. Diagnosis is made by physical examination with or without the addition of endoanal ultrasonography. Often examination under anesthesia and biopsy is necessary to make the diagnosis. When an individual presents with an enlarged inguinal mass, careful anal examination should be performed to evaluate the possibility of an anal primary.

Treatment

Treatment of these lesions in the literature is reported only anecdotally. Both abdominoperineal resection and Nigro protocol have been reported with varying results. Because of the rarity of these lesions, appropriate treatment remains speculative and prognosis is unknown.

Conclusions

Epidermoid carcinoma of the anus is a relatively rare entity. It often presents late due to patient anxiety and fear, as well as missed diagnosis by physicians. High risk groups include individuals practicing anal receptive intercourse, HIV positive individuals and patients who are immunosuppressed either secondary to chemotherapy or after allograft transplant. Screening should be considered for high risk groups. Anal cytology, similar to that for cervical screening, is currently recommended.

All individuals with anal and perianal complaints should undergo a thorough examination and suspicious lesions should be biopsied. If lesions are identified early, prognosis is good and sphincter function can often be maintained. Nigro protocol utilizing radiation therapy combined with mitomycin C and 5-fluorouracil has dramatically improved the prognosis as well as the functional outcome of individuals with epidermoid carcinoma of the anal canal.

Selected Readings

1. Cleary RK, Schaldenbrand JD, Fowler JJ et al. Perianal Bowen's disease and anal intraepithelial neoplasia. Dis Colon Rectum 1999; 42:945-951.
   This article provides an overview of perianal Bowen's disease, including diagnosis and management.
2. Fuchshuber PR, Rodriguez-Bigas M, Weber T et al. Anal canal and perianal epidermoid cancers. J Am Coll Surg 1997; 185:494-505.

3. Gordon PA, Nivatvongs S. Perianal and anal canal neoplasms. Principles and Practice of Surgery for the Colon, Rectum, and Anus, Second Ed. St. Louis: Quality Med Pub Inc, 1999; 447-471.

4. Klas JV, Rothenberger DA, Wong WD et al. Malignant tumors of the anal canal. Cancer 1999; 85:1686-1693.

    References 2, 3, 4. These articles provide state of the art information regarding epidemiology, diagnosis, and non surgical and surgical management of epidermoid carcinoma of the anal margin and canal.

5. Ky A, Sohn N, Weinstein MA et al. Carcinoma arising in anorectal fistulas of Crohn's disease. Dis Colon Rectum 1998; 41:992-96.

    This article summarizes the association of anal Crohn's disease and squamous cell cancer of the anal canal, including diagnosis and management.

6. Palefsky JM. Anal squamous intraepithelial lesions: relation to HIV and human papillomavirus infection. JAIDS 1999; 21:S42-S48.

7. Pfister H. The role of human papillomavirus in anogenital cancer. Ob Gyn Cl North Am 1996; 23:579-595.

    References 6, 7. These 2 articles summarize the relationship between HIV and anal neoplasia. They include recommendations regarding both screening and treatment.

8. Thibault C, Sagar P, Nivatvongs S et al. Anorectal melanoma—An incurable disease? Dis Colon Rectum 1997; 40:661-68.

    This is an overview of current information available on anorectal melanoma including treatment recommendations.

9

CHAPTER 10

Liver Cancer

Joseph A. Kuhn, Todd M. McCarty, and Robert M. Goldstein

Scope of the Problem

Hepatocellular carcinoma represents the most common tumor in males world-wide. In the Far East and Africa, this tumor has a reported incidence of 150 per 100,000 population. In the United States, hepatocellular carcinoma represents less than 2% of all tumors and has an incidence of 1-4 cases per 100,000 population. However, the incidence of hepatocellular carcinoma is increasing both in the United States and in other countries, mainly as a consequence of hepatitis C.

Risk Factors

There is a male to female ratio of approximately 5:1 in Asia and 2:1 in the United States. Several studies have shown a 250:1 relative risk for the development of hepatoma in asymptomatic HbsAg carriers as compared to noninfected controls.

Approximately 75% of hepatocellular tumors arise in cirrhotic livers. The risk of hepatoma is higher in cirrhosis associated with viral hepatitis compared to alcoholic cirrhosis or other causes of cirrhosis. Historically in Western countries, alcoholic liver disease has been the highest risk factor. This has changed over the past 10 years as hepatitis C-induced liver disease has increased in prevalence to become the most important risk factor for developing hepatoma. Other possible risk factors include aflatoxin, hemochromatosis, hepatic venous obstruction (Budd-Chiari), androgens, estrogens, and alpha-1 antitrypsin deficiency.

Screening

There are no firmly established guidelines for optimal screening of higher risk individuals. It is generally accepted that any patient with hepatitis B, hepatitis C, or alcohol-induced liver disease should be screened in a serial fashion. Due to the high male to female ratio, it is appropriate to consider screening any male patient with cirrhosis. The risk for hepatoma increases by 1-4% per year that an individual has cirrhosis. Therefore, patients with long-standing cirrhosis should definitely be screened.

The appropriate screening test is not as clear as the need for screening. The efficacy of any screening test depends on the ability to detect a tumor at an early enough time to influence the outcome of the patient. The most sensitive test is gadolinium-enhanced MRI. However, the cost of this test limits widespread applicability as a screening test. Ultrasound is the most common diagnostic screening tool due to a lower cost and widespread availability, with a sensitivity of 70% and a specificity of 93%. However, the nodularity of a cirrhotic liver and the ability of a

Surgical Oncology, edited by David N. Krag. ©2000 Landes Bioscience.

hepatoma to blend into the background of a cirrhotic liver limit ultrasound. CT scan offers no significant improvement over ultrasound.

Alpha-fetoprotein is produced by 40-80% of hepatocellular carcinomas. False positive values may be seen in patients with acute or chronic hepatitis, germ cell tumors, or normal pregnancy. The other available assay is that for des-y-carboxy prothrombin protein induced by a vitamin K abnormality (PIVKA-2), which is increased in as many as 91% of patients with hepatoma.[1] However, this level may also be elevated with chronic active hepatitis or vitamin K deficiency.

Depending on the number of risk factors, these patients may be monitored with a serologic test for alpha-fetoprotein every 3-12 months and liver ultrasound every 6-12 months. For very high-risk patients or patients with severe nodularity, a gadolinium-enhanced MRI would be most appropriate.

Diagnosis

The cirrhotic liver is eventually composed of numerous regenerating nodules as a component of the underlying fibrotic process. There is a molecular progression to dysplastic nodules and then to hepatocellular carcinoma. Based on needle biopsy results, it is often difficult to distinguish this continuum from hyperplastic or regenerating nodule to dysplastic nodule to hepatoma. The clinician must follow these nodules with a variety of radiological tools. Radiological evaluation may include ultrasound, CT scan, or contrast-enhanced MRI. Often, the initial test may be an ultrasound, based on a screening exam in a high-risk individual. However, ultrasound is limited by the difficulty in differentiating a regenerating nodule from a dysplastic nodule or a hepatocellular carcinoma. Following these lesions with ultrasound requires close attention to the size of each nodule.

A CT scan may be done with dynamic infusion of contrast, with helical images, or with portography based on mesenteric injection of contrast. Although nodules can generally be seen, the CT scan is also limited by the inability to distinguish hyperplastic nodules from hepatoma (Fig. 10.1A). The CT scan is also limited by the decreased ability to demonstrate vascular invasion, which is an important criteria of resectability and staging. CT with arterial portography is limited in the cirrhotic liver by abnormal flow characteristics resulting in a high incidence of false positive findings and flow artifact areas.

Positron emission tomography (PET) has been shown to be very sensitive for the detection of adenocarcinoma and lymphoma. However, early experience with hepatoma has not been promising due to the low metabolic activity and uptake of the radiolabeled glucose product. As newer targeting agents become available, the use of PET for staging and diagnosis may be reassessed.

Magnetic resonance imaging (MRI) is generally considered to be the ideal imaging technique for optimal lesion identification, and is typically performed with gadolinium infusion (Fig. 10.1B). Ideally, the MRI should be done with "breath-holding" technique and rapid image acquisition. Older techniques with averaging for breathing result in blurred images that were difficult to interpret. With attention to these imaging techniques, MRI has been shown to be the best test for assessing major vascular invasion and for differentiating nodular hyperplasia from hepatoma.

Biopsy is not routinely necessary in cases of a clearly abnormal mass in a cirrhotic patient with or without an elevated alfa-fetoprotein level. When suspicious or

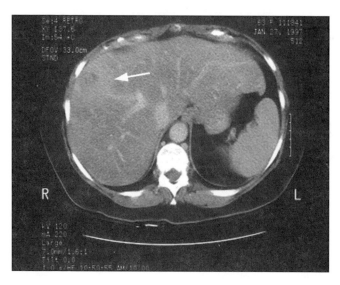

Fig.10.1A. This contrast enhanced CT scan correctly visualized a large hepatoma, but also identified possible posterior lesions that were found to be hyperplastic nodules.

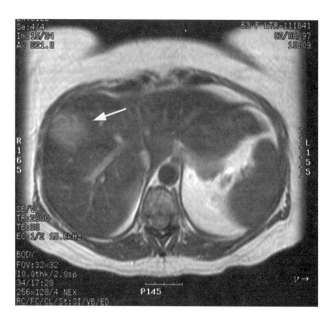

Fig. 10.1B. Gadolinium enhanced MRI in the same patient may be more effective for characterization of the large hepatoma and surrounding parenchyma.

indeterminate nodules are visualized, needle biopsy may be considered when management may be influenced. However, it is important to note that the pathologist often has a difficult time differentiating a regenerating nodule from a well-differentiated hepatoma. A negative biopsy from a highly suspicious lesion should not preclude operative management.

Laparotomy or laparoscopy may occasionally be indicated in patients with nodular cirrhosis and coagulopathy in order to obtain a biopsy of indeterminate suspicious nodules. However, the presence of a severely cirrhotic and nodular liver makes intraoperative ultrasound a difficult task due to the problems obtaining good contact of the ultrasound probe and the nodular surface.

Preoperative Evaluation

The radiological tests, which help to establish the diagnosis, also help to determine the treatment alternatives. Specific attention must be given to the number, size, and vascular relationships of the liver tumors. Tumors located near the bile duct bifurcation or major bile ducts should be identified since this could preclude ablative regimens. Evidence of extrahepatic disease must be sought.

The degree of hepatic reserve must be estimated, based on available tests of liver function including prothrombin time, albumin, bilirubin, liver enzymes, and the presence of ascites, varices, or encephalopathy. The Childs-Pugh classification is helpful, but can underestimate the degree of liver dysfunction. The vast majority of liver tumors in cirrhotic patients are not resectable due to parenchymal dysfunction. Alternate methods of tumor ablation must be considered in the majority of patients with cirrhosis and limited hepatic reserve.

Sophisticed tests of hepatic reserve such as the aminopyrine breath test, measures of serum bile acid concentrations, and indocyanine green clearance have been difficult to standardize and are not routinely available. Ultimately, the determination of underlying hepatic function is based on clinical judgement.

Staging

TNM Staging
Primary Tumor (T)
TX   Primary tumor cannot be assessed
T0   No evidence of a primary tumor
T1   Solitary tumor < or = 1 cm in greatest dimension without vascular invasion
T2   Solitary tumor < or = 2 cm in greatest dimension with vascular invasion
     Multiple tumors limited to one lobe, none > 2 cm, without vascular invasion
     Solitary tumor > 2 cm, without vascular invasion
T3   Solitary tumor > 2cm, with vascular invasion
     Multiple tumors limited to one lobe, any > 2 cm with or without vascular invasion
T4   Multiple tumors in more than one lobe, or tumors involving a major branch of portal or hepatic veins

Regional lymph Nodes (N)
Nx   Nodes cannot be assessed
N0   No regional lymph node metastases
N1   Regional lymph node metastases

Distant Metastases (M)
MX   Metastases cannot be assessed
M0   No distant metastases
M1   Distant metastases

Stage Grouping

Stage	
Stage I	T1, N0, M0
Stage II	T2, N0, M0
Stage III	T1, N1, M0
	T2, N1, M0
	T3, N0, M0
Stage IVA	T4, any N, M0
Stage IVB	any T, any N, M1

Treatment

Liver Anatomy

Knowledge of liver anatomy is essential prior to liver resection or ablation. The Bismuth-Couinad nomenclature divided the liver into 8 segments, beginning in the caudate lobe as segment 1, and then proceeding in a clockwise fashion, beginning with the left lateral lobe superior segment as segment 2, the left lateral inferior segment as segment 3, and ending with the right anterior segment as segment 8. (Fig. 10.2) Recent textbooks have begun to favor use of specific anatomic terminology rather than the Couinad classification. The anatomic terminology divides the left lateral lobe into superior and inferior segments. The right lobe is divided into four quadrants, including superior posterior, superior anterior, inferior posterior, and inferior anterior (Fig. 10.3). The interlobar fissure divides the right and left lobes. The falciform ligament divides the medial and lateral segments of the left hepatic lobe. Both systems base the divisions along vascular planes and are therefore essential to understand when determining the resectability of a hepatic tumor.

Surgical Resection

In the noncirrhotic patient, surgical resection is generally considered the preferred approach. The type of resection may include an anatomic segmental resection, a nonanatomic wedge resection, total lobectomy, or trisegmentectomy. An extended right lobectomy would include a portion of the medial segment of the left hepatic lobe. An extended left lobectomy would include the left hepatic lobe and some portion of the right hepatic lobe. The caudate lobe may be removed alone, or may be removed as part of a left or right hepatic lobectomy.

The technique of liver tissue separation may involve use of a Harmonic scalpel/scissors, the Cavitron Ultrasonic Aspirator (CUSA), blunt techniques with a scalpel blade or finger fracture, an unsheathed suction catheter, or simple electrocautery. The CUSA works by providing a rapidly vibrating instrument with simultaneous suction and irrigation. It is ideal for visualizing intrahepatic vessels prior to division. It does not provide coagulation during transection. Vessels are clipped, tied, or cauterized as they are visualized. The process can be more time-consuming and may ultimately be associated with more than expected blood losses due to the time required.

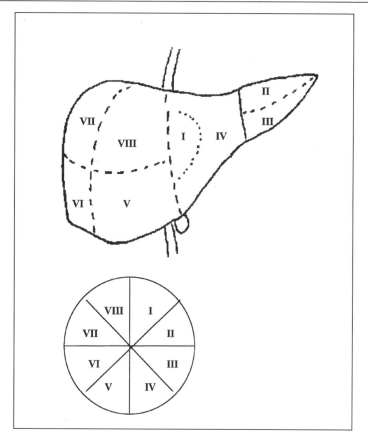

Fig. 10.2. Representation of Bismuth-Couinad nomenclature for liver segmental anatomy.

The Harmonic scalpel/scissors is composed of an instrument that uses ultrasonic vibration, which leads to coagulation and tissue separation. This generally allows for less bleeding due to the ultrasonic coagulation. However, this device provides a charred tissue edge, which generally precludes visualization of the large vessel anatomy prior to tissue transection. Therefore, the Harmonic scalpel/scissors is better suited for nonanatomic wedge resections, which do not require visualization of large vascular structures.

Some surgeons use blunt dissection with scalpel, clamp, or finger. This is felt to be a more rapid approach and may result in equivalent blood loss (when compared to the ultrasonic devices) due to the shorter resection time. In cases of massive bleeding or severe operative laceration, a rapid blunt tissue separation may be most appropriate.

A Pringle maneuver, which involves vascular inflow occlusion, is frequently used regardless of which surgical technique is used for hepatic resection. Complete inflow

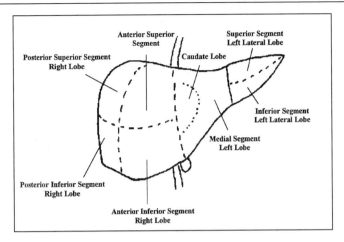

Fig. 10.3. Representation of liver segmental anatomy based on simplified anatomic terminology.

occlusion can safely be done for 45 minutes and in some patients as long as 60 minutes. Intermittent clamping or selective occlusion of the artery or vein may prolong this time.

Controlled hypotension with a low central venous pressure may be used as an adjunct to the Pringle maneuver in selected cases. In very complex cases, complete vascular exclusion and veno-veno bypass with or without liver hypothermia may be indicated.

Technical issues must also include a decision about whether to utilize a cell-saver for aspiration of operative blood loss. In spite of the concern about tumor cell dissemination, many surgeons would consider use of a cell saver in cases of a major hepatic resection when there is no chance of tumor transection or direct tumor spill.

Tumor Ablation Techniques

These techniques should generally be considered only in circumstances in which the tumor cannot be resected by the standard techniques. The indications for tumor ablation includes bi-lobar tumors, multiple tumors, and insufficient hepatic reserve (cirrhosis), and severe comorbid illness.

Radiofrequency Thermal Ablation

The equipment for thermal ablation has largely relied on radiofrequency waves with resultant high temperature, which leads to tissue coagulation and cell death (Fig. 10.4). The advantage of this equipment is related to the small size of the trocars, the absence of adjacent tissue damage around the shaft of the trocar and the possibility to perform the procedure percutaneously or laparoscopically in selected cases. In cirrhotic livers, the use of the small trocars is particularly important. The small trocar size decreases the risk of bleeding from the trocar site or cracking of the liver surface. There is limited clinical data with short follow-up (12-15 months), which shows an approximate 2-10% recurrence of the ablated lesion.[2,3] A probe

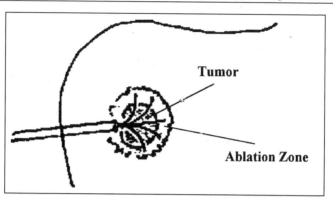

Fig 10.4. Radiofrequency ablation probe with retractible tines and a 12 gauge needle. This creates a trapezoidal zone of ablation.

with 3.5 cm diameter tynes will create a pyramidal zone of ablation measuring approximately 3.5 x 3 cm. A probe with 5 cm tynes with average temperatures of 100-110°C will create a lesion approximately 5 cm in diameter in 30 minutes. For a 5 cm tumor, the probe must be repositioned multiple times so that a 7 cm zone of necrosis is achieved. Repositioning must be done with ultrasound guidance and is somewhat difficult due to the gaseous changes which occur following an ablation cycle. There are no absolute limits about the number of lesions which might be ablated. However, the location of the lesions must be clear from the main bile ducts in order to avoid bile duct injury and stricture. The main limitation of radiofrequency ablation is related to the relative difficulty of intraoperative assessment of the extent of tissue destruction and the short follow-up from existing reports.

Cryo-ablation

The equipment for cryo-ablation relies on liquid Argon or Nitrogen. The probes are designed to deliver a temperature of −170-190 C to the tip. Tissue necrosis occurs by direct freeze injury, microvascular thrombosis, and crystal formation with cell death during thawing. A typical 5 mm probe would lead to a 3.5-4 cm iceball. An 8-10 mm probe would lead to a 6 cm iceball. The probes are often placed in combination in order to achieve an adequate zone of ablation at one time (Fig. 10.5). Two freeze-thaw cycles are optimal and typically requires about 25-30 minutes. The lesion is usually thawed under ultrasound guidance in order to allow the outer 1 cm (which may only reach −10-20°C) to thaw prior to a repeat freezing. Preclinical experience shows a more effective tissue necrosis at these temperatures with a second freeze cycle. The primary limitation of cryo-ablation for hepatoma is related to larger probe sizes and higher incidence of hepatic cracking and bleeding in a firm, cirrhotic liver.

Alcohol Injection

Absolute ethanol has been used for local tissue lysis for over 50 years. Ethanol-ablation is ideally suited for hepatocellular carcinoma due to the generally softer

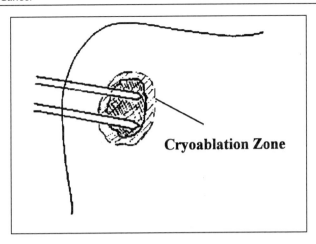

Cryoablation Zone

Fig. 10.5. Cryoablation probes and the zone of ablation/freezing with anticipated 1 cm margin around the tumor.

consistency of the tumor, which limits the tendency of the ethanol to leak into the surrounding parenchyma. Ethanol injection is generally considered a palliative tool in patients with insufficient hepatic reserve and unresectable lesions based on vascular invasion or extrahepatic involvement. Although generally used percutaneously by interventional radiologists, ethanol injection may also be considered at the time of surgical exploration when the tumor is found to be unresectable or otherwise not a candidate for thermal or cryo-ablation. Tumors larger than 2.5-3.0 cm are considered too large for alcohol injection.

Liver Transplantation

Initial enthusiasm for transplantation in patients with hepatoma has steadily diminished due to the limited availability of donor livers and the high incidence of systemic recurrence following transplantation and immunosuppression. Adverse factors which increase the likelihood of hepatocellular carcinoma recurrence after liver transplantation include tumor size > 5 cm in diameter, presence of major vascular invasion, presence of positive lymph nodes, and tumors of higher grade. However, in selected patients with either a single tumor or no more than three tumors (none greater than 3 cm) a 4 year survival of 75% and a recurrence rate of 8% has been reported. The only absolute contraindications include major vascular involvement or extrahepatic disease, and each transplant center must define their own specific policy or philosophy. In actual practice, the primary limiting feature is related to the waiting time required for availability of a suitable donor liver. During this wait, many patients will progress and no longer be candidates for transplantation. In many centers, patients will receive interim treatment with thermal ablation or chemoembolization while waiting for transplant consideration.

10

Chemotherapy

Adjuvant Chemotherapy following Resection
There is no established chemotherapeutic regimen that has demonstrated a survival benefit following resection or transplantation.

Systemic Chemotherapy for Advanced Disease
Intravenous multi-agent chemotherapy regimens have been associated with 10-20% response rates, but have shown no significant effect on overall survival. The most active systemic agents are doxorubicin and cisplatin.

Regional Chemotherapy with Chemoembolization
In contrast to the poor response rates for systemic chemotherapy, a large number of reports have shown promising response rates of 40-68% for a variety of agents which may be administered through the hepatic artery as a bolus injection, often with some agent such as lipiodol, microspheres, or starch which allows for arteriolar embolization. These agents include doxorubicin, cisplatin, mitomycin C, and neocarzinostatin.[4] Side effects include fever (95%), abdominal pain (60%), anorexia (20%), and ascites (20%).

Outcome
Five-year overall survival rates following resection, cryosurgery, or thermal ablation have been reported at approximately 40-70% for stage I or II disease. Patients with vascular invasion (Stage III) who undergo resection may be associated with a 20-40% 3-year survival. Although transplantation has been reported with a 40-60% 3-year survival, this approach is no longer immediately available due to the limited organ availability and low incidence of disease free survival. This group of patients is often treated with intra-arterial chemotherapy before or after surgery under controlled clinical trials.

Posttreatment Surveillance
Follow-up should include serial measures of alpha-fetoprotein following curative treatment. Ultrasound or MRI offers a reasonable method for annual radiological surveillance.

Special Considerations: Other Types of Primary Hepatic Malignancy

Fibrolamellar Hepatocellular Carcinoma
This variant is not typically associated with hepatitis B, hepatitis C, or elevated AFP. Neuroendocrine differentiation has been identified histologically. A CT scan may show pathognomonic intratumoral calcification similar to that seen with focal nodular hyperplasia. Some reports have concluded that this variant may be associated with improved long-term survival after hepatic resection.

Epithelioid Hemangioendothelioma
This low grade, malignant, soft tissue tumor is of endothelial origin. It is seen in patients from 19-86 years of age and may be related to vinyl chloride exposure.

Sarcomas

Angiosarcoma (malignant hemangioendothelioma) is the most common hepatic sarcoma, but remains a very rare liver tumor. It is generally seen in elderly males. Most patients die within 6 months of diagnosis, with few patients surviving more than 1 to 3 years following complete resection.

Cholangiocarcinoma

There are approximately 2500 cases of bile duct malignancy diagnosed each year in the US. Bile duct cancer may be classified into three anatomic groups:

1. intrahepatic,
2. perihilar, and
3. extrahepatic or distal.

The peri-hilar tumors are most common (70%). Risk factors includes liver flukes, choledochal cysts, hepatolithiasis, and sclerosing cholangitis. An anomalous junction of the pancreatic and bile duct junction outside the wall of the duodenum (common channel greater that 10 mm) has been shown to be associated with an increased incidence of bile duct and gallbladder cancer.

Pathology

More than 95% are adenocarcinoma, with other histologic types including squamous, carcinoid, mucoepidermoid, and granular cell tumors. The perihilar tumors have been classified by Bismuth. Type I tumors involve the common hepatic duct. Type II tumors are located at the hepatic duct bifurcation but do not involve the secondary branches on the right or left. Type IIIa tumors involve the right secondary branches. Type IIIb tumors invade the left secondary branches, and Type IV tumors involve both the right and left secondary intrahepatic ducts.

10

Diagnosis

CT scan, ultrasound, or MRI generally identify tumor location, which demonstrate a mass or the level of biliary obstruction. These studies also characterize the extent of involvement around the portal vein and hepatic artery. Percutaneous or endoscopic cholangiography is the ideal method for defining the extent of disease in the secondary hepatic ducts, for obtaining brushings, and for placement of stents which allow biliary decompression and facilitate operative localization of the biliary ducts.

Staging

Primary Tumor	
T1	Tumor invades mucosa or muscle layer
T2	Tumor invades periductal tissue
T3	Tumor invades adjacent structures
Regional Nodes	
N0	No regional lymph node metastases
N1	Metastases in regional lymph nodes
Metastasis	
M0	No distant metastasis
M1	Distant metastasis

Stage Grouping	
Stage I	T1, N0, M0
Stage II	T2, N0, M0
Stage III	T1, N1, M0
	T2, N1, M0
Stage IVa	T3, N0-1, M0
Stage IVb	T1-3, N0-1, M1

Treatment

Surgery

The type of surgery depends on the site and extent of disease of the primary tumor. For intrahepatic cholangiocarcinoma, the surgical approach is similar to a primary hepatocellular carcinoma. For perihilar cholangiocarcinoma which is localized to the bile duct bifurcation without involvement of the secondary bile ducts (Bismuth type II and I), a local resection of the tumor and adjacent nodes should be considered if the portal vein and hepatic artery are not involved. When secondary bile ducts to the right or left are identified (Bismuth IIIa or IIIb); resection of the involved hepatic lobe may also be planned.

Adjuvant Therapy

There are no clinical trials which convincingly show any survival benefit to adjustment chemotherapy.

Palliation

Self-expanding metallic stents are indicated in patients with metastatic or unresectable tumors. Numerous reports have suggested that radiation therapy may improve survival for patients with unresectable cholangiocarcinoma. Some techniques to enhance the effect of radiation include intraoperative radiotherapy, brachytherapy with iridium 192 or cobalt 60 via percutaneous or endoscopic stents, or charged particle therapy. Prospective studies involving combination chemotherapy and radiation therapy are necessary in order to determine the value of this treatment.

Hepatic transplantation has previously been used in patients with unresectable intrahepatic and perihilar cholangiocarcinoma. Poor outcome and limited donor availability have rendered this treatment impractical.

Outcome

For resectable intrahepatic cholangiocarcinoma, median survival has varied between 18 and 30 months, and the 5-year survival rate has been reported from 35-40%.[5] For patients with resectable perihilar cholangiocarcinoma, median survival is 18-24 months and the 5-year survival rate is approximately 10-30%.[6] Patients with distal cholangiocarcinoma have a median survival of 32-38 months and a 5-year survival rate of 35-45% following resection. Median survival for unresectable cholangiocarcinoma within the liver or perihilar regions ranges between 5-8 months despite multimodality radiation chemotherapy.

Liver Metastases
Scope of the Problem

A wide variety of cancers will metastasize to the liver. Approximately 50% of patients with colorectal carcinoma or gastric carcinoma will develop liver metastases. Nearly 75% of patients with pancreatic carcinoma will develop liver metastases. Endocrine tumors frequently metastasize to the liver, including carcinoid, gastrinoma, and insulinoma. Patients with breast cancer, melanoma, renal cell carcinoma, and sarcoma may also metastasize to the liver. In the absence of other sites of metastases, the surgeon may consider ablation or resection in select cases.

Screening

Patients with node-positive colorectal carcinoma are considered to be at high risk and should undergo serial measurement of CEA at 3-6 month intervals. In patients with node-negative colorectal carcinoma, the benefit of serial CEA measurement is somewhat more controversial, but is still generally practiced at less frequent intervals. A CT scan is often performed on node-positive colorectal cancer patients on an annual basis for approximately 3-5 years. Early detection may allow for potentially curative treatment prior to the time when abnormal liver enzymes are detected or symptoms develop from liver metastases.

Diagnosis

Radiologic evaluation of the liver has improved dramatically in the past 10 years and tumors as small as 3-4 mm are detectable. Prior survival studies based on presumptive removal of a solitary lesion may be inaccurate when compared to current experience with improved radiological imaging which more accurately reports small nonpalpable disease.

Current CT scans are performed with helical reconstruction, which has improved imaging compared to nonhelical CT. The limitations of CT scan are largely related to the difficulty with identifying subcentimeter lesions which may significantly influence the type of treatment offered. Small metastases are often difficult to differentiate from benign lesions.

CT scan with arterial portography (CTAP) was popularized in 1990 as a technique which could identify the sub-centimeter lesions. Wide acceptance of this technique was limited due to the cost and potential morbidity of simultaneous arteriography. Flow artifacts and false positive rates of 10-15% have markedly decreased the utilization of this technique.

Magnetic resonance imaging (MRI) with a breath-holding and rapid-acquisition technique allows for images that are clear and more readily interpretable. Imaging of metastases is significantly improved with infusion of a contrast agent which is taken up by Kupfer cells. With a contrast agent the specificity for diagnosis of metastases is 98%. Iron oxide MRI also offers improved anatomic information about the precise location of each tumor and the proximity to major vascular and biliary structures. Recent studies have shown that the Iron oxide MRI can alter the clinical management in up to 60% of patients who present with colorectal liver metastases, generally through the identification of additional sub-centimeter tumors. The changes in management included: avoiding unnecessary laparotomy or changing the type of treatment from resection to ablation or placement of an infusion pump.

Positron emission tomography (PET) has been approved for preoperative imaging in patients with liver metastases. Although liver imaging is possible, the primary benefit of a PET scan is related to extrahepatic imaging in the lungs, bones, and abdomen. Several studies have reported accuracy between 90-95%. Additional studies have shown that the PET scan can impact clinical management in approximately 35-65% of patients with liver metastases from colorectal carcinoma.[8] The presence of extra-hepatic disease could preclude attempts at resection or ablation.

Preoperative Evaluation

The presence of extrahepatic disease is generally considered to be a contraindication for resection of liver metastases. The staging of the lungs is generally achieved with a CT scan. Most centers utilize an abdominal and pelvic CT scan for intra-abdominal staging. However, experience with PET scans in patients with liver metastases has shown a significantly higher sensitivity for extrahepatic disease. As experience is gained, the PET scan may become the optimal screening test for the extrahepatic abdomen and chest prior to exploration for colorectal liver metastases.

Monoclonal antibodies, which are labeled with indium 111 or technetium 99, have been proposed as a technique for imaging of the extrahepatic abdomen and chest. These agents are limited to tumors with CEA or TAG 72 expression. The intense normal uptake of these agents in the liver or kidney makes the nuclear medicine image less than ideal. In addition, there is a small incidence of false positive uptake in a draining lymph node due to antigen shedding which has further limited the widespread application of this technique. When these radiolabeled antibodies are used, a gamma detection probe may be used intraoperatively to identify "hot spots" which are seen preoperatively on the nuclear medicine scan. Some centers have investigated Iodine-labeled antibodies, which do not allow for an external image due to the lower energy of this isotope and limited tissue penetration. Rather, a gamma detection probe can be used to identify the areas of antibody uptake. As newer agents become available, interest in tumor specific imaging will increase.

In addition to the assessment for extrahepatic disease, preoperative imaging should also fully characterize the number, size, and vascular relationships of the liver metastases. It is generally desirable to be prepared preoperatively for all apparent tumor deposits and bilobar disease. The test with the highest sensitivity and specificity for subcentimeter metastases is an iron oxide MRI. However, some surgeons will take the patient to surgery based on the preoperative CT scan and rely on intra-operative ultrasound to identify sub-centimeter metastases and the vascular relationships of the metastases.

Patients should be assessed for the presence of hepatitis or cirrhosis. The degree of hepatic reserve must be estimated. This is based on tests of liver function including prothrombin time, albumin, bilirubin, liver enzymes, and the presence of ascites, varices, or encephalopathy. For patients with colorectal hepatic metastases the colon should be assessed for the possibility of a second primary tumor.

Treatment

Surgical Resection

Segmental resection, nonanatomic segmental resection, total lobectomy, or extended trisegmentectomy may be considered in order to remove the tumor or tumor deposits. Knowledge of liver anatomy is essential prior to liver resection.

The first step in the hepatic resection is complete mobilization of the left triangular ligaments and falciform ligament, which allows the liver to be mobilized towards the left. This subsequently allows for division of the right triangular ligaments.

Intraoperative ultrasound should be a standard component of any exploration for liver metastases. Although the ultrasound may be useful to identify new tumor deposits within the liver, it is ideally suited to identify and precisely localize the metastases that were visualized preoperatively. A sonographic guided core biopsy may occasionally be necessary when the diagnosis of a particular lesion is uncertain.

The technique of segmental resection may involve use of a Harmonic scalpel/scissors or the CUSA. Blunt techniques include a finger fracture technique, scalpel blade technique, or suction dissector technique.

Formal lobectomy requires occlusion or ligation of the portal vein branch and hepatic arterial branch associated with the affected lobe. For a right lobectomy, the liver should be fully mobilized, including small venous branches coursing directly into the vena cava. The right hepatic vein should ideally be ligated. A vascular stapling device may also be used to divide the right hepatic vein. Tissue separation through the interlobar fissure can be accomplished with an ultrasonic dissector (CUSA), which allows visualization and direct ligation of large or vascular structures. For a left hepatic lobectomy, the portal venous branch and hepatic artery branch should be ligated. The left hepatic vein may be intrahepatic, and may be difficult to ligate prior to tissue transection. In either case, there may still be significant bleeding during liver transection, despite initial ligation of the associated inflow vessels. When necessary, a Pringle maneuver with a Penrose drain around the entire porta hepatis may be utilized safely for approximately 45 minutes.

Tumor Ablation Techniques

These techniques should generally be considered only in circumstances in which the tumor cannot be resected by the standard techniques. The indications for tumor ablation include bilobar tumors, multiple tumors, extrahepatic metastases, insufficient hepatic reserve and severe comorbid medical conditions.

Cryoablation

The mechanism and technical aspects of cryo-ablation are outlined in the section on primary hepatic tumors. The probes are often placed in combination in order to achieve an adequate zone of ablation at one time. Tumors up to 8 cm (ablation zone = 10 cm) have been treated successfully. Argon-based cryosurgery devices may be used laparoscopically since the freezing is limited to the tip of the probe. Complications of cryosurgery include postoperative bleeding (1-5%), renal insufficiency (1-3%), and cryo-site abscess (1%). Postoperative care requires urinary alkalinization and osmotic diuresis, along with attention to the platelet count and prothrombin time. Current literature has shown an approximate 5-10% recurrence at an ablated lesion.[9] The location of the lesions must be clear from the main bile ducts in order to avoid bile duct injury and stricture.

Thermal Ablation

The mechanism and technical aspects of thermal ablation are outlined in the section on primary hepatic tumors. Potential advantages include the small trocar size that leads to a smaller risk of trocar site bleeding. Larger lesions (greater than

4 cm) are more cumbersome to treat with this modality due to the inability to use more that one probe at a time on a given lesion. For these larger tumors, thermal ablation requires sequential placement in 4-8 quadrants in order to obtain a complete ablation, including a 1-cm margin in all directions. Sequential placement is somewhat difficult due to gaseous changes in the ablation zone. There is limited clinical data, which shows an approximate 2-10% recurrence at the ablation site in patients with colorectal metastases.[2,3] The location of the lesions must be clear from the main bile ducts in order to avoid bile duct injury and stricture.

Adjuvant Chemotherapy Following Liver Resection or Ablation

Adjuvant Systemic Chemotherapy

Following liver resection or ablation, there is no single agent or combination of agents which have demonstrated a survival benefit in an adjuvant setting. In spite of this, many patients who have not previously received 5-FU chemotherapy will be offered this regimen. Folinic acid (5-FU) is an antimetabolite which works by preventing DNA methylation of tumor cells. Patients who have already received 5-FU may be offered Camptosar (CPT 11) or a continuous infusion of 5-FU. Ideally, these patients should be studied in a prospective clinical trial.

Adjuvant Regional Chemotherapy

Decreased recurrence of liver metastases following infusion of floxuridine (FUDR) through the hepatic artery using an infusion pump following resection of liver metastases for colorectal carcinoma has recently been reported.[11] A similar decrease in hepatic recurrences has been reported following infusion of FUDR through the portal vein, which is felt to represent the primary blood supply of microscopic deposits. Based on these studies, it is possible that some high-risk patients could be offered this mode of adjuvant therapy. However, additional experience in randomized clinical trials would be most appropriate.[12]

Chemotherapy for Unresectable Liver Metastases

Systemic Chemotherapy

First line chemotherapy for advanced colorectal carcinoma has included 5-FU, with an average response rate of 20 to 25% and a median response duration of 6 months. Combination therapy with other agents rarely achieves response rates higher than 30%. Camptosar is considered a second line regimen for advanced colorectal cancer with response rates of approximately 25% and median response duration of 6 months.

Regional Chemotherapy

The hepatic artery is the dominant source of blood supply for macroscopic tumor deposits. The liver during the first pass extracts chemotherapeutic drugs that are ideally suited for regional infusion. Floxuridine (5-FUDR) is almost completely extracted (94-99%) during the first pass through the liver. There are seven randomized trials which showed a significantly higher response rate for intrahepatic infusion (48-62%) compared to systemic chemotherapy (0-38%). Two prospective randomized trials (National Cancer Institute study and Memorial Sloan Kettering study) have shown a modest median survival benefit only in patients without evidence of portal node metastases or other sites of extrahepatic disease. Most of the studies were compromised by small numbers, patients who never got a particular treatment, and patients who were allowed to crossover from the systemic arm to the intrahepatic arm. There has been no study which demonstrated a benefit in patients with nodal or other extrahepatic disease. The decision to consider use of an infusional pump for regional chemotherapy is often based on institutional factors, patient factors, and surgeon preference.

The risks of infusional FUDR are related to direct hepatotoxicity or adjacent perfusion into the gallbladder, bile ducts, or duodenum. Hence, proper catheter placement requires removal of the gallbladder and ligation of extrahepatic branches of the right hepatic artery and small branches along the hepatic artery proper. Coadministration of dexamethasone may be considered as well.

Outcome

Five-year disease-free survival following resection of hepatic colorectal metastases have been reported between 5-45%, depending on the margin of resection, number of tumors, disease-free interval from the time of colectomy, and presence of nodal disease at the time of colectomy. The Hepatic Tumor Registry offers the largest multi-institutional experience with hepatic resection for colorectal liver metastases. The best 5-year disease-free and overall survival (25% and 45%) was seen in patients with solitary lesions, > 1 cm margins, > 1 year disease free interval, and negative lymph nodes. However, even in patients with 3 lesions, clear margins, < 1 year disease free interval, or positive nodes, the disease-free survival (10-20%) and overall survival (20-30%) compares quite favorably to the 5 year survival in patients treated with chemotherapy or no therapy (0-1%).[13] Recent experience with liver resection also confirms the importance of the above prognostic factors and emphasizes the value of hepatic resection when disease is limited to the liver.[14]

Long term data for thermal-ablation is not available. Results following cryosurgery have been fairly similar to the results following resection when similar groups with multiple, bilobar liver tumors are compared.[15]

Posttreatment Surveillance

Follow-up may include serial measures of CEA at approximately 6-month intervals following curative treatment. CT scan or iron oxide MRI may be used on an annual basis for detection of recurrent liver metastases. Early detection is warranted since some of these patients may be candidates for repeat resection or ablation.

Selected Readings

1. Carr Bi, Flickinger JC, Lotze MT. Hepatobiliary cancers. In: Cancer: Principles and Practice of Oncology, 5th edition. Philadelphia: Lippincott 1997; 1087-1107.
    This textbook is one of the most comprehensive sources of current multimodality infomration for nearly all tumors.
2. Pearson AS. Intraoperative radiofrequency ablation or cryoablation for hepatic malignancies. Am J Surg 1999; 178:592-599.
    A good review of technical details for this novel technique.

3. Jiao LR, Hansen PD, Havlik R et al. Clinical short term results of radiofrequency ablation in primary and secondary liver tumors. Am J Surg 1999; 177:303-6.
    This article highlights the limitations of short term results of radiofrequency ablation.

4. Berger DH, Carrasco CH, Hohn DC et al. Hepatic artery chemoembolization or embolization for primary and metastatic tumors: post-treatment management and complications. J Surg Oncol 1995; 60:116-121.
    This is a valuable article for understanding the various agents and outcomes after chemoembolization.

5. Schlinkert RT, Nagorney DM, Van Heerden JA et al. Intrahepatic cholangiocarcinoma: Clinical aspects, pathology and treatment. Hepatobiliary Surgery 1991; 5:95.
    This article provides a classification and treatment guide for these uncommon tumors.

6. Pitt HA, Grochow LB, Abrams RA. Cholangiocarcinoma: In: Cancer, Principles and Practice of Oncology 5th Edition. Devita VT, ed. Philadelphia: Lippincott-Raven 1997; 1119-1127.
    This textbook chapter provides an excellent review of these challenging tumors of the bile duct.

7. Schultz JF, Bell JD, Goldstein RM et al. Hepatic tumor imaging using iron oxide MRI: Comparison with computed tomography, clinical impact, and cost analysis. Ann Surg Onc 1999; 6:691-698.
    This article is based on the newest techniques for MRI imaging and contrast agents.

8. Delbeke D et al. Staging recurrent metastatic colorectal carcinoma with PET. J Nucl Med, 1997.
    This is the first of several articles which characterize the potential benefits of this new imaging modality.

9. McCarty TM, Kuhn JA. Cryotherapy for liver tumors. Oncology 1998; 12:979-87.
    This review article provides and understanding of the indications, technique, and complications of cryosurgery.

10. Fong Y, Blumgart LH. Hepatic colorectal metastasis: Current status of surgical therapy. Oncology 1998; 12:1489-1499.
    This is a review article with recent outcomes for surgical removal of liver metastases.

11. Kemeny N, Huang Y, Cohen AM et al. Hepatic arterial infusion of chemotherapy after resection of hepatic metastases from colorectal cancer. NEJM 1999; 341:2039-48.
    This is the principal article that demonstrates the value for regional adjuvant therapy of liver metastases.

12. Kemeny N, Fata F. Arterial, portal, or systemic chemotherapy for patients with hepatic metastasis of colorectal carcinoma. J Hepatobiliary Pancreat Surg 1999; 6:39-49.
    This review article summarizes many of the risks, benefits, and alternatives for regional therapy for liver metastases.

13. Hughes KS, Simon R, Songhorabodi S et al. Resection of the liver for colorectal carcinoma metastases: A multi-institutional study of patterns of recurrence. Surgery 1988; 103:278-284.
    This is a valuable article which provided the first multicenter database emphasizing the value of surgical resection of liver metastases.

14. Fong Y, Cohen AM, Fortner JG et al. Liver resection for colorectal metastases. J Clin Oncol 1997; 15:938-46.
    This is an article which outlines current success with liver resection.

15. Seifert JK, Junginger T, Morris DL. A collective review of the world literature on hepatic cryotherapy. J R Coll Surg Edinb. 1998; 43:141-154.
    This is a good review article for hepatic cryotherapy.

CHAPTER 11

Cancers of the Pancreas and Biliary Tract

James C. Hebert

Cancer of the Pancreas

Scope of Problem

Cancer of the pancreas is a highly lethal disease with the worst prognosis among all the common malignancies. Although it accounts for only 2% of new cancer cases per year, it accounts for 5% of all cancer deaths per year and is currently the fourth leading cause of cancer death of both men and women in the United States (Table 11.1). More than 28,000 Americans will be diagnosed in the year 2000 with cancer of the pancreas and an almost equal number will die of this disease. Worldwide, pancreatic cancer ranks 13th in incidence, but is the 9th most common cause of cancer death in both sexes with an estimated 168,000 deaths per year. Less than 19% of patients diagnosed with pancreatic cancer will survive one year after diagnosis. The five-year survival rate is about 4%.

Cancers of the pancreas can arise from any cell type (Table 11.2). The majority of pancreatic cancers are adenocarcinomas that arise from the pancreatic duct. The smallest ducts are lined with cuboidal epithelium. As the ducts enlarge and progress toward the duodenum, the epithelium becomes columnar with more mucin-secreting cells. More than 60% of pancreatic cancers occur in the head of the gland. While tumors arising from the islet cells of the endocrine pancreas have a better survival rate, they account for fewer than 2% of all pancreatic cancers. Cystic tumors and mucinous cystadenocarcinomas are less common, but they too have a higher survival following resection if there is no evidence of metastatic disease. Tumors arising from the acinar cells of the pancreas are quite rare.

Over the past 20 years the rate of pancreatic cancer has declined in men while the rate among women has remained stable. Although it was felt that men were more likely to develop cancer of the pancreas than women, recent estimates suggest that incidence and mortality rates are almost equal, with women slightly outnumbering men in recent years. African Americans have a higher incidence and mortality rate for pancreatic cancer compared to white and Asian Americans. Discrepancies due to race in this country are felt to be associated with lifestyle-related risk factors such as cigarette smoking and diet, as the risk of developing pancreatic cancer in both Africans and Asians is lower than that of African Americans and Asian Americans.

The risk for developing pancreatic cancer increases after the age of 50. It is rare in patients 30-50 years of age. Most patients are diagnosed between the ages of 60 and 80 years.

Surgical Oncology, edited by David N. Krag. ©2000 Landes Bioscience.

Table 11.1. Pancreatic cancer statistics

2% of all new cancers
5% of all cancer deaths
4th leading cause of cancer death
4% 5-year survival

Table 11.2. Pancreatic cancer types

Duct Cell Carcinoma 90%
    Adenocarcinoma (most common)
    Giant cell carcinoma
    Adenosquamous carcinoma
    Cystadenocarcinoma (serous and mucinous)
    Mucinous carcinoma
    Signet ring carcinoma
Acinar Cell Cancer 1%
Others 9%
    Papillary-cystic neoplasm
    Pancreaticoblastoma
    Mixed-ductal/islet cell carcinoma
    Undifferentiated
    Connective tissue tumors
    Islet cell tumors

Risk Factors for Development of the Disease

Cigarette smoking has consistently been identified as a risk factor for developing pancreatic cancer (Table 11.3). Up to 25% of all cases of pancreatic cancer may be attributable to cigarette smoking. The mechanism for this increase in risk is unknown, but laboratory animals exposed to carcinogens in cigarette smoke have shown susceptibility to developing pancreatic tumors. Recent increases in pancreatic cancer rates among women have followed increased rates of smoking by women.

In some studies, a diet high in meat and fat has been associated with an increased risk for pancreatic cancer. In animal models, increased amounts of dietary unsaturated fats lead to pancreatic carcinogenesis. The way meats are cooked may lead to carcinogenic substances in the fat, or the fat may act to promote the effects of other carcinogens. In contrast, diets high in fruits, vegetables and fiber are associated with a decreased risk of developing pancreatic cancer. Older studies found some correlation with coffee and alcohol consumption with the risk for developing pancreatic cancer, but recent studies have not confirmed these findings.

Many patients with pancreatic cancer have diabetes mellitus. Diabetes is not a cause of pancreatic cancer. Although studies suggest that patients with longstanding diabetes are not at increased risk, patients over 50 years of age who develop diabetes with a negative family history and a lean body mass may be at an increased risk for having pancreatic cancer. Abnormal glucose tolerance tests associated with pancreatic cancer may be related to substances elaborated by the pancreatic cancers. Islet amyloid polypeptide levels, which are increased in patients with pancreatic cancer, can reduce insulin sensitivity.

Table 11.3. Risk factors for developing pancreatic cancer

Environmental Factors

Definite	Cigarette smoking
Possible	Excess dietary fats
	Exposure to DDT and nonchlorinated solvents

Host Factors

Chronic pancreatitis
Genetic predisposition
BRCA 2 mutations
Hereditary pancreatitis
Gardner's syndrome
Familial and atypical multiple mole melanoma (FAMMM) syndrome

Chronic pancreatitis has been associated with an increased risk for pancreatic cancer. However, most patients with pancreatitis never develop pancreatic cancer. Some studies suggest that the association of pancreatic cancer with pancreatitis is likely due to other associated risk factors such as cigarette smoking. In rare cases, some patients with familial forms of chronic pancreatitis are at high risk for developing cancer.

Patients who have had gastric resection to treat ulcers may have an increased risk for developing pancreatic cancer. Less consistently, some studies have identified that women who have undergone cholecystectomy are possibly at higher risk for later developing pancreatic cancer.

There appears to be some occupational risk for developing pancreatic cancer for those working in aluminum mining, the metal industries, the tanning industry and machine repair. Exposure to pesticides such as DDT, certain dyes and certain chemicals such as beta-naphthylamine and benzidine are thought to increase the risk of developing pancreatic cancer.

Between 5 and 10% of pancreatic cancer cases may be related to an inherited risk.

Recent research has demonstrated that many of the risk factors affect the DNA of pancreatic cells resulting in carcinogenesis. Mutations involving the K-ras gene are the most common abnormalities identified with a prevalence of 70-100%, higher than any other cancer type.

K-ras mutations are seen in ductal cells before the presence of invasive cancer and are often seen in chronic pancreatitis. This may be an early event in the pathogenesis of the disease that, when followed by other gene abnormalities, may lead to the development of an invasive tumor. Subsequent mutations in tumor suppressor genes may lead to deregulation of cell growth. A number of mutated tumor-suppressor genes have been identified, the most common being P16, P53 and DPC4 (deleted in pancreatic carcinoma locus 4). Men and women with BRCA2 mutations have a 10 to 20 times higher risk for developing pancreatic cancer than the general population. Patients with close relatives who have had breast or ovarian cancers may have BRCA2 mutations. Other inherited gene abnormalities associated with an increased risk of pancreatic cancer include hereditary pancreatitis, MEN1 (multiple

11

endocrine neoplasia type 1), Gardner's syndrome, nonpolyposis colorectal cancer (Lynch II variant) and familial and atypical multiple mole melanoma (FAMMM) syndrome.

Other than smoking cessation and a diet that is high in fresh fruits, vegetables, fiber and low in fat, there are no specific guidelines for preventing cancer of the pancreas.

Screening for Problem

Currently, there are no specific tests available to screen the general population for pancreatic cancer. Since the tests available to identify pancreatic cancer are relatively expensive and nonspecific, only patients at high risk should have an investigation for pancreatic cancer. To date the only group in which this seems useful, other than those with genetic risks discussed above, are patients with new-onset diabetes who are over 50 years old, have lean body mass and no other reason is found for their diabetes. In these individuals, high-resolution CT scanning may be of value in identifying early tumors. Patients with vague abdominal symptoms with no obvious diagnosis should also undergo CT.

Methods of Diagnosis

The symptoms of pancreatic cancer are quite nonspecific (Table 11.4). Delays in diagnosis are not uncommon. Weight loss appears to be the most consistent finding. Despite a normal appetite, patients demonstrate a significant weight loss prior to other symptoms. Since more than 60% of pancreatic cancers appear in the head of the gland, jaundice is often a presenting symptom. Dark urine and light, clay-colored stools are associated with the jaundice and often, when the jaundice is quite profound, the patients complain of pruritus. Diarrhea due to biliary and/or pancreatic duct obstruction can occur. Many patients also present early with vague abdominal pain and generalized dyspepsia.

Patients with abdominal pain are often treated symptomatically because of the mild nature of the early symptoms. Patients with persistent vague abdominal symptoms should be investigated for an occult malignancy including pancreatic cancer. More advanced cancers, particularly involving the body and tail of the pancreas, present with characteristic midepigastric pain that radiates to the central portion of the back. This pain is usually more severe when the patient is supine and is relieved when the patient is sitting up and leaning forward. This type of pain suggests that there is perineural invasion and splanchnic nerve compression. Patients with this type of pain are unlikely to be cured by resection. Vomiting suggests that the tumor has invaded the duodenum or other viscus, causing a complete or partial bowel obstruction. Patients may also present with a hypercoagulable state and peripheral phlebitis.

The classic finding of painless jaundice associated with a palpable gallbladder (Courvoisier's sign) is present in only about 25% of patients with pancreatic cancer. Occasionally there may be an enlarged liver due to biliary congestion.

In advanced disease there may be enlarged lymph nodes, particularly in the left supraclavicular fossa (Virchow's node). Similarly, on occasion, an umbilical lymph node may be identified (Sr. Mary Joseph's node) as a sign of advanced disease. Patients may present with a mass on rectal exam (Blumer's shelf), representing pelvic

Table 11.4. Signs and symptoms of pancreatic cancer

Early

Weight loss
Dyspepsia
Jaundice (cancers of the head)
Courvoisier's sign (painless jaundice, palpable gallbladder) < 25%
Late

Pain-midepigastric and mid-back
Palpable abdominal mass
Ascites
Palpable lymph nodes (L supraclavicular, umbilical)
Vomiting

metastases. On occasion, patients may have advanced disease and ascites. Occult blood in the stool may be present due to tumor invasion of the duodenum, stomach or colon.

There are no highly specific blood tests to determine the presence of a pancreatic cancer. A number of tumor-associated antigens have been identified as being associated with adenocarcinoma of the pancreas. Of these, CA19-9 appears to be the most sensitive tumor marker for differentiating an inflammatory mass from a tumor. CA19-9 is also a useful marker to follow after resection of tumors. CA19-9 is not elevated in small tumors and appears to rise as the tumor volume increases. It can be elevated in 10% of patients with benign diseases of the pancreas, liver, and bile ducts. Recently, large elevations of CA19-9 have been identified in patients with simple splenic cysts. Elevations have also been noted in patients with ovarian and other tumors. Abdominal ultrasonography and computerized tomography are both likely to demonstrate a mass in the pancreas. In patients with obstructive jaundice, ultrasonography can be performed first to rule out the level of obstruction. If a mass is identified, it should be evaluated for resectability. If no mass is identified on either ultrasonography or CT scanning, then endoscopic retrograde cholangio-pancreatography (ERCP) may be useful in identifying and potentially biopsying a tumor. ERCP also has the potential advantage of differentiating pancreatic cancer from other periampullary tumors such as tumors of the distal common bile duct and ampullary and duodenal tumors.

Patients who have an identifiable mass in the pancreas and serum CA19-9 elevation greater than 120 μ/mL have a positive predictive value of 100% for pancreatic cancer. Some patients cannot synthesize CA19-9 and values may be lower than normal in some patients with small tumors. Normal CA19-9 levels can also occur with some tumor types. Patients who are acceptable surgical candidates and have a suspicion for pancreatic cancer based on these preliminary studies should be evaluated for resectability. If the tumor is resectable and the patient is a good operative risk, then there is no need to attempt to biopsy the mass. If the patient is a poor surgical risk or the tumor is unresectable, then CT-guided fine-needle biopsy should be performed to confirm the diagnosis. If the patient is jaundiced and has pruritus, an attempt at ERCP biopsy with placement of a palliative stent to establish bile drainage is a very useful procedure while awaiting further evaluation. Although external biliary drainage prior to resection may lead to more infections and metabolic complications, internal drainage usually is well tolerated. Occasionally, severe acute pancreatitis can complicate ERCP and compromise resectability. For this reason, ERCP with stenting in patients with resectable tumors should be performed only to alleviate symptoms when there are delays to definitive resection.

Preoperative Evaluation

Many cancers of the pancreas are not resectable at the time of presentation. Surgery, however, remains the only chance of cure for patients with pancreatic cancer. Assessing individuals for curative resection is key to successful treatment. High-resolution, thin section computerized tomography using helical scanning during early phases of intravenous contrast administration is the cornerstone to accurate assessment of the resectability of tumors. When patients are allergic to contrast dye, magnetic resonance imaging with fast scanning and suppressed sequences, both with and without contrast enhancement, allows for similar evaluation. Use of these advanced imaging techniques in recent years has obviated the need for angiography to assess tumor invasion of the portal or the superior mesenteric veins. Patients are deemed suitable for resection by CT or MR imaging if there is:

1. absence of any extra-pancreatic disease;
2. no evidence of involvement of the superior mesenteric vein and portal vein confluence; and
3. no evidence of direct tumor extension to the celiac or superior mesenteric arteries.

CT and MR imaging are unable to detect small, peritoneal metastases and small liver metastases. While CT or MR can accurately assess nodal enlargement, they can not reliably determine whether nodal enlargement is due to tumor metastases.

Recently, endoscopic ultrasonography (EUS) has been used to help identify vascular invasion of the portal vein/superior mesenteric vein (PV/SMV) confluence, to identify lymph nodes, and to biopsy nodes and masses within the pancreas. Because of the proximity of the pancreas to the lateral edge of the PV/SMV complex where the uncinate wraps posteriorly, this area is difficult to visualize by any technique. Only evidence of significant compression or invasion of the vein should suggest that the tumor is not resectable. EUS has limited availability and is quite operator-dependent. It does not seem to offer any more information than high-quality helical CT or MR imaging at this time.

The use of laparoscopic evaluation prior to laparotomy has helped identify small tumor implants along the serosal and visceral peritoneum. Some investigators have argued that the addition of laparoscopic ultrasonography at the time of laparoscopic evaluation has further improved the sensitivity for detecting extra-pancreatic disease, particularly small liver metastases and lymph node metastases. Likewise, there may be some advantage for assessing the tumor-PV/SMV interface. For the most part, if high-resolution helical CT scan or MR imaging using proper technique is obtained preoperatively, laparoscopic ultrasonography does not seem to have much impact. Peritoneal lavage for cytology at the time of laparoscopy is recommended by some. Positive cytology portends a very poor prognosis and some authors would not then proceed with resection even if this were the only finding. This is controversial.

CT-guided, fine-needle aspiration cytology and biopsy are useful in making a tissue diagnosis of pancreatic cancer. Occasionally ductal brushings or biopsy specimens obtained during ERCP are helpful. Because the sensitivity is of questionable value, the decision to proceed with resection does not hinge on a tissue diagnosis of cancer if, indeed, the patient is an acceptable surgical candidate and has no other risk factors for undergoing curative surgery and the preoperative clinical assessment is that of a pancreatic malignancy.

In general, tumors of the body and tail present late and often are unresectable. Approximately 70% of all patients with pancreatic cancer have unresectable disease identified at the time of their initial CT scanning. Of the remaining 30% of patients, one third will have unresectable disease as determined by further preoperative evaluation and laparoscopic assessment. Only about 20% of all patients with pancreatic cancer present with resectable disease.

Staging

Staging for pancreatic cancer has had little effect on survival. Accurate staging will become more important as new therapies are devised. The American Joint Committee on Cancer (AJCC) has described staging by TNM classification (Table 11.5).

Treatment Options

An approach to patients with suspected pancreatic cancer is presented in Figure 11.1. Resectable tumors of the head of the pancreas are best treated by pancreaticoduodenectomy. Some surgeons prefer to do a pylorus-preserving procedure when possible. The advantages in some series appear to be decreased time and blood loss. Disadvantages of this technique over standard pancreaticoduodenectomy appear to be related to decreases in gastric emptying time following pylorus preservation, although this has not been everyone's experience. Operative mortality should be less than 5%. Tumors in the body and tail of the pancreas are treated with distal resection including splenectomy. Occasionally a total pancreatectomy is performed for multifocal disease, but these patients often do not fare well. Some surgeons prefer resection of the PV/SMV complex due to the proximity of tumors. There is no convincing data, however, that this improves survivability beyond standard resection. Ongoing studies of more radical resections have not convincingly shown improvement in survival. Complications from these more radical resections tend to be higher than with standard resections.

Patients, with unresectable disease, who present with jaundice are best treated by endoscopic stenting of the bile duct. Permanent metal wall stents are available for long-lasting palliation. Patients who are found to be unresectable at the time of open surgical exploration should undergo palliative biliary-enteric bypass. Because these individuals often have smaller tumors, a significant proportion may survive long enough to develop duodenal obstruction. For this reason, patients who undergo surgical biliary-enteric bypass should have a prophylactic gastrojejunostomy at the time of biliary bypass. Unresectable patients who have had prior stenting can undergo gastrojejunostomy at a later time if they develop duodenal obstruction. Laparoscopic techniques have increased the options for palliative bypass. Chemical splanchnicectomy with 50% alcohol at the time of surgery is useful to relieve cancer-related pain and may be effective for up to six months.

Table 11.5. American Joint Committee on Cancer (AJCC) staging for exocrine pancreatic cancer

TNM Definitions

Primary Tumor (T)
    TX: Primary tumor cannot be assessed
    T0: No evidence of primary tumor
    Tis: In situ carcinoma
    T1: Tumor ≤ 2 cm; limited to pancreas
    T2: Tumor > 2 cm; limited to pancreas
    T3: Tumor extends directly into duodenum, bile duct or
          peripancreatic tissues
    T4: Tumor extends directly into stomach, spleen, colon or adjacent
          large vessel
Regional Lymph Nodes (N)
    NX: Regional lymph nodes cannot be assessed
    N0: No regional lymph node metastasis
    N1: Regional lymph node metastasis
Distant Metastasis (M)
    MX: Distant metastasis cannot be assessed
    M0: No distant metastasis
    M1: Distant metastasis
AJCC Stage Groupings
    Stage O: Tis, N0, M0
    Stage I: T1 or T2, N0, M0
    Stage II: T3, N0, M0
    Stage III: T1, T2 or T3, N1, M0
    Stage IVA: T4, any N, M0
    Stage IVB: Any T, Any N, M1

11

Patients who have had resections with curative intent should be evaluated for combined therapy with 5-fluorouracil and radiation therapy. Patients with tumors 2-3 cm in diameter with negative resection margins and no nodal involvement have reported 5-year survivals up to 40%.

Unresectable patients, with locally-advanced pancreatic cancer without metastatic disease should be considered for combined radiation and chemotherapy. Studies have shown potential advantages of combined therapy versus radiation or chemotherapy alone with either single or combination drug protocols. Occasionally, patients with locally-advanced pancreatic cancer precluding resection can be down-staged following combined therapy. Whether there is any survival benefit with this approach is unclear.

For patients with advanced disease, there is no definite survival advantage from undergoing combined radiation/chemotherapy. Recently, however, gemcitabine (2',2'-difluoro-2'-deoxycytidine), a nucleoside analogue, has been found to offer a slight increase in survival times when compared to 5-FU. Furthermore, it appears that gemcitabine treatment results in improved symptom control and better performance status. Patients with advanced disease should be considered for enrollment into clinical trials with these and newer agents.

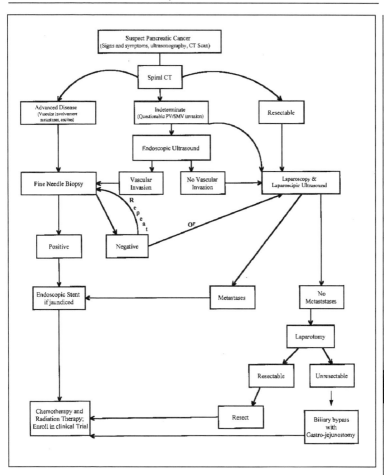

Fig. 11.1. Approach to evaluation and treatment of patients with suspected pancreatic cancer.

Approximately 80% of patients with advanced disease will develop pain. Chemical splanchnicectomy with 50% alcohol is effective in about 60% of patients with pain. Celiac plexus block can be performed at the time of surgery or percutaneously with CT guidance. In experienced hands, EUS-guided blocks may be superior to CT-guided blocks. Many patients have some or complete pain relief, but usually pain recurs after a few months and then becomes persistent. Pharmacologic treatment using a variety of nonopioid and opioid drugs is the cornerstone of pain management. Radiation therapy and chemotherapy may control pain in some instances.

Table 11.6. Factors associated with prolonged survival of patients with pancreatic cancer

Tumors < 2-3 cm in diameter
No lymph node metastases
Well-differentiated histology
No microvascular invasion
Diploid tumors
Less than 2 units in the perioperative period

Outcome of Treatment

Resection of pancreatic cancers has resulted only in a 4% 5 year survival for all patients. A variety of factors account for prolonged survival (Table 11.6). For patients with small cancers, no lymph node metastasis and no evidence of extension of tumor beyond the pancreas, the survival for 5 years following resection may be as high as 20%. The DNA content of pancreatic cancer cells appears to be a significant determinant of survival as well, with diploid tumors having a higher median survival compared to aneuploid tumors. Assessing lymph nodes, other tissue and cytology specimens for markers other than histology (such as mutated K-ras) may allow for more sensitive determination of micrometastases.

Posttreatment Surveillance and Special Considerations

Patients with advanced disease and patients who have undergone resection with chemotherapy and radiation should be considered as having pancreatic exocrine insufficiency. Pancreatic enzyme replacement therapy is important for all patients with pancreatic cancer to maintain nutrition. CA19-9 levels should be monitored following resection for curative intent. Recurrent pain or elevations of CA19-9 are indications that tumors have recurred.

Cancer of the Gallbladder and Extrahepatic Biliary Tract

Scope of Problem

Cancers of the gallbladder and biliary tract are uncommon. In the United States, it is estimated that 6,900 individuals will develop gallbladder and extrahepatic bile duct cancers with an estimated 3,400 individuals dying of these diseases in the year 2000. Superficial gallbladder cancers are often discovered on pathological examination following cholecystectomy for other reasons. In this situation, patients are usually cured without further therapy. Patients presenting with invasive gallbladder cancer usually cannot be cured of their disease. Cancers arising in the extrahepatic bile duct have varying prognoses depending on the site of origin. Tumors of the distal bile duct can be resected 25-30% of the time, while tumors in the proximal bile duct are resectable only occasionally. Although women develop gallbladder cancer more frequently than men, the female to male ratio of 2–3:1, cancer of the bile ducts occurs more commonly in males. The age of onset of gallbladder cancers is usually in the mid-60s, while the age range for diagnosis of carcinoma of the bile ducts is slightly lower, between 50 and 70 years. While gallbladder cancer is less common in African Americans, it is 5-6 times more common in Southwest Native Americans, Mexicans, and Hispanics living in the United States and Native Americans in Alaska.

Table 11.7. Gallbladder cancer histology

Adenocarcinoma-85%
      Papillary (best prognosis)
      Mucinous
      Clear cell
      Signet-ring cell
      Small cell (oat cell)
Squamous-5%
      Adenosquamous
Undifferrentiated-10%

Most gallbladder cancers are well-differentiated adenocarcinomas (Table 11.7). Occasionally squamous cell cancers are found and approximately 10% are highly anaplastic neoplasms. The spread of gallbladder cancer is by lymphatic and venous drainage from the gallbladder. There is usually direct extension into the liver and to periportal lymph nodes and can invade local structures, particularly the common bile duct, causing biliary obstruction.

Adenocarcinomas are also the most common type of extrahepatic bile duct cancers, although there are a few, rare mesenchymal tumors that can also occur (Table 11.8). The disease can be multifocal. Often, there is extensive fibrosis associated with the tumors that makes diagnosis difficult by small fragment or needle biopsy.

Risk Factors

The majority of patients with gallbladder cancers have associated cholelithiasis (Table 11.9). Conversely, less than 1% of patients with cholelithiasis will ever develop gallbladder cancer. It is postulated that chronic inflammation secondary to gallstones somehow leads to the development of cancer. Chronic typhoid carriers have an increased risk also thought to be due to chronic inflammation. Calcification of the gallbladder wall, porcelain gallbladder, is associated with gallbladder cancer in up to 60% of patients with this condition. Although there may be some association with bile duct cancers in cholelithiasis, 50-75% of these patients will not have associated gallstones.

Certain biliary tract infections are associated with a higher incidence of bile duct cancer and these include infestation with *Clonorchis sinensis* and chronic typhoid. Patients with congenital bile duct abnormalities including choledochal cysts, Caroli's disease and hepatic fibrosis have a higher incidence of bile duct cancers than the general population. Patients with ulcerative colitis and associated sclerosing cholangitis have a markedly increased risk for developing bile duct cancer. Bile duct cancers in these individuals tend to be more aggressive, and the risk for developing these tumors is not decreased following total proctocolectomy for ulcerative colitis.

Screening for the Problem

There are no specific screening tests available to screen the general population for gallbladder cancer or cancers of the extrahepatic bile ducts. Patients at high risk, such as those with histories of choledochal cyst or other congenital abnormalities associated with bile duct tumors, and patients with inflammatory bowel disease

11

Table 11.8. Bile duct cancer histology

Adenocarcinoma (most common)
    Types similar to gallbladder cancer
Malignant mesenchymal tumors (rare)
    Embryonal rhabdomyosarcoma
    Leiomyosarcoma
    Malignant fibrous histiocytoma

Table 11.9. Risk factors for gallbladder and bile duct cancer

Gallbladder Cancer
    "Porcelain" gallbladder
    Gallstones
Bile Duct Cancer
    Infections
        Clonorchis sinensis
        Chronic typhoid carriers
    Congenital
        Choledochal cysts
        Caroli's disease
Ulcerative colitis

---

should be monitored closely. There are no recommendations for screening CT scans in these populations, however.

Patients with porcelain gallbladders on x-ray should be evaluated for gallbladder cancer and undergo cholecystectomy.

Methods of Diagnosis

Patients with gallbladder cancer present with signs and symptoms that are indistinguishable from those of cholelithiasis (Table 11.10). Small tumors are often found when patients have concomitant symptoms from their gallstones. More-advanced cancers often present with obstructive jaundice and are often associated with other, nonspecific signs such as weight loss, anorexia and malaise. Occasionally, patients can present with fever and severe pain. Tumors that obstruct the cystic duct can cause a clinical picture indistinguishable from acute cholecystitis. Tumors invading the common bile duct will often present as jaundice; and sometimes, patients have associated ascending cholangitis.

Jaundice is associated with pain, fever from ascending cholangitis or acute cholecystitis due to obstruction of the cystic duct, and pruritus when the jaundice becomes more profound. Patients with bile duct cancers usually present with jaundice.

Ultrasonography will identify a mass in the gallbladder and will demonstrate the level of bile duct dilatation. CT will demonstrate large masses, however, small masses in the bile duct may not be demonstrable with CT. With biliary obstruction, it is necessary to examine the bile ducts either using endoscopic retrograde cholangiopancreatography (ERCP) or transhepatic cholangiography (PTC). For tumors in the proximal bile ducts, PTC is usually preferred. It will allow better visualization of the proximal extent of the tumor as well as provide better access to

Table 11.10. Signs and symptoms of gallbladder and bile duct cancers

Jaundice
Pain
Fever
Weight loss
Anorexia
Malaise
Pruritus

facilitate biliary drainage. ERCP is preferred for tumors of the distal bile duct. With current expertise, both methods allow placement of internal drainage catheters, allow for sampling of the tumor for cytologic diagnosis and direct biopsy. ERCP may have some advantage over PTC secondary to the decreased risk of bile leak due to liver puncture necessary for PTC. Celiac angiography has been useful to determine involvement of the major vascular structures such as the portal vein and hepatic arteries. Newer techniques such as intra-bile duct ultrasonography have provided more detail for more accurate preoperative staging. However, this technique is not widely available and because of the low incidence of this problem, its value for prolonging survival has yet to be demonstrated.

Tumor markers are often normal with gallbladder cancers and bile duct cancers, however, there may be elevations in carcinoembryonic antigen (CEA) and/or CA19-9.

Preoperative Evaluation

Since many of these patients are elderly, a careful assessment of other comorbid conditions should be explored prior to surgical therapy for these tumors. Extensive resections may be contraindicated in patients who are poor surgical candidates.

Prior to surgical resection, it is necessary to have an accurate understanding of the tumor location and the extent of local invasion, as well as the presence of metastatic disease. Tumors of the proximal bile ducts and the hilum (Klatskin tumors) require extensive resections and offer little hope for cure. Patients with tumors of the middle and distal bile duct can sometimes be resected with a chance for cure. While distal metastatic disease is uncommon, intraabdominal metastases, often with small, peritoneal metastases, can occur; and at this point, the disease is incurable. High resolution CR or MR, as with pancreatic cancer discussed previously, provides the most useful staging information. Visualization of the proximal and distal extent of the tumor with ERCP or PTC is likewise essential. Laparoscopy should be performed in all patients prior to laparotomy to rule out peritoneal metastases.

Staging

The American Joint Committee on Cancer has described cancer staging for gallbladder cancer (Table 11.11) and extrahepatic bile duct cancers (Table 11.12).

Treatment Options

For patients with invasive but resectable disease of the gallbladder, most would recommend a wedge resection of the liver around the gallbladder fossa with a 2-3 cm margin. Lymph node dissection includes the nodes from the hilum to the distal common bile duct as well as nodes along the hepatic artery and the celiac axis.

11

Table 11.11. American Joint Committee on Cancer (AJCC) staging for gallbladder cancer

---

TNM Definitions

Primary Tumor (T)
  TX: Primary tumor cannot be assessed
  T0:   No evidence of primary tumor
  Tis: Carcinoma in situ
  T1:   Tumor invades lamina propria or muscle layer
  T1a: Tumor invades lamina propria
  T1b: Tumor invades the muscle layer
  T2:   Tumor invades the perimuscular connective tissue; no extension
        beyond the serosa or into the liver
  T3:   Tumor perforates the serosa (visceral peritoneum) or directly invades
        one adjacent organ, or both (extension 2 cm or less into the liver)
  T4:   Tumor extends more than 2 cm into the liver, and/or into two or
        more adjacent organs (stomach, duodenum, colon, pancreas,
        omentum, extrahepatic bile ducts.)
Regional Lymph Nodes (N)
  NX: Regional lymph nodes cannot be assessed
  N0:   No regional lymph node metastasis
  N1:   Metastasis in cystic duct, pericholedochal, and/or hilar lymph nodes
        (i.e., in the hepatoduodenal ligament)
  N2:   Metastasis in peripancreatic (head only), periduodenal, periportal,
        celiac, and/or superior mesenteric lymph nodes
Distant Metastasis (M)
  MX: Distant metastasis cannot be assessed
  M0:   No distant metastases
  M1:   Distant metastases present
AJCC Stage Groupings
  Stage 0:   Tis, N0, M0
  Stage I: T1, N0, M0
  Stage II: T2, N0, M0
  Stage III: T1, N1, M0 or T2, N1, M0 or T3, N0, M0 or T3, N1, M0
  Stage IVA: T4, N0 or N1, M0
  Stage IVB: Any T, N2, M0
               Any T, Any N, M1

Tumors of the distal gallbladder and cystic duct should include resection of the middle portion of the common bile duct.

For patients with symptomatic cholelithiasis undergoing laparoscopic cholecystectomy, conversion to open cholecystectomy should be performed before manipulation of the gallbladder cancer. Tumor implantation at the trocar sites and intraabdominal dissemination have been reported following laparoscopic removal of gallbladder cancers. There is no evidence to support the roles of chemotherapy and radiation in the treatment of gallbladder cancer.

Proximal bile duct cancers at the hilum of the liver have an extremely poor prognosis, much like gallbladder cancer. While there is some evidence to support the value of radical resections in terms of improved 5-year survival compared to palliative measures, very few patients are suitable candidates for these aggressive resections.

Table 11.12. American joint committee on cancer (AJCC) staging for bile duct cancer

TNM Definitions

Primary Tumor (T)
- TX: Primary tumor cannot be assessed
- T0: No evidence of primary tumor
- Tis: Carcinoma in situ
- T1: Tumor invades subepithelial connective tissue or fibromuscular layer
- T1a: Tumor invades subepithelial connective tissue
- T1b: Tumor invades fibromuscular layer
- T2: Tumor invades perifibromuscular connective tissue
- T3: Tumor invades adjacent structures: liver, pancreas, duodenum, gallbladder, colon, stomach

Regional Lymph Nodes (N)
- NX: Regional lymph nodes cannot be assessed
- N0: No regional lymph node metastasis
- N1: Metastasis in cystic duct, pericholedochal and/or hilar lymph nodes (i.e., in the hepatoduodenal ligament)
- N2: Metastasis in peripancreatic (head only), periduodenal, periportal, celiac, and/or superior mesenteric and/or posterior pancreaticoduodenal lymph nodes

Distant Metastasis (M)
- MX: Presence of distant metastasis cannot be assessed
- M0: No distant metastasis
- M1: Distant metastasis

AJCC Stage Groupings
- Stage 0:   Tis, N0, M0
- Stage I: T1, N0, M0
- Stage II: T2, N0, M0
- Stage III: T1 or T2, N1 or N2, M0
- Stage IVA: T3, any N, M0
- Stage IVB: Any T, Any N, M1

11

For patients whose tumors are potentially resectable, hepatic lobectomy is necessary for potential cure. For tumors extending along the right hepatic ducts or invading the right hepatic vasculature, extended right hepatic lobectomy is necessary. Similarly, left hepatic lobectomies or extended left hepatic lobectomies are necessary for lesions extending into the left lobe of the liver. All patients should have stenting of the biliary anastomosis because of the high incidence of postoperative strictures and recurrent tumors. While hepatic transplantation has been performed for these tumors, results are poor and it is not recommended.

For small tumors of the mid-portion of the bile duct, resection with reanastomosis might be possible. However, reconstruction with a biliary enteric anastomosis, preferably a Roux-en-Y hepaticojejunostomy or other biliary enteric anastomosis is necessary. Lesions involving the distal third of the bile duct require pancreaticoduodenectomy for resection.

External beam radiation therapy has been used in association with surgical resection for bile duct cancers. External beam irradiation, as well as intraoperative radiation or local irradiation with 192 IR wires, seems to have some benefit, although the number of reported cases is too small to make any conclusive judgements. There are no effective chemotherapeutic regimens reported.

The majority of patients with gallbladder and bile duct tumors require palliative therapy. For patients with bile duct obstruction, biliary drainage can be accomplished either with nonsurgical stenting or with surgical bypass. Stents can be placed across obstructing tumors with ERCP or percutaneous transhepatic approaches. Endoprostheses are preferred since they are better tolerated by patients and have fewer resulting complications. Usually unilateral stent placement is adequate, although some patients may require bilateral stent placements to control biliary infection for symptoms such as pruritus. Patients with advanced bile duct tumors that have obvious metastases or unresectable disease should undergo either endoscopic or percutaneous stenting. Patients who are explored for potential resection and are found intraoperatively to be unresectable should undergo surgical drainage as indicated by the location of the tumor. U-tubes have been used in the past for palliation. These involve passing a silastic tube through the abdominal wall and the liver into the proximal bile duct, traversing the tumor and exiting through the distal bile duct through the abdominal wall. Most surgeons prefer to construct their Roux-en-Y hepaticojejunostomy and have the U-tube exit through the Roux-en-Y limb prior to passing the tube through the abdominal wall. This will minimize leak into the abdomen. The advantage of a U-tube is continued access to the tumor to facilitate biliary drainage. Newer endoprostheses, including metal wall stents, have been more widely used recently and seem better tolerated by patients.

Outcome of Treatment

Patients with mucosal cancers or cancers invading only the submucosa often can be cured by simple cholecystectomy. When there is muscular invasion or beyond, survival decreases drastically to less than 15%. The majority of cases of gallbladder cancer, however, are unresectable Stage III and IV disease. Patients with proximal bile duct cancers can rarely be resected. Even when resection is performed, cure is not often achieved. Tumors of the distal bile duct that are localized tend to have much better survival rates. Periampullary bile duct cancers have a much better survival rate compared to similarly staged pancreatic cancers.

With wide resection around the gallbladder fossa associated with regional lymphadenectomy, radical or extended cholecystectomy, survival times have increased slightly. However, five-year survival rates have not improved. Similarly, hepatic lobectomies and other radical resections including an associated bile duct resection and pancreaticoduodenectomy have not led to increased 5-year survivals. There is, however, increased perioperative morbidity and mortality associated with radical procedures.

Posttreatment Surveillance and Special Considerations

Many patients will have recurrences after resection. There are no specific tests available to determine early recurrences. Recurrent gallbladder and bile duct tumors are always unresectable and patients should receive palliative care when appropriate.

Selected Readings

1. Hruban RH, Petersen GM, Ha PK et al. Genetics of pancreatic cancer. Surg Oncol Clin N Am 1998; 7:1-23.
   Overview of the genetics of pancreatic cancer.

2. Gold EB, Goldin SB. Epidemiology of and risk factors for pancreatic cancer. Surg Oncol Clin N Am 1998; 7:67-91.
   Review article of epidemiology and risk factors.

3. Burris HA 3rd, Moore MJ, Andersen J et al. Improvements in survival and clinical benefit with gemcitabine as first-line therapy for patients with advanced pancreatic cancer : A randomized trial. J Clin Oncol 1997; 15:2403-2413.
   Key paper describing trial forming basis for important therapy.

4. Yeo CJ, Cameron JL, Lillemoe KD et al. Pancreaticoduodenectomy for adenocarcinoma of the pancreas. J Clin Oncol 1997; 15:928-937.
   Key article describing the status of surgical resection.

5. Diehl S, Lehmann KJ, Sadick M et al. Pancreatic cancer: value of dual phase helical CT in assessing resectability. Radiology 1998; 206:373-378.
   Article describes current status of diagnostic imaging.

6. Hunstad DA, Norton JA. Management of pancreatic carcinoma. Surg Oncol 1995; 4:61-74.
   Major review article.

7. Chao T, Greager JA. Primary carcinoma of the gallbladder. J Surg Oncol 1991; 46:215-221.
   Review article of gallbladder cancer.

8. Bismuth H, Nakache R, Diamond T. Management strategies in resection for hilar cholangiocarcinoma. Ann Surg 1992; 215:31-38.
   Major review article describing management and therapy.

9. Hejna M, Pruckmayer M, Raderer M. The role of chemotherapy and radiation in the management of biliary cancer: a review of the literature. Eur J Cancer 1998; 34:977-986.
   Major review of the current status of adjuvant therapy for biliary cancer.

11

CHAPTER 12

Lung Cancer 2000

Frederic W. Grannis Jr.

Scope of Problem: Incidence and Mortality

Lung cancer (LC) is the most important unsolved problem in surgical oncology. 180,000 Americans will develop LC in the year 2000; 160,000 will die. Treatment of the disease has limited effectiveness; only 14% of patients attain five-year survival. LC is the number one cause of cancer mortality in the U.S. in both men and women. LC is also the main cause of cancer mortality in Western Europe, and its incidence is rapidly increasing in Third World nations. Global annual mortality will exceed three million during the next century. There has been very little improvement in survival during the past 20 years.

LC is unique in that its cause is known. It is theoretically preventable, by the single measure of primary prevention of cigarette smoking. Unfortunately, the issue is complicated by the fact that the etiologic agent is highly addictive, and is aggressively marketed by a rich and deeply entrenched industry, which is protected by powerful legal and political allies. Furthermore, it is a grim joke that our government actually taxes the victims and their physicians to provide subsidies of many millions of dollars to those who grow the etiologic agent (Fig. 12.1)! There are currently 1.1 billion cigarette smokers on earth. In the U.S. there are approximately 50 million smokers and another 50 million ex-smokers, who are at risk. Smoking cessation (SC) interventions therefore represent another potential strategy to reduce mortality, but current smoking cessation therapy, including group counseling, nicotine replacement and bupropion, have only modest success, and are not widely applied. The risk of LC is approximately 2.5% per decade in smokers. Risk diminishes by half after ten years of abstinence, but probably never returns to that of a never-smoker.

History and Epidemiology

LC was a curiosity in the nineteenth century. Only 350 cases were reported in the medical literature as late as Adler's review in 1912. There was a sudden increase in incidence in the 1920s. Almost all cases were in men, and a striking association with cigarette smoking was soon noted. Ernst Wynder, a medical student working with Evarts Graham, collected retrospective data documenting a strong correlation of the disease with cigarette smoking. This finding was confirmed in large retrospective studies performed by Hammond and Horn. Auerbach demonstrated sequential pathologic changes in the bronchial mucosa in smokers, but not nonsmokers, in two large postmortem studies. Smoking was associated with bronchial hyperplasia, dysplasia, carcinoma in situ and invasive bronchogenic carcinoma. Similar changes

Surgical Oncology, edited by David N. Krag. ©2000 Landes Bioscience.

Fig. 12.1. Tobacco industry executives swear that nicotine is not addictive, before a Congressional committee. © 2000 American Lung Association. Reproduced with permission.

were noted in experimental animals in a famous study of beagle dogs exposed to tobacco smoke.

Molecular Biology

LC arises as the result of prolonged, repetitive exposure of the airway to an aerosolized broth of carcinogens, including polycyclic hydrocarbons and nitrosamines, produced during the combustion of tobacco. Mutations in a number of oncogenes and tumor suppressor genes accumulate over years of repetitive exposure, with the result that LC typically occurs in persons who have been heavy smokers for 10 or more years. LC is uncommon before age 40 and reaches a peak incidence in the 60s and 70s. Although LC in men who have never smoked is very unusual, approximately one quarter of cases in women occur in never-smokers. There is now good evidence that women who have never smoked, but are exposed to environmental tobacco smoke (ETS), who have null glutathione S-transferase m 1 (GSTM1) status, have a dose-dependent increased risk of developing adenocarcinoma of the lung.

Aneuploidy is common in bronchogenic carcinomas with frequent deletions and transpositions of chromosomal material. The pathologic changes noted by Auerbach have been demonstrated to correlate with sequential mutations in bronchial mucosal cells.

Cell Types and Tumor Biology

Cell type characterization by pathologists is important in the determination of prognosis and treatment. Bronchogenic carcinomas account for more than 95% of all primary lung malignancies. These tumors can be separated according to biologic behavior into two major categories, small cell lung cancer (SCLC) comprising about

12

20% of bronchogenic carcinomas and nonsmall cell lung cancer (NSCLC) approximately 80%. Using modern pathologic techniques of immunohistochemistry and electron microscopy, precise differentiation should be possible in almost all cases.

Treatment of SCLC differs markedly from that of the multiple sub-categories of NSCLC. NSCLC is further subdivided into adenocarcinoma, squamous carcinoma, large cell undifferentiated carcinoma and bronchioloalveolar carcinoma, in approximate order of relative frequency.

LC kills by destruction of lung tissue, airway obstruction, local invasion of chest wall, spine and central mediastinal structures, and by distant metastasis.

Although these tumors are grouped together because their clinical behavior is generally less aggressive than SCLC, there are individual differences. Squamous cancers tend to arise centrally in large airways, and may cavitate when they arise peripherally. Adenocarcinomas tend to arise in the periphery and have a higher incidence of lymphatic metastasis than squamous tumors of comparable size. Large cell undifferentiated cancers usually present as large peripheral masses, and frequently have more rapid doubling times. Bronchioloalveolar cancers arise peripherally and may show a different mode of progression, spreading like a pneumonia to surrounding and contralateral lung tissue. Apart from these differences, each NSCLC cell type typically grows rapidly, with a doubling time of 30-90 days. When tumor size reaches 1 cm, spread to hilar and then mediastinal nodes (MN) is first observed. With increase in size of the primary tumor, and lymphatic metastasis, there is an increasing incidence of distant metastasis. LC shows a predilection for metastasis to brain, bone and adrenal in approximate order of avidity. All cell types may cause obstruction of proximal airways complicated by hemoptysis, atelectasis and obstructive pneumonia as the disease progresses.

SCLC is biologically distinct. It arises from the Kulschitzky cells of the bronchial mucosa, usually arising in the submucosa of central bronchi. Doubling time is typically very rapid. Extensive nodal and systemic spread is usually present at clinical presentation. SCLC is infrequently diagnosed in early stages, and is seldom curable by surgical resection alone. Large cell and neuroendocrine variants of SCLC have intermediate behavior.

Carcinoid tumors comprise approximately 5% of lung malignancies. They are subdivided pathologically into typical and atypical varieties. Carcinoid tumor presents as a round, smooth, red submucosal mass in the central airway or less often, as a peripheral nodule, and seldom metastasizes to lymph nodes or distant sites. The more aggressive atypical variant is characterized by cellular atypia, areas of necrosis and more frequent metastasis.

Mucoepidermoid and adenoid cystic carcinomas are uncommon tumors that arise centrally in the trachea and main bronchi. Other rare primary lung malignant tumors include sarcomas, primary pulmonary lymphoma, blastoma and lymphoepithelioma.

Clinical Presentation

The most important item in clinical evaluation is the history of smoking. LC is uncommon in the never-smoker. It is not sufficient to ask if the patient smokes. We must ask if he has ever smoked. Because more than half of all smokers in the U.S. have now quit, LC is now actually more frequent in ex-smokers. Patients who have

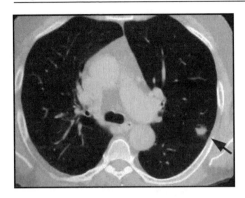

Fig. 12.2. CT scan asymptomatic pulmonary nodule. Stage IA bronchioloalveolar cancer.[1]

had a prior head and neck or LC have risks as high as 20%/10 years of developing a second LC.

Bronchogenic carcinoma is asymptomatic during most of its natural history, with the exception of cough, which is typically ignored in the presence of tobacco-induced chronic bronchitis. When cough is persistent and worsening, or hemoptysis or dyspnea secondary to airway obstruction occurs, disease is often locally advanced or metastatic. Headache, a focal neurological symptom or localized bone pain are common initial findings, signaling distant metastasis. The crucial lesson is that if physicians wait for the patient to present with symptoms of LC, the usual result is disease incurable by current therapy. In happier circumstances a small asymptomatic pulmonary density is discovered in an early stage, fortuitously, on a chest roentgenogram (CXR) done for an unrelated problem or because of a deliberate screening attempt (Fig. 12.2).

Tumors growing in the central airways produce cough and hemoptysis early, sometimes before development of a mass on CXR. Although tumors can be detected early by sputum cytology, trials of cytological screening have proved disappointing. CXRs are not sensitive in the detection of early tumors in the central airways. CXRs may also miss peripheral lesions smaller than 1 cm in diameter and apical, paramediastinal and retrocardiac masses. Bronchoscopic examination of patients with persistent cough or hemoptysis in the presence of a normal CXR is controversial; neoplasm is found in approximately 2% of such cases. LC is painless until far into its natural history. Chest pain is commonly a reflection of parietal pleural invasion or pleural effusion, denoting a locally advanced tumor.

Since most patients with LC also have some degree of chronic obstructive lung disease (COPD), shortness of breath is common. Central airway growth is frequently late in the course of disease and can cause stridor and dyspnea. Extrinsic compression by lymph node metastases or phrenic and recurrent laryngeal nerve palsy secondary to mediastinal invasion are other causes of dyspnea.

Initial distant metastases are to brain and bone. Therefore a careful history of bone pain and neurologic symptoms, as well as a careful neurologic examination are in order. Persistent headache or even subtle disturbance of balance or vision warrants careful evaluation and MRI imaging (Fig. 12.3). Palpation of cervical lymph nodes and liver will occasionally discover metastasis.

12

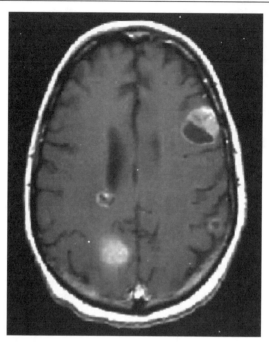

Fig. 12.3. MRI of the brain demonstrates multiple metastases in a NSCLC patient with mild headache.

Systemic tumor effects can be produced by nonmetastatic, paraneoplastic syndromes and must be differentiated from metastasis. For example, severe, rapidly progressive bone and joint pain in a patient with digital clubbing may denote hypertrophic pulmonary osteoarthropathy associated with primary LC.

From a practical standpoint, the most important clinical presentation of LC is that of a pulmonary radiographic density. The modern decrease in pulmonary nodules caused by tuberculosis and other infectious disease results in an increased percentage of new nodules that are neoplasms. A new radiographic nodule in a smoker or exsmoker is LC in approximately 70% of cases. This percentage is higher in masses larger than 3 cm in diameter. Unless the mass in question has been present without growth for two or more years, or there is unequivocally benign calcification, such a pulmonary density must be considered a LC until proven otherwise.

The most common error in LC management is failure to aggressively pursue the diagnosis of a pulmonary density. It is important to be cognizant of the fact that the average doubling time of a LC approximates 60 days. Therefore, untoward delay in workup might allow one or more doublings in tumor volume, with a potential resultant upstage of the cancer and decrease in the possibility of cure. Timely workup and treatment is essential to proper management.

Table 12.1. Radiographic features of lung cancers

Radiographic Appearance	Predicted Cell Type
Small peripheral nodule	Adenocarcinoma
Large peripheral nodule	Large cell undifferentiated carcinoma
Cavitary peripheral nodule	Squamous carcinoma
Hilar mass with Atelectasis	Squamous carcinoma
Hilar mass without Atelectasis	Small cell carcinoma
Diffuse infiltrate	Bronchioloalveolar carcinoma

Tissue diagnosis is desirable but not essential prior to surgical treatment of LC. Bronchoscopy will return a tissue diagnosis by biopsy or cytologic samples obtained by brushing or washing in most central tumors and in approximately half of peripheral masses. Transthoracic needle biopsy has a higher yield in the diagnosis of peripheral nodules, but has significant disadvantages. Even in the hands of experienced radiologists there is a significant incidence of false-negative biopsy, particularly in the case of small lesions. Furthermore, penetration of the lesion exposes the patient to risks of pneumothorax and hemoptysis as well as potential needle track spread of tumor to lung, pleura or soft tissues. Surgical resection of the nodule may be the best option in many cases.

Based on knowledge of tumor biology, Dines observed that one can often predict the cell type of a LC by its appearance on a CXR or computerized tomogram of the chest (CT).

Diagnostic Workup

The goals of the workup are to be as certain as possible that distant metastases are not present, that the tumor will be resectable at the time of surgery, and that the patient will survive the operation with sufficient pulmonary reserve to maintain a reasonable quality of life.

CT of the chest and upper abdomen is the keystone of LC assessment. CT allows reasonably accurate determination of T and N stage factors, and helps rule out adrenal and liver metastasis (Table 12.2). Laboratory tests are seldom abnormal in LC. Serum calcium may be elevated in the absence of bone metastasis secondary to paraneoplastic production of a parathormone like protein by the neoplasm. Carcinoembryonic antigen (CEA) can be mildly elevated in smokers, but levels higher than 5 µg/ml suggest nodal or distant metastasis. Electrolyte abnormalities may suggest Cushing's syndrome or inappropriate serum antidiuretic hormone production by SCLC.

If history and physical examination or laboratory studies suggest metastases, or CT scan findings indicate Stage III disease, a search for distant metastases using bone scan and brain MRI scan is indicated. Positron emission tomography (PET) scans graphically portray increased metabolic uptake of [18F]fluorodeoxyglucose by neoplastic tissue. Early studies suggest that PET scanning may be more valuable than contemporary methodologies in identifying nodal and distant metastasis. My current practice is to order preop PET scans on patients with elevated CEA or cN2 or cT3-4 status on CT scan (Fig. 12.4). Biopsy should confirm positive test results before the patient is considered unresectable. If pleural effusion is present,

12

Fig. 12.4. PET scan shows uptake of [18F]fluoro-
deoxyglucose in a lung mass and mediastinal nodes.

Table 12.2. CT scan warning signs of possible unresectability

Pleural effusion
Atelectasis
Tumor abuts hilum
Tumor abuts mediastinum
Loss of normal fat planes
Large tumor mass
Large mediastinal lymph nodes

Despite application of best available technology, approximately 8% of patients are
found to be unresectable during exploratory thoracotomy.

Table 12.3. NSCLC resections City of Hope Cancer Center 1987-1996

Pulmonary resection and MND	123
Pulmonary resection without MND	34
Limited resections	20
Unresectable at thoracotomy	15
Total cases	194

---

thoracentesis should be performed. If pleural fluid cytology is negative, preliminary
thoracoscopy with negative visual and cytologic examination of the pleura should
be confirmed prior to thoracotomy. Bronchoscopic examination prior to thoracotomy determines the location and extent of endobronchial disease and rules out
synchronous tumors.

It is important to recognize the shortcomings of preoperative CT in both under
and over estimation of T and N clinical staging. Even experienced radiologists and
surgeons are frequently unable to differentiate tumor that abuts rather than invades
mediastinal structures.

Some experts recommend that mediastinoscopy (MSC) be performed in all cases
of NSCLC; others advocate MSC in all cases with nodes larger than 1.0 cm on CT
scan. CT, PET and MSC, alone or in combination, can all miss N2 disease. Data on
PET is less extensive, but indicates that this test may have higher sensitivity than

CT. Nodes in stations R3, 8 and 9 and L 5, 6, 8 and 9 cannot be sampled by MSC, and sampling error may result in a false negative result when there is minimal disease in R or L stations 2, 4, or 7. The best test for detection of N2 disease, with respect to sensitivity, specificity and accuracy, is a complete and systematic mediastinal node dissection (MND) done at the time of thoracotomy. The author has accordingly chosen an alternative strategy. In the case of resectable right-sided tumors, where the presence of contralateral left sided N3 disease is uncommon, preliminary MSC is not performed, since all right-sided mediastinal nodes (MN) will be routinely resected at the time of surgery. Since left to right nodal metastasis is common, preliminary MSC to rule out contralateral N3 nodes is important. Current methodologies will not completely solve this problem, but judicious preliminary thoracoscopy prior to planned thoracotomy, in appropriate cases of LC with clinical or CT findings predictive of unresectability, might prevent futile thoracotomy.

Once staging studies have demonstrated that the patient is a candidate for surgical resection, the next important consideration is whether the patient has physiological reserves sufficient to sustain the requisite loss of pulmonary function. A major problem in treating LC lies in the fact that cigarette smoking causes multiple other diseases that may complicate the surgical treatment of the lung neoplasm. Chronic obstructive lung disease (COPD) is almost always present and may be severe enough to preclude lobectomy or pneumonectomy. Cigarette smoking is also a major risk factor for carotid and coronary artery disease. Basic workup consists of a careful history and physical examination with particular attention to exercise capacity, sputum production, and history of asthma and COPD. Pulmonary function studies (PFTs) and arterial blood gas determination are critical. Chronic hypercarbia or hypoxemia severe enough to require oxygen supplementation are contraindications to major pulmonary resection in most instances. The predicted postoperative forced expiratory volume in one second (FEV1), rather than preoperative FEV1, determines the safety of pulmonary resection. It is important to determine by bronchoscopic examination whether bronchi are obstructed by tumor. In calculating predicted FEV1, we have found the following formula to be reasonably accurate.

12

Predicted postoperative FEV1 = Preop FEV1 x # of lung segments remaining postop (anticipated) / # of unobstructed lung segments before surgery

If predicted postoperative FEV1 is less than 1.0 L, the risk of surgery and/or long term respiratory insufficiency is increased. If less than 0.7 L, surgery is definitely contraindicated, according to almost all authorities. McKenna has shown that some small peripheral tumors can be resected in conjunction with pulmonary volume reduction surgery even in the face of very low FEV1. Significant decrease in diffusing capacity for carbon monoxide may presage major postoperative difficulties even in the face of otherwise acceptable PFTs. In difficult situations further helpful information can be obtained by quantitative ventilation perfusion lung scans. If the lung tissue to be resected is found to have minimal or absent perfusion, it is nonfunctional and may be resected even in the face of low predicted FEV1. Exercise PFTs suggest increased morbidity if maximal oxygen consumption is less than 15 ml/kg/min and high mortality if below 10 ml/kg/min. Other major risk factors including congestive heart failure, pulmonary hypertension and myocardial ischemia require individualization. For example, preoperative cardiac catheterization and angioplasty might increase safety of pulmonary resection in patients with severe coronary artery disease. In the final analysis it is important to understand that the decision to deny surgical treatment on the basis of increased risk is equivalent to a decision not to treat for cure. Most patients with LC that cannot be resected will not be salvaged by subsequent RT (RT) and/or chemotherapy (ChT). This justifies an aggressive attempt by the surgeon to resect all resectable LC. On the other hand, thoracotomy that results in a failure to resect for cure only worsens an already bad situation, and curative surgery that leaves the patient ventilator dependent is a true catastrophe.

Principles of Surgical Treatment

Treatment of LC is very complex, with multiple variables of cell type, stage (see Table 12.4) and general medical condition that must be considered in clinical decision making. Although comprehensive guidelines for treatment have been formulated, in the actual delivery of care there are wide variations, particularly in patients with stage III disease. Surgical resection is by far the most efficacious treatment for LC, and all patients in stages I and II should have surgical resection if it can be done with acceptable risk. Many patients in surgical stage IIIA, and some in stages IIIB and IV, are surgical candidates. Pulmonary resection in properly selected patients has a mortality rate of less than 2% for limited resection or lobectomy and 5-10% for pneumonectomy. The higher morbidity and mortality associated with pneumonectomy can sometime be averted with lobe sparing, bronchoplastic, "sleeve" resections. If necessary, partial pericardiectomy with intrapericardial ligation or stapling of proximal pulmonary artery and veins will allow safe resection of tumors with central hilar invasion.

Like other solid organ neoplasms, bronchogenic carcinoma is best treated by en bloc, negative margin (R0) pulmonary resection with dissection of draining lymphatic tissues. Small peripheral tumors are technically resectable by wedge or segmental resections, but local recurrence is possible if in-transit tumor is present in the lobar remnant, or if the tumor has spread to unresected hilar or MN. The current standard is open pulmonary resection of a lobe, lobes or an entire lung, with individual dissection and ligation of pulmonary arteries and veins and secure bronchial closure with either suture or staple techniques (Table 12.3). Pneumonectomy should not be performed without tissue diagnosis.

NSCLC typically metastasizes first to lymph nodes in the pulmonary hilum and then on into MLN, but "skip" metastases into MLN occurs in approximately 30% of instances. Patterns of nodal metastases are different for each lobe of the lung. Left lower lobe (LLL) tumors commonly metastasize to the subcarinal lymph nodes (station L7), then cross over to the right (station R7) before extending up into the right paratracheal lymph nodes (station R4). Left upper lobe tumors commonly metastasize to the left aorto-pulmonary window and para-aortic nodes (stations L 5 and 6). Right lung tumors rarely metastasize into left-sided lymph nodes. Right upper lobe (RUL) tumors typically metastasize to subcarinal nodes (station R7) then on to R4 and R2. RUL tumors typically metastasize to R4 and R2. Right middle lobe tumors can follow either pattern. Lower lobe tumors occasionally metastasize into nodal stations 9 and 8. Nodal metastasis is rare in tumors smaller than 1 cm in diameter.

Table 12.4. NSCLC staging in brief

T IS	Carcinoma in situ
T1	Primary tumor 3 cm. or less not invading visceral pleura or involving a main bronchus.
T2	Primary tumor larger than 3 cm., or invades visceral pleura or involves main bronchus, but more than 2 cm. distal to carina.
T3	Primary tumor invades chest wall, diaphragm, parietal or mediastinal pleura, parietal pericardium, or main bronchus less than 2 cm. distal to carina, but not causing atelectasis or obstructive pneumonia of the entire lung.
T4	Primary tumor invades carina, trachea, great vessel, heart, mediastinum, esophagus or vertebra, or is associated with a satellite lesion, malignant pleural effusion, or atelectasis or obstructive pneumonia involving the entire lung.
N0	No nodal metastasis
N1	Metastasis to hilar lymph node(s)
N2	Metastasis to ipsilateral mediastinal lymph node(s)
N3	Metastasis to contralateral hilar or mediastinal node(s), or supra clavicular node(s).
M0	No distant metastasis
M1	Distant organ metastasis or second malignant tumor in another lobe of ipsilateral or contralateral lung or a discontinuous lesion in chest wall or diaphragm.
Stage 0	Carcinoma in situ only.
Stage IA	T1N0M0
IB	T2N0M0
Stage IIA	T1N1M0
IIB	T2N1M0, T3N0M0
Stage IIIA	T3N1MO, T1-3N2M0
IIIB	T4N0-3M0, T1-3N3M0
Stage IV	M1, any T, any N

12

Tumors larger than 1 cm (adenocarcinoma > squamous carcinoma) have a small incidence of N2 nodal metastasis that increases progressively with tumor size.

A complete and systematic mediastinal node dissection (MND) is not technically demanding, can be performed in 30-45 minutes, and adds little morbidity to the procedure. Because the technique has not often been described in thoracic surgical texts, I will describe my technique in some detail. The technique was derived from previous operations described by Brock, Cahan, Higginson and Weinberg (Fig. 12.5).

Double lumen or Univent tubes are used routinely to allow single lung ventilation. Exposure is through a generous posterolateral thoracotomy incision. The inferior pulmonary ligament is transected upward to the inferior pulmonary vein, and all

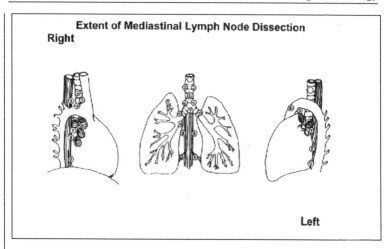

Extent of Mediastinal Lymph Node Dissection
Right

Left

Fig. 12.5. Extent of right and left mediastinal lymph node dissections for NSCLC.[2]

MLN in stations 9 and 8 are dissected. The periesophageal nodes are dissected upward, and all MLN in the subcarinal area (R7), extending a short distance onto the contralateral main bronchus (L7), are dissected. A sizable bronchial artery may be encountered in this region and must be carefully secured. One must be careful not to injure the posterior tracheal or bronchial wall at the locations where the tracheal and main bronchial balloons are inflated. On the right, the azygos vein is ligated and divided, in order to provide safe access to the precarinal nodes, and a longitudinal incision is carried up along the course of the vagus nerve. A second incision is carried along the posterior margin of the superior vena cava (SVC). The SVC is mobilized upward and to the left. Small venous branches enter the SVC directly in some patients; these branches must be carefully identified and secured. All lymphatic tissues are dissected between the anterior tracheal wall, the ascending aorta on the left, the SVC anteriorly, the pulmonary artery below and the subclavian artery above (stations R4 and L4 and the lowest nodes in stations R2 and L2). We emphasize that the nodes to the left of the tracheal midline are accessible, and are resected en bloc as far to the left as the underside of the aortic arch. The MLN around the vagus nerve and esophagus posteriorly are cleared, and nodal tissue anterior to the phrenic nerve (station R3) is resected. On the left, the dissection of stations L9, L8, L7 and L3 is identical. Above, incisions are made posterior to the phrenic nerve and along the course of the vagus. All MLN between the nerves, overlying the pulmonary artery and aortic arch are resected en-bloc (nodal stations 6 and 5).

We have adopted a conservative approach in not routinely extending dissection by division of the ligamentum arteriosum on the left or above the right subclavian artery in order to avoid recurrent nerve injuries. If, however, palpably abnormal MLN extend into the area of the ligamentum, the vagus and recurrent nerves are dissected, the ligamentum is transected, and all nodes in this area are resected. If the phrenic, vagus or recurrent nerves are invaded by tumor, they are resected en bloc.

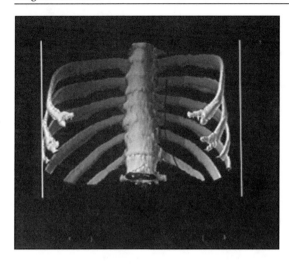

Fig. 12.6. Continuous extrapleural block. The percutaneous catheter (black) is inserted in a pocket (grey) created by dissection of the posterior parietal pleura.[3]

Typical carcinoid tumors are curable by local resection of pulmonary tissue or bronchus with narrow margins of resection and without nodal dissection, while atypical carcinoid should be treated like NSCLC.

Postoperative Care

Complications following thoracotomy should be anticipated in the operating room. A meticulous search for bleeding and lymphatic leaks should be performed with repair. The bronchial stump and inflated lung should be examined for air leak with repair by suture, staple or fibrin glue as appropriate to avoid prolonged postoperative air leak or bronchopleural fistula. Potential lobar torsion can be averted by appropriate two-point fixation of lung tissue. Cardiac herniation is prevented by closure of pericardial defects with 1 mm thick polytetrafluorethylene (PTFE, Gore-Tex).

Pain following thoracotomy is the major cause of morbidity and mortality and improved pain management results in less postsurgical morbidity and mortality. Transection of skin, muscle and pleura, retraction of muscles, ligaments, and intercostal nerves, pleural irritation by chest tubes and fractured ribs all contribute to severe pain following thoracotomy. If pain is not adequately controlled, a series of events including shallow respiration, chest wall splinting and avoidance of cough and deep breathing result in ventilation-perfusion mismatch, hypoxemia, atelectasis and pneumonitis. The end result can be prolonged ICU and hospital stay or even respiratory failure and death. Traditional treatment with parenteral opioids on demand provides inadequate pain relief and superimposes morbidity secondary to the sedative and respiratory depressant side effects of narcotics. Our preference is for continuous extrapleural block with a 1% lidocaine infusion at 1 mg/kg/hr delivered through a plastic catheter placed in a posterior pocket created by dissection of the posterior parietal pleura (Fig. 12.6). This technique, combined with patient-controlled analgesia with morphine sulfate results in effective reduction of pain.

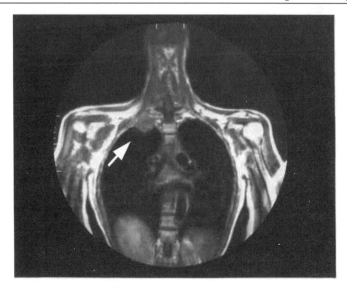

Fig. 12.7. MRI scan of NSCLC patient with a right superior sulcus tumor.

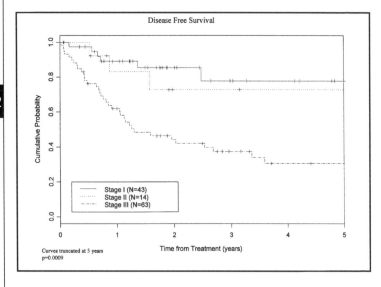

Fig. 12.8. Five year disease-free survival (Kaplan-Meier) of patients following pulmonary resection, MND and postoperative radiation therapy for patients in stages II and III at City of Hope Cancer Center (1986-1997).

Transcutaneous oximetry provides early warning of deteriorating pulmonary function. If pulmonary physiotherapy or nasotracheal suctioning do not result in improvement, bronchoscopic clearance of secretions is better applied early than late.

Atrial dysrhythmias are very common, with increased incidence following pneumonectomy and in elderly patients.

Radiation Oncology

Radiation therapy (RT) can cure approximately 15-20% of stage I and 5% of stage IIIA NSCLC in patients unwilling or unsuited to undergo surgical resection. RT is commonly given preoperatively in the treatment of NSCLC invading into the superior sulcus, so-called Pancoast tumor (Fig. 12.7). External beam radiation has been demonstrated to reduce local recurrence in squamous carcinoma. It is currently controversial whether postoperative RT is valuable following surgical resection of N1 and N2 LC. The author's experience at City of Hope National Medical Center over the past decade with adjuvant RT (ART) in patients with N1 or N2 disease following complete nodal dissection has resulted in 70% disease free survival in N1 patients and 30% disease free survival in N2 disease without adjuvant ChT (AChT) (Fig. 12.8).

Following confirmation by permanent section, patients with N1 or N2 disease or microscopically positive margins receive ART for a total of 5040 cGy, administered in anterior and posterior opposed fields. These results suggest that such adjuvant treatment results in survival similar to that in Stage I patients. RT is often valuable in the palliation of symptoms secondary to airway obstruction and bone and brain metastasis. Endobrachy-RT and intraoperative-RT are valuable in selected circumstances.

Treatment Results

Treatment results are dependent upon the stage in which the disease is discovered. As many as one-half of LC patients are not suitable for surgical treatment upon first diagnosis because of distant metastasis, advanced local disease or severe impairment of cardiopulmonary function by associated tobacco-related disease. Although treatment results with surgery alone in Stage I patients are good, it is clear from a very large data base from many surgical series that multimodality treatment is needed in more advanced disease.

LC staging is revised periodically to reflect accumulating data on treatment of subsets of patients. The most recent revision was published in 1997. New changes in staging reflect the realization that size greater than 3 cm in diameter is associated with increased chances of recurrence. Stage IB tumors are larger than 3 cm in diameter or have other unfavorable features such as invasion of visceral pleura or hilar proximity (T2). Stage II tumors have hilar node metastasis (N1) and are also subdivided into A and B.

Realization that the prognosis of T3N0M0 tumors have a better prognosis than T1-2N2M0 has resulted in a reclassification of this group into IIB from it's prior IIIA stage classification.

I consider two new features of the 1997 classification to be of questionable wisdom. The presence of "satellite" tumor within the same lobe as the primary tumor is classified as T4 and therefore stage IIIB. In general, stage IIIB tumors are not thought

12

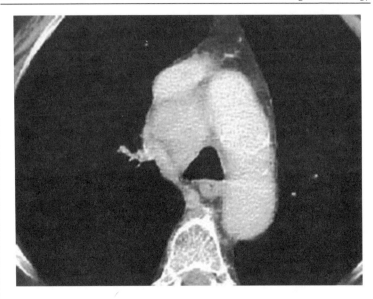

Fig. 12.9. Six months following lobectomy, without MND, for NSCLC, this CT scan demonstrates local recurrence in station R4 mediastinal lymph nodes.

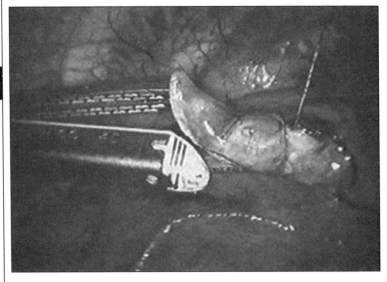

Fig. 12.10. Thoracoscopic view of VAT wedge resection following CT guided needle localization of a small peripheral mass in a patient with severe COPD. An endoscopic GIA stapler is being used to resect the neoplasm.

Table 12.5 Is limited resection of clinical stage IA adenocarcinoma safe?

Tumor size in mm.	N0	N1	N2	Total
< 10	9	0	0	9
11-20	84	9 (8.5%)	12 (11.3%)	106
21-30	96	9 (7.4%)	14 (11.5%)	122
>31	61	10 (10.4%)	25 (26%)	96
Total	250 (75%)	28 (8.4%)	51 (15.3%)	333

to be suitable candidates for surgical resection and are treated by combination RT and ChT. Unlike other T4 tumors, the tumor and a satellite can usually be resected simply and safely by lobectomy. Adjuvant therapy can be given after surgery where appropriate. Multiple surgical series show five-year survival greater than 20% in patients with satellite lesions. In my opinion, there is insufficient data regarding the results of treatment of patients with potential satellite lesions to withhold potentially curative treatment.

Intrapulmonary ipsilateral or contralateral LC is classified as M1, and therefore stage IV. Patients with two lesions, either contralateral or ipsilateral, in different lobes often have separate primary tumors as reflected by different cell types and subsequent long term survival. There is very little information in the medical literature on distant metastasis from LC back into the lungs. The existence of second or multiple cancers is entirely consistent with our knowledge of the field carcinogenesis effect of tobacco smoke and such lesions, both metachronous and synchronous, are quite common. In my personal experience with several thousand cases of LC, I have seen few cases where a NSCLC has metastasized to lung in the absence of metastases to other organs. It is my practice to consider that patients with two simultaneous and distinct lung masses, either ipsilateral or contralateral, have two separate primary LCs until proven otherwise. The chance of cure of both lesions is a mathematical product of the survival percentage of each separate lesion. The presence of synchronous second or multiple primary LCs has an incidence of approximately 2% of all cases of NSCLC. The incidence of a second metachronous primary LC is at least 20% in patients who survive 10 years.

Stage I

Only 15% of patients with NSCLC currently achieve five-year survival; almost all such long term survival follows surgical resection in patients with stage I disease. Stage I NSCLC can be cured in a majority of cases with surgery alone. The crucial problem continues to be that surgeons only see a very small percentage of patients referred in stage I. Standard treatment, with lobectomy where possible, and complete MND will cure the majority of these patients. Failure to dissect MN risks undetected N2 metastasis. In such cases, leaving local recurrence followed by dissemination and death will predictably follow (Fig. 12.9).

One recent advance in the treatment of early stage LC is limited resection performed open or by video assisted or thoracoscopic techniques (VAT) in patients with poor lung function. Patients with limited respiratory reserve are at increased risk for perioperative respiratory complications. Frequently, impairment in pulmonary

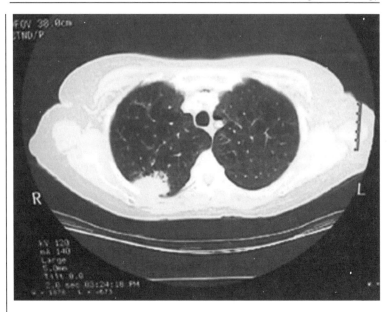

Fig. 12.11. CT scan in a patient with T3N0M0 Stage IIB NSCLC, confirmed at thoracotomy.

function is the reason not to proceed with thoracotomy and lung resection in the presence of otherwise clear indications. Recent experience with the use of thoracoscopic procedures in benign lung disorders, especially emphysema, confirms that limited thoracoscopic lung resections can be performed safely in this setting. Accordingly, criteria for an elective limited VAT resection include T1 lesions in peripheral location, without gross lymph node involvement, that can be removed with negative margins (Fig. 12.10). Better understanding of PFTs and limits of resection now allow resection of small peripheral LCs in patients with poor pulmonary function via open segmental resection, thoracoscopic wedge resection, or a combination of reduction pneumoplasty with wedge resection in carefully selected patients. Unfortunately, this progress has a counterproductive flip side, since some surgeons are performing imprudent limited resections, open or VAT, in patients who are not prohibitive candidates for resection. Increase in local recurrence and decreased survival can be predicted from available data.

Table 12.5 shows data derived from experience in Japanese surgical centers demonstrating that despite presurgical staging indicating no lymph node involvement, patients with lesions larger than 1 cm. in diameter have a significant risk of hilar and MN metastasis.

Lesser resections, i.e., segmental and wedge resections, represent a compromise of oncologic surgical principles and carry a significant risk of local recurrence. Reoperation or RT can occasionally salvage patients with local recurrence of NSCLC, but the chance of long term survival is very poor. Such resections should, accordingly, only be performed in the face of increased risk of death or disability with standard resection. Techniques of oncologically sound, video-assisted lobectomy or pneumonectomy and MND are currently being developed in a number of centers, but survival data is not yet available.

Although SCLC is seldom discovered in early stage, SCLC in stages I and II has proven curable in approximately 30-40% of cases following surgical resection followed by AChT.

RT can achieve cure in approximately 15-20% of patients with small peripheral LCs who are not candidates for surgical resection. Photodynamic therapy using hematoporphyrin derivatives and a tuneable dye laser may be curative in carefully selected patients with in situ or small localized central tumors.

Stage II

Standard therapy for T1-2 N1 M0 Stage II disease is lobectomy or pneumonectomy. Since the risk of mediastinal metastasis is higher in N1 patients, MND is indicated. Most centers do not recommend postoperative adjuvant therapy. Our preference at City of Hope Cancer Center has been for surgical resection including MND followed by RT in patients with stage II disease. This program has resulted in five -year disease-free survival (Kaplan-Meier) of 77.7% in stage I patients and 71.8% in stage II patients (Fig. 12.8).

Multiple reports have documented survival in approximately 40% in patients with T3 chest wall invasion. Patients with such T3 stage II B disease should undergo en-bloc, margin-negative resection of the invaded ribs together with pulmonary resection (Fig. 12.11). The chest wall is reconstructed using prosthetic materials. Our preference is for 2.0 mm thick PTFE patch.

Tumors invading the chest wall in the region of the superior sulcus represent a special subset. Following the initial report of Shaw, most centers attempt to shrink the tumor with preoperative RT before proceeding with en bloc surgical resection of the upper lobe of lung, chest wall, portions of vertebra and autonomic and brachial nerves as indicated. Survival of approximately 40% has been reported from a number of centers. A current trial investigates the efficacy of adding neoadjuvant ChT (NChT) to RT. Results are not yet available.

Stage IIIA

More than 40,000 patients present in Stage III in the U. S. each year. Although formerly there was considerable pessimism regarding the treatment of Stage III NSCLC, it has now been shown in multiple surgical series from the U.S., Europe and Japan that many patients can be salvaged with surgical resection alone, including MND, and that survival is probably improved by multimodality therapy. Most centers recommend NChT followed by surgical resection. This is based upon two 1994 reports of randomized prospective trials from the U.S. (Roth) and Spain (Rosell) that demonstrated statistically significant improvement in survival over patients treated with surgery alone. Recent updates of these trials, however, indicate that the disease free survival difference in N2 patients in the Roth study is not statistically significant, and that survival in the surgery-only arm of this series equals that in the neoadjuvant arm of the Rosell series. Although NChT will result in a partial response in one-half to two-thirds of patients and a complete response in approxi-

12

Fig. 12.12. CT scan in a patient with T2N2M0 Stage IIIA NSCLC treated with pneumonectomy, MND and adjuvant postoperative radiation therapy.

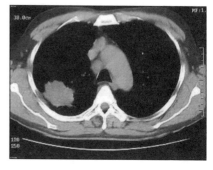

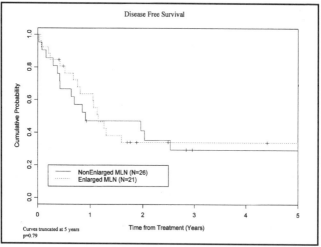

Fig. 12.13. Five year disease-free survival (Kaplan-Meier) of patients following pulmonary resection, MND and postoperative radiation therapy for patients in stages IIIA N2 NSCLC; cN0 vs. cN2 survival differences do not reach statistical significance. City of Hope Cancer Center (1986-1997).

mately 10%, posttherapy surgery is technically difficult and morbidity and mortality are increased.

It is important to emphasize that stage IIIA NSCLC is a complex group with many subsets, including cN0 proven N2 by immunohistochemistry techniques only, cN0 proven N2 by postresection pathologic examination of MN dissected at the time of resection, cN2 disease resected by MND and "bulky" cN2 disease that can be reliably predicted to be unresectable by an experienced surgeon. The first three groups listed above are all amenable to a complete and systematic MND (Fig. 12.12).

The author has chosen a surgical approach in such patients. Resection including routine systematic MND has been employed at the City of Hope National Medical Center in the management of nonsmall cell LC between 1987 and 1996. ART (5040 cGy to the mediastinum via posterior and anterior opposed fields) was routinely recommended to patients with N2 involvement. ChT was given only for salvage in

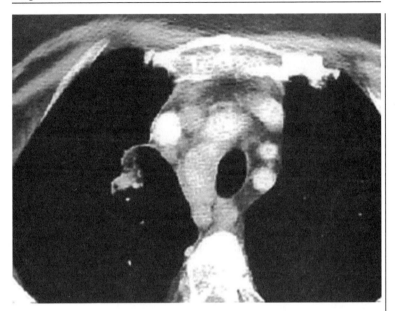

Fig. 12.14. CT scan in a patient with T2N2MO Stage IIIA NSCLC shows multiple enlarged mediastinal lymph nodes in station R2; not considered resectable for cure.

case of recurrence. Actual five-year disease free survival in 51 patients with N2 stage IIIA disease was 30% (see Fig. 12.13).

Survival in patients with N2 disease treated in this fashion appears to be equivalent to that of patients treated with NChT followed by surgical resection. Although this retrospective data suggests a trend toward survival advantage in surgically resected stage IIIA N2 patients who receive ART without ChT, this has not been confirmed in randomized prospective studies. More recently, the Mayo Clinic group has reported survival exceeding 30% at four years in patients with completely resected Stage IIIA N2 NSCLC with MND and ART. Prospective studies comparing pulmonary resection with MND and ART with NChT followed by pulmonary resection and MND have not been done. Results in patients with T3 N2 disease are very poor.

Although one would intuitively expect that the same ChT regimens that are being advocated for the neoadjuvant treatment of IIIA would also improve survival in these patients when given postoperatively, there is no series which demonstrates such efficacy. Our current protocol at City of Hope Cancer Center is a single arm prospective study of surgical resection followed by postoperative adjuvant cisplatin and gemcitabine and RT. It is hoped that adjuvant ChT will prove feasible and might further increase long term survival.

It is important to emphasize that patients with N2 disease must be carefully selected before advising surgical resection. Preoperative CT scans will allow recognition of patients with "bulky" MN metastases with large tumor burden in nodes, multiple levels of metastasis, contralateral involvement, and obliteration of normal tissue planes (Fig. 12.14). Such cases will almost certainly not be completely resect-

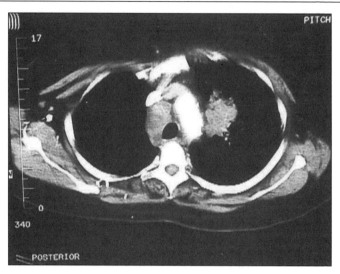

Fig. 12.15. CT scan in a patient with T2N3MO Stage IIIB NSCLC shows a large left upper lobe mass and markedly enlarged mediastinal lymph nodes in station R4.

able. Patients with such "bulky", "clinically apparent" or "technically unresectable" N2 disease may benefit from NChT or combination NChT and RT, but it must be recognized that morbidity and mortality of such treatment is high. Furthermore, surgery is very difficult in these patients who have fibrous obliteration of normal dissection planes. For patients with bulky N2 disease, best evidence would seem to indicate that NChT, or NChT and RT, followed by an attempt at surgical resection is reasonable. Results are not yet available on a recently completed prospective randomized trail comparing such treatment with ChT and RT alone.

Stage IIIB

Few patients with stage IIIB disease are cured by current treatment protocols. T4 tumors with local invasion of vital structures are generally not amenable to surgical resection, although carefully selected patients can be resected by sleeve pneumonectomy or SVC resection and reconstruction. It is important to be cognizant of the fact that a CT scan showing that a tumor abuts a vital structure, the aorta for example, does not indicate authoritatively that the structure is invaded. In some cases the question of invasion can only be resolved during exploration. The decision as to whether a NSCLC is resectable is frequently complicated and should be made by an experienced thoracic surgeon rather than a medical oncologist.

Pleural and pericardial effusions are common complications of locally advanced LC that produce considerable morbidity and represent surgically incurable T4, stage IIIB disease. Palliation of malignant pleural effusion can be attained by talc pleurodesis (by chest tube or VAT) or by placement of a PleurX catheter. Surgical palliation in case of malignant pericardial effusion has been approached by a number of different techniques. The author's preference is for thoracoscopic partial pericardiectomy.

Bilateral MND is technically possible in the surgical treatment of N3 IIIB disease by either median sternotomy or "clam-shell" thoracotomy approaches, but experience is limited to a small number of cases in a few centers (Fig. 12.15). A recent report by Eberhart suggests that some NSCLC patients in Stage III B may be salvaged by radiation therapy, surgical resection, and standard ChT regimens.

Stage IV

Occasional patients with solitary metastasis to brain or adrenal gland can be salvaged by surgical resection; otherwise, the aim of treatment is palliative. Palliative surgical interventions are sometimes in order to prevent or treat pathologic fractures in weight-bearing bones or spinal cord compression. There have been major improvements in the palliation of patients with progressive airway obstruction using neodymium-YAG laser, photodynamic therapy and stents. Endobrachy-RT using new technology can now be rapidly and simply performed on an outpatient basis to extend the palliative benefits of laser ablation. Complex new technology facilitates delivery of precise "radiosurgery" RT for brain metastasis.

There is currently no curative ChT for NSCLC in patients with distant metastasis. Five-year survival in SCLC is less than 5%. Cisplatinum based ChT can offer NSCLC patients with good performance status some modest gains in the duration and quality of life. ChT regimens for SCLC offer predictable prolongation of survival and improved quality of life. Multiple new ChT options are currently under investigation, without any clear prospect for a curative therapy in the near future.

Posttreatment Follow-Up

Managed care groups are exerting pressure on thoracic surgeons to turn over postresectional follow-up of LC to primary care physicians. The rationale is that these doctors can do an equally effective job at a lower cost. A recent study from M.D. Anderson Hospital has concluded that follow-up of LC patients after surgical treatment is ineffective. I strongly disagree. There are major potential advantages to postsurgical follow-up by surgeons, including a reduction in unnecessary and ineffective studies. Although most patients with recurrent LC present with either distant metastasis or a combination of local and distant recurrence, approximately 20-30% recur locally and are potentially salvageable by re-resection or RT. Perhaps more important is the realization that at least 25% of survivors will develop a second neoplasm related to tobacco smoke, most frequently LC, but not uncommonly in the tongue, pharynx, larynx, esophagus, pancreas, kidney or bladder. Screening for such second cancers by history, physical examination and annual CT scan is a crucially important aspect of follow-up care.

Routine lab studies have a very low yield and should be ordered only as guided by pertinent positive findings on history and physical examination. Hypercalcemia can represent a life threatening paraneoplastic complication that requires urgent treatment. CEA is sometimes the first and only sign of recurrence. The chance of recurrence is highest in the first two years after surgery. During this time period the patient is seen in the office at four-month intervals and at six-month intervals thereafter. At each clinic visit a history and physical examination is performed. In addition, a CXR or CT scan is obtained with particular attention to detecting recurrence of LC, detecting second cancers in the lung and upper aerodigestive tract, and detecting long term problems related to the surgical procedure. Counseling and assisting the patient in smoking cessation is an important feature of follow-up care. Approximately 30% of my patients continue to smoke after treatment of a cancer caused by tobacco smoke.

If patients experience symptoms suggestive of recurrent or new LC, they are seen between regularly scheduled visits. Specialized imaging techniques such as brain MRI and bone scan are done only on suspicion of disease based on history and physical examination. Early recognition of bone and brain metastasis followed by appropriate referral and intervention can result in improvements in quality of life.

Prospects for Improvement

Tobacco Control

"If you want to beat malaria, you have to understand the malaria mosquitoes that spread malaria. If you want to prevent LC, you have to study the tobacco industry to see how it spreads smoking." Stanton Glantz

1999 has been a very eventful year in the area of public health promotion by tobacco control. A Master Settlement Agreement was signed between a small group of state attorney-generals, private litigators and the tobacco industry, without input from public health experts. Although the tobacco industry will have to pay upwards of $200 billion in damages, this weak settlement failed to address crucial public health measures such as FDA control of tobacco products or an absolute ban on tobacco advertising. We can expect the tobacco industry to continue to seek immunity with the help of their supporters in Congress.

Most LC could potentially be prevented by prevention of initiation of cigarette smoking by young people. Initiation of cigarette smoking seldom occurs after age 21. Even if primary prevention were completely successful, however, substantial reduction in LC mortality would not be achieved for at least 20-30 years.

Most adult smokers in the U.S. would like to quit. Unaided quit attempts have a disappointingly low success rate. A large body of experimental evidence, much of it from tobacco industry labs, has proven that nicotine is a highly addictive substance. This addictiveness is the primary obstacle to smoking cessation. Randomized prospective studies have proven that nicotine replacement therapy and bupropion are effective modalities in smoking cessation, individually and in combination. Despite the potential benefit of smoking cessation interventions in the reduction of tobacco-related morbidity and mortality, such therapy is not covered by private health insurance entities or by Medicare and Medicaid. This shortsighted policy must be changed.

Political Reform

"Members of Congress are as addicted to large campaign contributions as smokers are to nicotine." Ann McBride

Stanton Glantz and his colleagues at the University of California-San Francisco have published a remarkable series of papers demonstrating a statistically significant correlation between the voting behavior of state and national political office holders and the amount of money they accept in campaign contributions from tobacco industry sources. This has had a chilling effect on preventing passage of effective

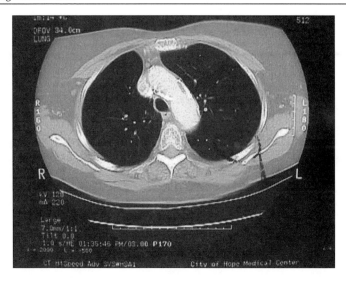

Fig. 12.16. 9 mm NSCLC in the left lower lobe discovered on screening CT scan in a patient with a prior contralateral NSCLC.

tobacco control legislation. Campaign contribution reform and tobacco control have become major bipartisan issues in the 2000 presidential election campaign.

Legal Activism

Because political efforts at effective tobacco control have been hamstrung by the political problem outlined above, legal actions assume increasing importance. Although cigarettes are the only consumer product sold in the United States that are damaging to the health of consumers when used as intended by the manufacturer, all product liability suits against the tobacco industry had failed until the mid 1990s. Since that time, corporate insiders have leaked internal corporate memoranda and further documents have been released by court order. An enormous archive of tobacco industry internal documents now is available to the public health and legal communities. This information has helped to make lawsuits against tobacco companies increasingly successful. Legal action has a powerful potential to force public health concessions from the tobacco industry. All patients with tobacco related diseases should be informed by their physicians that tobacco products have been detrimental to their health.

Chemoprevention

Research into chemoprevention, to determine whether it is possible to reduce the risk of LC in smokers and ex-smokers, is underway. Most trials have studied cis-retinoic acid and results have been disappointing.

Screening

"A decision not to screen for lung cancer is tantamount to a decision not to treat for cure." John C. McDougall M.D.

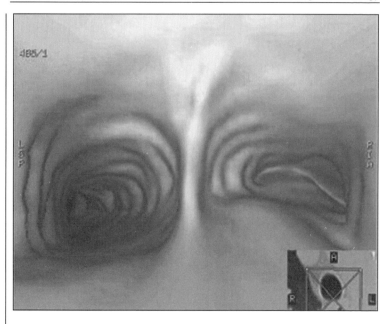

Fig. 12.17. "Virtual bronchoscopy" image of the distal trachea and carina created by digital reconstruction of spiral CT images.[4]

Improvement in survival has been achieved in colon, breast and prostate cancer following widespread application of mass screening for these diseases. There is currently no such screening for LC. Although only 15% of NSCLC in the U.S. is discovered in stage I, 85% of all five-year survivors are from this group of patients. The implication is clear that mass screening should improve survival. Three large prospective trials in the early 1970s demonstrated that screening with CXR resulted in discovery of LCs in stage I in 40% of cases. This screening did not, however, result in a reduction in population mortality, and no policy of mass screening has been applied in the U.S.. In Japan, mass screening with CXR has resulted in a gratifying increase in the number of early cases discovered.

This year, Henschke reported that screening with low-dose, noncontrast, spiral computerized tomograms found 27 cases of LC on initial screen in a group of 1000 patients who had smoked for at least 10 years; 85% of cases were Stage I. This critically important study strongly suggests that mass screening for LC is now feasible and offers an opportunity to drastically reduce the mortality from NSCLC in the early years of the new century (Figs. 12.16 and 12.17).

Surgical Research

Although LC is very common, there is a remarkable dearth of quality research studies. This is particularly true of surgical therapy trials. Public research dollars are scanty with only $800 per LC death as compared with $7500 per breast cancer death. Most clinical research currently involves ChT trials funded by drug companies. Such studies have had a very minimal impact on overall survival. Thoracic surgeons will need to commit themselves to more involvement in LC prospective clinical trials. The American College of Surgeons' Oncology Group (ACOSOG) may provide a foundation for this effort and currently has three new LC trials in accrual.

The full texts of three new NSCLC prospective surgical research trials are available at URL http://www.acosog.org/. Trials include: Z0030: randomized trial of mediastinal lymph node sampling versus complete lymphadenectomy during pulmonary resection in the patient with N0 or N1 (less than hilar) nonsmall cell carcinoma; Z0040: a prospective study of the prognostic significance of occult metastases in the patient with resectable nonsmall cell lung carcinoma, and Z0050: the utility of positron emission tomography (PET) in staging of patients with potentially operable nonsmall cell lung carcinoma.

Basic Science Research

Rapid progress is being made in the understanding of the biological processes that contribute to the development of LC. Mutations in a number of tumor suppressor genes and proto-oncogenes, including RAS, MYC, p53, and RB have been identified. Connections are beginning to emerge from research laboratories with a rapidly developing understanding of the molecular mechanisms disrupted in cancer cells including growth factors and their receptors, cell-cycle proteins, intracellular signaling cascades, telomerase, apoptosis, DNA repair, vasculoneogenesis, tissue matrix integrity and many other molecular cellular functions. A "smoking gun" direct link between carcinogens in tobacco smoke and specific point mutations at codons 248 and 273 in the p53 gene, identical to mutations found in clinical cases of LC, has been reported by Denissenko et al. Bennett's recent study demonstrated a dose-dependent correlation between environmental tobacco exposure and LC in female never-smokers with null GSTM1 gene status. In 1983 Whang-Peng made the important discovery that there is a specific and almost universal transposition or deletion of the short arm of chromosome 3, with loss of DNA in the region 3p14-23 in small cell lung cancer (SCLC). Biologists at City of Hope Cancer Center and elsewhere are currently closing in on a putative gene at 3p21.3 that is deleted in almost all cases of SCLC, and many cases of NSCLC. Each such discovery will open up a potential chink in the armor of LC, toward which new diagnostic tests and treatments can be directed. Monoclonal antibodies, ribozymes, antisense constructs and other gene-therapy modalities are currently in development in university and pharmaceutical industry labs, and early experimental clinical trials are in progress. We can expect exciting new molecular applications in screening, prognostication, and treatment of LC in the first decade of the new millenium.

12

Conclusion

LC is the most important neoplasm in the U.S. in 2000 in terms of mortality and is well on its way to becoming number one world wide. New developments in tobacco control, smoking cessation, chemoprevention, early detection and treatment offer hope that the mortality from this dread disease will be markedly diminished in the early years of the new century.

Acknowledgments

Image courtesy of Mark Levinson, M.D.; Line drawing courtesy of Elizabeth Shaghnessy, M.D.; 3-dimensional reconstruction of CT scan courtesy of Howard Marx, M.D.; Virtual bronchoscopic image courtesy of Howard Marx, M.D.

Selected Readings

1. Henschke CI, McCauley DI, Yankelevitz DF et al. Early Lung Cancer Action Project: Overall design and findings from baseline screening. Lancet July 10, 1999; 354:99-105.
    Henschke reported that screening with low-dose, noncontrast, spiral-computerized tomograms found 27 cases of LC on initial screen, in a group of 1000 patients who had smoked for at least ten years; 85% of cases were Stage I. This critically important study implies that mass screening for LC is now feasible and offers an opportunity to dramatically reduce the mortality from NSCLC.

2. Mountain CF. Revision in the international staging system for staging lung cancer. Chest 1997; 111:1710-17.
    Accurate staging lies at the core of effective management of LC. The 1997 revisions are outlined in this article.

3. Ettinger DS (Nonsmall-Cell Lung Cancer Practice Guidelines Panel Chairman) Oncology 1996; 10 (Supplement):81-112.
    This guideline provides a remarkably comprehensive framework for the diagnostic workup and treatment of patients with NSCLC. A new guideline revision will be released this year. This guideline is also available on the Internet at URL http://www. cancernetwork.com/journals/Oncology/nc960811.htm

4. Izbicki JR, Passlick B, Pantel K et al. Effectiveness of radical systematic mediastinal lymphadenectomy in patients with resectable nonsmall cell lung cancer. Ann Surg 1998; 227:138-144.
    Results from a recently published randomized, prospective study of MND by Izbicki demonstrate that the addition of MND to standard pulmonary resection did not improve overall survival in patients with NSCLC as compared to controls at all disease stages at resection, but suggested that there was an improvement in survival among those patients with N1 or N2 disease, with involvement of only one lymph node.

5. Vansteenkiste JF, De Leyn PR, Deneffe GJ et al. Clinical prognostic factors in surgically treated stage IIIA-N2 nonsmall cell lung cancer: Analysis of the literature. Lung Cancer 1998; 19:3-13.
    This 1998 review of multiple surgical series by Vansteenkiste demonstrates that many selected patients with N2 disease (range 6-34%) can be salvaged by surgical resection.

6. The Lung Cancer Study Group. Effects of postoperative mediastinal radiation on completely resected stage II and stage III epidermoid cancer of the lung. N Engl J Med 1986; 315:1377-1381.
    This prospective, randomized trial demonstrated statistically significant reduction in local recurrence, but not survival with adjuvant postoperative RT.

12

7.   Sawyer TE, Bonner JA, Gould PM et al. The impact of surgical adjuvant thoracic radiation therapy for patients with nonsmall cell lung carcinoma with ipsilateral mediastinal node involvement. Cancer 1997;80:1399-1408.
      This retrospective data from the Mayo Clinic shows improved local control and actuarial survival in N2 patients treated with ART following MND, versus patients with surgical resection and MND alone (43% versus 22% four-year survival).

8A.   Roth JA, Fossella F, Komaki R et al. A randomized trial comparing perioperative chemotherapy and surgery with surgery alone in resectable stage IIIA nonsmall-cell lung cancer. J Natl Cancer Inst 1994; 86:673-680.

8B.   Roth JA, Atkinson EN, Fossella F et al. Long-term follow-up of patients enrolled in a randomized trial comparing perioperative chemotherapy and surgery with surgery alone in resectable stage IIIA nonsmall-cell lung cancer. Lung Cancer 1998; 21:1-6.
      This 1994 article from M.D. Anderson Cancer Center provides the best evidence for the use of preoperative chemotherapy in the treatment of LC. Careful review of the 1998 follow-up article demonstrates that the disease-free five-year survival for stage IIIA N2 NSCLC is much lower than predicted, and does not reach statistical significance compared to surgical resection alone.

9.   Fong KM, Sekido Y, Minna JD. Basic science review: Molecular Pathogenesis of lung cancer. J Thorac Cardiovasc Surg 1999; 118:1136-52.
      This recent review details current knowledge of the molecular alterations in LC cells and explores the clinical implications in terms of possible therapy with experimental biologic and gene therapy interventions.

10.   http://www.tobaccodocuments.org/index.cfm?page=Project_Overview
      Some of the most important discoveries of the past few years in the areas of epidemiology, primary prevention, nicotine addiction and tobacco control have come from a cache of millions of internal tobacco industry documents made public during court hearings and trials. These documents are accessible on the Internet at the URL listed above.

12

CHAPTER 13

Mediastinal Tumors

Bruce Jason Leavitt

Scope of Problem

The mediastinum is an anatomically distinct area of the body located within the center of the chest. The boundaries are the sternum anteriorly and the spine posteriorly. The thoracic inlet is the superior boundary and the diaphragm the inferior one. The lateral pericardium forms the lateral borders. Quite a wide variety of tumors exist in the mediastinum.

Approximately two thirds of all mediastinal tumors are benign and the rest are malignant. Mediastinal tumors are most common in young and middle-aged adults and occur with less frequency in the older population. However, a higher incidence of malignancy is noted in older patients with primary lesions of the mediastinum. The majority of tumors are discovered by radiography, most commonly a plain chest x-ray. Most lesions are benign and have no symptoms. Malignant lesions often produce symptoms such as substernal pain or cough. A clinician can determine the location and type of tumor by an accurate history and physical examination, chest x-ray, CT scan of the chest, and simple tumor marker studies. Once identified, proper treatment can be instituted.

Risk Factors

Most mediastinal tumors do not have known risk factors. Patients with Von Recklinghausen's disease have a high incidence of neurogenic tumors. For the most part, these tumors are neurofibromas of the posterior compartment. Tumors of the thymus gland have many associated syndromes, the most common of which is myasthenia gravis. Myasthenia gravis is present in 30-50% of patients with thymoma. Of all patients with myasthenia gravis, only 15% have a thymoma. Other conditions associated with thymoma include pure red cell aplasia, hypogammaglobulinemia, myotonic dystrophy, and systemic lupus.

Screening

Screening for mediastinal tumors has not been a generally accepted practice. Patients with atypical chest pain should be evaluated with a PA and lateral chest x-ray. A thoracic CT scan with intravenous contrast should follow any masses or abnormal mediastinal shadows. Patients with myasthenia gravis and a normal chest x-ray still should have a CT scan of the chest to evaluate for the possibility of a thymoma. Tumor markers, human chorionic gonadotropin (HCG) and alpha fetoprotein (AFP) should be drawn on all patients with anterior mediastinal masses.

Surgical Oncology, edited by David N. Krag. ©2000 Landes Bioscience.

Methods of Diagnosis

Methods of diagnosis depend on the size and location of the tumor. Tumors can be grouped according to an anatomic location within the mediastinum. The mediastinum is separated into three compartments: anterior; middle; and posterior (Fig. 13.1).

Anterior Mediastinum

The anterior compartment lies between the sternum and the anterior pericardium. The most cephalad aspect extends to the thoracic inlet and neck (Fig. 13.1). Tumors of this region include thymic tumors, lymphomas, germ cell tumors, and thyroid masses (both benign and malignant).

Thymomas and thymic carcinomas can often be identified on a CXR (Fig. 13.2), but can always be detected on a CT scan (Fig. 13.3). CT scan with contrast combined with transthoracic needle biopsy (FNA) can diagnose most of the above conditions. If the thymic tumor has been diagnosed by FNA, surgical excision is warranted. If the FNA is not suggestive of a thymic mass, and further tissue is needed to make the diagnosis, an anterolateral thoracotomy with biopsy (Chamberlain procedure) should be considered. Recently, thoracoscopy with biopsy of masses in the anterior mediastinum has been utilized as a diagnostic technique.

Germ cell tumors of the mediastinum are uncommon. There are both benign (teratoma, teratodermoid) and malignant germ cell tumors. Almost all occur in the anterior mediastinum. Most malignant germ cell tumors occur in adult males under the age of 35. The types of germ cell tumors are similar to those found in the testicle. Pure seminoma is the most common, followed by embryonal carcinoma, teratocarcinoma, endodermal sinus tumor, and choriocarcinoma. Germ cell tumors most commonly present with localized chest pain or pressure, although asymptomatic tumors often occur and are diagnosed via CXR. These seminomatous germ cell tumors are seen as smoothly marginated anterior mediastinal masses on CXR (Fig. 13.4) and CT scan (Fig. 13.5). Nonseminomatous germ cell tumors tend to be more lobular and extensive than pure seminomatous tumors (Figs. 13.6, 13.7). Well-formed teeth have rarely been identified within a teratoma by CXR. CT scans often show a fatty mass with varying amounts of denser tissue such as bone, teeth, or nonspecific calcifications. Beta HCG and AFP levels should always be measured if a mediastinal germ cell tumor is suspected. FNA is the initial diagnostic approach while video-assisted thoracic surgery and/or thoracotomy can be performed if the FNA is nondiagnostic. A full metastatic work-up should be performed with any patient who is suspected of having a germ cell tumor. Nonseminomatous germ cell tumors are very malignant and most always present with metastases.

Mediastinal thyroid masses are most often goiters that arise in the neck and descend into the anterior mediastinum through the thoracic inlet. Thyroid carcinomas that arise within the mediastinum are quite rare. Substernal extension of thyroid goiters is often seen as dense masses on CXR that are located in the most superior aspect of the chest (Fig. 13.8). CT imaging of these lesions often show an enhancing lesion arising in the neck and extending down the right side of the anterior mediastinum (Fig. 13.9). If the goiter is metabolically active, a radioactive iodine scan can be diagnostic. Metabolically inactive substernal goiters will not enhance with the administration of radioactive iodine.

13

Table 13.1. Mediastinal compartments

Mediastinal Compartment	Tumors and Cysts
Anterior	Thymomas
	Lymphomas
	Teratodermoids
	Germ cell tumors
	Substernal thyroid goiters
Middle	Cysts (pericardial, bronchial, enteric)
	Lymphomas
Posterior	Neurogenic tumors (neurofibroma,
	neurolemmoma, neurosarcoma)

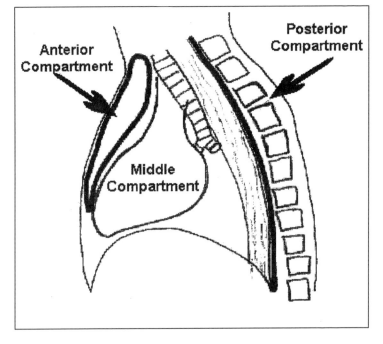

Fig. 13.1. Mediastinal compartments

Middle Mediastinum

The middle mediastinum contains the pericardium and its contents, as well as the hilar structures of the lung (Fig. 13.1). The most common masses in this area are cysts. Tumors are less common with lymphomas being the most prevalent. The typical radiologic presentation of a lymphoma on CXR is that of a lobular, often bilateral hilar mass (Fig. 13.10). CT scan is often impressive for a large central and anterior lobular mediastinal mass with differing densities within the tumor

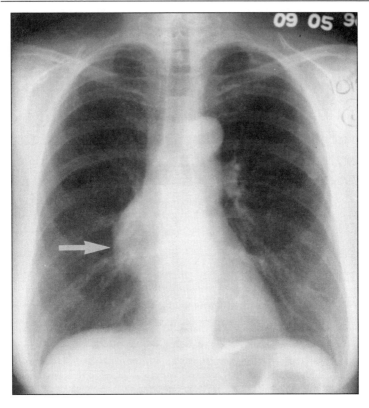

Fig. 13.2. Thymoma: Frontal CXR shows a right mediastinal mass

(Fig. 13.11). Certain mesynchymal neoplasms, and metastases from other primary cancers can occur in the middle mediastinum.

Lymphomas can be difficult to differentiate from other lymph node pathology such as sarcoidosis and benign lymphoid hyperplasia. Because of the major vascular organs and vessels in this area, FNA should be recommended with caution. This procedure should be performed by an experienced invasive thoracic Radiologist. Mediastinoscopy and biopsy provide the standard of care for lymph node biopsies in this region. Mediastinoscopy involves a 4-cm incision just above the sternal notch. A surgical plane is developed anterior to the trachea. This plane is carried down to the carina. Lymph nodes in the anterior mediastinum can be approached via this procedure. Posteriorly located masses can be safely approached by FNA. Video-assisted thoracic surgery or thoracotomies are usually reserved for tumors that cannot be biopsied by other methods.

Posterior Mediastinum

This compartment is bounded anteriorly by the pericardium and posteriorly by the spine and ribs. It extends from the thoracic inlet inferiorly to the diaphragm

13

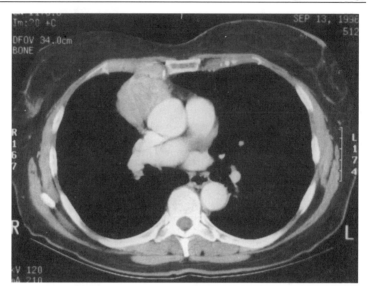

Fig. 13.3. Thymoma: CT scan shows a soft tissue mass in the right anterior mediastinum

(Fig. 13.1). Various cysts are present in this region, as well as neurogenic tumors. Neurogenic tumors arise from the sympathetic and intercostal nerves. Most of these neurogenic tumors are benign. Presenting symptoms include regional pain caused by compression of the surrounding structures or neurologic radiculopathy. Subtle CXR findings often include a smooth, homogenous posterior mass (Fig. 13.12). CT scans of a neurofibroma show a smooth mass adjacent to the thoracic vertebral body (Fig. 13.13). Tumors of the intercostal nerves are located in the costovertebral sulcus or sometimes more laterally along the course of the nerve. Some tumors can invade through the intervertebral foramen and grow into the spinal canal causing neurologic spinal cord symptoms. Magnetic resonance imaging (MRI) of the spine would be indicated if the CT scan suggests the above scenario. Most tumors can be identified safely in this region by FNA.

Preoperative Evaluation

A preoperative evaluation should be done on all patients who are considered candidates for surgery. For mediastinoscopy, only a simple preoperative work-up is needed since this is a potentially minimally invasive procedure that can be performed under local anesthesia with intravenous conscious sedation. Many surgeons prefer to perform this surgery with patients under general anesthesia. These patients can be discharged on the same day as the procedure.

Anterior thoracotomy with biopsy is a major procedure and should include a full preoperative work-up including pulmonary function tests (PFTs), electrocardiogram (ECG), type and screen and routine laboratory tests. Younger patients in good health may not need PFTs or an ECG.

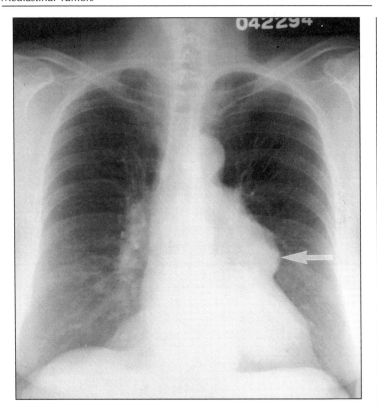

Fig. 13.4. Seminoma: CXR shows a left-sided mediastinal mass

Posterior thoracotomy and median sternotomy are the most invasive procedures utilized for resection of mediastinal tumors. These approaches require a full preoperative evaluation, including all of the above studies as well as an anesthesia consultation. A hospital stay of several days is routine following these more extensive procedures.

Staging

Anatomic location and the presence or absence of invasion of adjacent structures most commonly stages thymic tumors (Table 13.2). It is often stated that the surgeon determines the malignancy of the thymoma, not the pathologist.

A staging system for germ cell tumors has recently been developed. The staging is based on whether or not the tumor has invaded adjacent mediastinal structures and the presence or absence of metastases (Table 13.3).

Thymic carcinomas are aggressive tumors. Squamous cell carcinoma and lymphoepitheliocarcinoma are two of the most common types of thymic carcinomas. Thymic carcinoid, thymolipoma, and non-neoplastic thymic cysts rarely occur in the thymus.

13

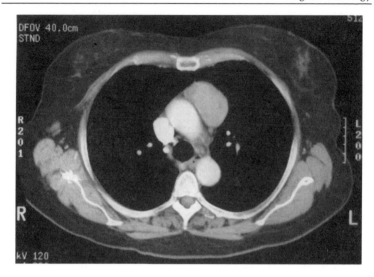

Fig. 13.5. Seminoma: CT scan shows a smooth mass in the left anterior mediastinum

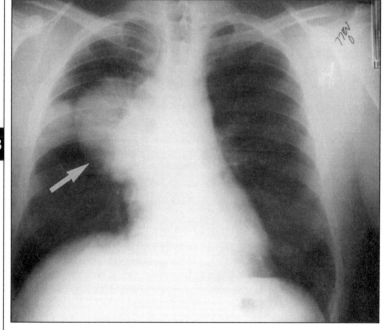

Fig. 13.6. Choriocarcinoma: CXR shows a lobulated right mediastinal mass

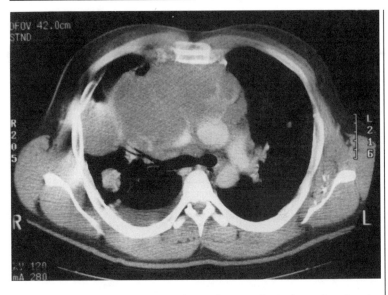

Fig. 13.7. Choriocarcinoma: CT scan shows a large mediastinal mass invading the superior vena cava

Fig. 13.8. Sub-sternal thyroid goiter: CXR shows a large, dense anterior and superior mediastinal mass

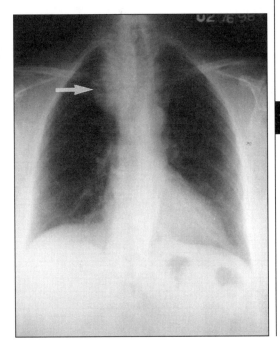

13

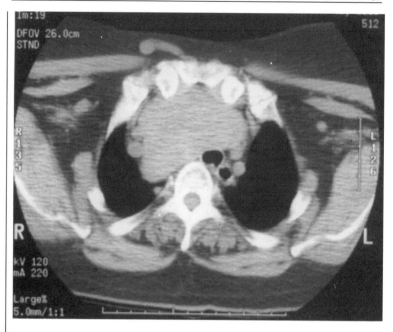

Fig. 13.9. Sub-sternal thyroid goiter: CT scan shows a right-sided dense anterior mediastinal mass

13

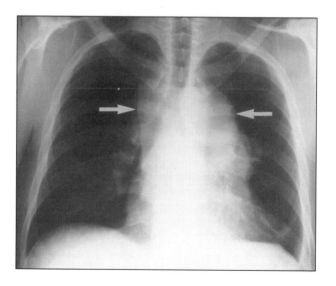

Fig. 13.10. Lymphoma: CXR shows a lobulated bilateral mediastinal mass

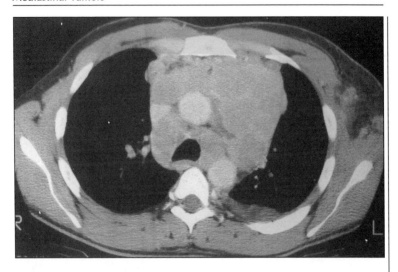

Fig. 13.11. Lymphoma: CT scans shows an extensive, lobulated anterior and middle mediastinal mass

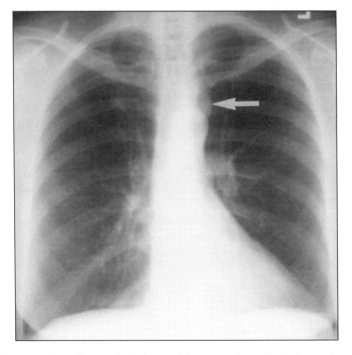

Fig. 13.12. Neurofibroma: CXR shows a left supraaortic mediastinal mass (arrow)

13

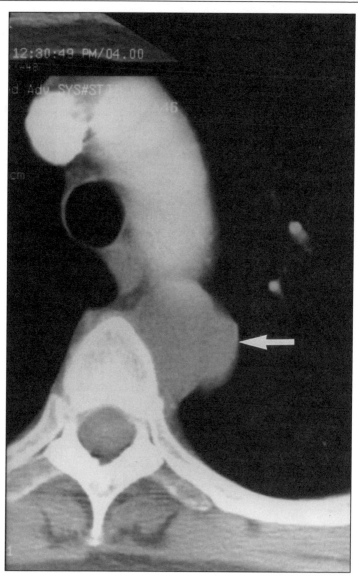

Fig. 13.13. Neurofibroma: CT scan shows a left paravertebral mass (arrow)

Table 13.2. Current staging of thymomas

Stage I	No capsular invasion
Stage II	Invasion through capsule, but not into adjacent structures
Stage III	Direct invasion into adjacent structures
Stage IVA	Pleura or pericardial metastases
Stage IVB	Distant metastases

Lymphomas occur in two forms: Hodgkin's and non-Hodgkin's lymphoma. The staging for Hodgkin's lymphoma most commonly used is the Ann Arbor staging (Table 13.4). Histologic criteria, rather than anatomic staging best classify non-Hodgkin's lymphoma.

Treatment Options

Most thymomas should be surgically resected. The standard approach is through a median sternotomy, but video-assisted thoracic surgery with thymectomy has been successfully performed. Stage I and II thymomas should be totally resected. Stage III tumors should undergo radical resection. Pericardium, lung, and diaphragm have all been resected along with the thymic tumor. The phrenic nerves should be preserved in most cases, unless an ipsilateral pneumonectomy is performed. If a total resection is not possible, subtotal resection is indicated. Surgical clips should be placed at the margins to mark the extent of resection. These patients can have improved survival with postoperative radiation and chemotherapy.

For all Stage II and III resected thymomas, postoperative radiation therapy is indicated. Chemotherapy has not been clearly shown to improve survival. Patients with Stage IV disease should have combined radiation therapy and chemotherapy. Survival in these patients is quite poor. However, some long-term survivors have occurred with aggressive therapy. Only very small and isolated germ cell tumors should be resected. Most of these tumors present with bulky disease and often have metastatic spread. This tumor is extremely sensitive to radiation therapy and therefore postoperative therapy should be done in all surgically resected patients. Patients with advanced disease should undergo radiation therapy and chemotherapy. In these cases surgery plays almost no role. Patients with bulky disease that have recurrent isolated tumor after aggressive chemotherapy and radiation therapy, may be candidates for resectional surgery. Patients with nonseminomatous germ cell tumors are generally not treated by resection due to the aggressiveness of the disease. Chemotherapy is the treatment of choice. This tumor is resistant to radiation therapy.

Substernal thyroid goiters should be surgically resected if the mass is responsible for any symptoms. Surgery is carried out via a generous neck incision with the operative field prepared for sternotomy, if needed. Since the blood supply arises from the neck vessels nearly all of these masses can be successfully removed through the neck incision alone.

Germ cell tumors are most commonly approached through a median sternotomy. Benign teratomas and dermoid cysts should be surgically removed. Pure seminomas of the mediastinum should be surgically resected through a median sternotomy if the extent of the tumor is such that it leads to complete removal. If excision is not possible, the only surgical option is a biopsy. Radiation therapy and chemotherapy

Table 13.3. Staging of mediastinal germ cell tumors

Stage I	Distinct tumor without direct or microscopic invasion of surrounding structures
Stage II	Tumor within the mediastinum with direct and/or microscopic invasion of surrounding structures
Stage IIIA	Tumor with intrathoracic metastases
Stage IIIB	Tumor with extrathoracic metastases

Table 13.4. Staging of Hodgkin's Lymphoma

Stage I	Single node region or lymph organ
Stage II	Two or more regions on the same side of the diaphragm
Stage III	Node regions on both sides of the diaphragm
Stage IV	Diffuse involvement of one or more extranodal organ

are the hallmarks of treatment for this condition. Nonseminomatous germ cell malignancies are not a surgical disease. A surgical biopsy is indicated when all other diagnostic avenues have failed.

Lymphoma should be diagnosed by FNA and surgery. The surgical approach is decided by the anatomic location of the suspected mediastinal lymphoma. Mediastinoscopy is the most common approach; however VATS with biopsy can be of use for masses in the hilar or aorto-pulmonary window regions. No surgical options other than biopsy are considered. Neurogenic tumors are always approached surgically. Tumors involving the intercostal nerves or sympathetic chain can be approached by thoracotomy or video-assisted thoracic approach.

Tumors that are "dumbbell" in configuration and are present in the spinal canal or intervertebral foramen should be approached by both thoracic and neurosurgical teams. Complete resection can usually be obtained. Neurofibroma, Neurolemoma, and ganglioneuroma are easily resected and have a good prognosis. Neurosarcoma occurs frequently in older patients and complete surgical resection is difficult.

Outcomes

Thymoma

Completely resected Stage I thymomas have a 95% or better five-year survival. Patients with larger tumors tend to have a poorer prognosis than those with smaller tumors. Patients with Stage II or greater disease that can be surgically resected have a greater survival than those who have residual thymoma. Overall, ten-year survival rates are shown in Table 13.5.

Germ Cell Tumors

Benign germ cell tumors (teratomas) are not malignant and have virtually 100% survival rate. In patients with seminoma, five-year survival ranges from 60-80% following chemotherapy and radiation therapy. Moran showed that patients over the age of 37 at the time of diagnosis of their seminoma had a 55% five year survival compared to an 80% five year survival in those patients under the age of 37.

Table 13.5. Ten year survival rates of thymoma by stage

Stage I	85-100%
Stage II	60-84%
Stage III	21-77%
Stage IV	26-47%

Nonseminomatous germ cell tumors have the worst prognosis of all tumors in this category. Prior to 1975, no long-term survivors were noted with this disease. With intensive chemotherapy, long term survival rates can vary between 13 and 40%.

Lymphomas

Patients with Hodgkin's disease have better survival than patients with non-Hodgkin's lymphoma. A greater than 90% survival is achieved with radiation therapy in patients with stage I or II Hodgkin's disease. Sixty percent or greater of patients with stage III Hodgkin's disease survive with chemotherapy and radiation therapy. Patients with non-Hodgkin's lymphoma often relapse. The goal of therapy in these patients is often one of palliation, using radiation therapy and chemotherapy.

Tumors of the Posterior Mediastinum

Eighty percent of tumors in this region are benign and therefore the majority of patients experience long term survival. Neurosarcoma is the most common malignant tumor of this region. Five year survival for patients treated for a malignant tumor of the posterior mediastinum is 36%.

Posttreatment Surveillance

Patients who undergo treatment for mediastinal tumors should be followed very closely by their surgeon, oncologist or primary care physician. Surgically resected thymomas should be followed with routine chest x-rays. CT scans of the chest should be done if there is any suggestion of recurrence of the tumor within the anterior mediastinum.

Mediastinal germ cell tumors should be followed carefully as well. Benign teratoma's should be followed with serial chest x-rays, while follow-up for seminomas and nonseminomatous germ cell tumors should include not only radiologic evaluation, but also beta HCG and AFP levels. If a patient has successful treatment of this disease, the tumor markers fall. Elevation of the tumor markers most commonly indicates recurrence of the disease.

Posterior mediastinal tumors are usually benign and should be followed with serial chest x-rays in the postoperative period.

Selected Readings

1. Sabiston DA, Spencer FC. Surgery of the Chest, Seventh edition. WB Saunders Company, 1999.
   Excellent review chapter on mediastial tumors. This text has been a standard of reference for cardiac and thoracic surgery for many years.
2. Shields TW. Mediastinal Surgery. Lea & Febiger, 1991.
   Authoritative review of mediastinal tumors including pathology, presentation, current surgical therapy and outcomes. The text is one of the best for mediastinal disorders.

3.   Haskell CM. Cancer Treatment, Fourth edition. WB Sanders Company, 1995.
This is a review text involving oncologic treatment of most malignant conditions.

4.   Shields TW. General Thoracic Surgery, Fourth edition. Williams & Wilkins, 1994.
Superb chapter review of mediastianl tumors in a general thoracic surgery text. Dr. Shield's textbook on general thoracic surgery is one of the most authoritative texts in this surgical area.

5.   Whooley BP, Urshcel JD, Antkowiak JG et al. Primary tumors of the mediastinum. J Surg Oncol 1999; 95-99.
General review article on primary tumors of the mediastinum in a recent journal.

6.   Bacha EA, Chapelier AR, Macchiarini P et al. Surgery for invasive primary mediastinal tumors. Ann Thor Surg 1998; 66:234-9.
Recently published French series involving surgery of invasive mediastinal tumors

7.   Kaga K, Nishiumi N, Iwasake M et al. Thoracoscopic diagnosis and treatment of mediastinal masses; usefulness of the two windows method. J Cardio Surg 1999; 40(1): 157-160.
This paper discusses thoracoscopic diagnosis of mediastinal tumors.

8.   Moran CA, Suster S. Primary germ cell tumors of the mediastinum. I Analysis of 322 cases with special emphasis on teratomatous lesions and a proposal for histopathologic classification and clinical staging. Cancer 1997; 80(4):681-690.
A large series of primary germ cell tumors of the mediastinum. This series focused on teratomas.

9.   Moran CA, Suster S, Przygodzki RM et al. Primary germ cell tumors of the mediastinum. II. Mediastinal seminomas a clinicopathologic and immunohistochemical study of 120 cases. Cancer 1997; 80 (4):681-690.
A large series of primary germ cell tumors of the mediastinum. This series focused on seminomas.

10.   Moran CA, Suster S, Koss MN. Primary germ cell tumors of the mediastinum. III. Yolk sas tumor, embryonal carcinoma, choriocarcinoma, and combined nonstratomatous germ cell tumors of the mediastinum. Cancer 1997; 80(4):699-707.
A large series of primary germ cell tumors of the mediastinum. This series focused on embryonal carcinoma, choriocarcinoma and non-seminomatous germ cell tumors.

11.   Feo CF, Chironi G, Porcu A et al. Videothoracoscopic removal of a mediastinal teratoma. Amer Surg 1997; 63(5):459-461.
Case report of a thoracoscopic (VATS) removal of a mediastinal teratoma.

12.   Bousamra M, Haasler GB, Patterson GA et al. A comparative study of thoracoscopic vs open removal of benign neurogenic mediastinal tumors. Chest 1996; 109(6):1461-1465.
Comparative study of VATS versus open thoracotomy for resection of posterior compartment benign neurogenic tumors.

13.   Luketich JD, Ginsberg RJ. The current management of patients with mediastinal tumors. Adv in Surg 1996; 30:311-332.
Review manuscript of the treatment of mediastinal tumors.

14.   Fizazi K, Culine S, Droz JP et al. Primary mediastinal nonseminomatous germ cell tumors: results of modern therapy including cisplatin-based chemotherapy. J Clin Onc 1998; 16(2):725-732.
Recent review article on chemotherapy for nonseminomatous mediastial germ cell tumors.

15.   Strollo DC, Rosado de Christenson ML, Jett JR. Primary mediastinal tumors. Part I. Chest 1997; 112:511-522.
Review article of primary mediastial tumors.

16.   Strollo DC, Rosado de Christenson ML, Jett JR. Primary mediastinal tumors. Part II. Tumors of the middle and posterior mediastinum. Chest 1997; 112:1344-57.
Review article of primary mediastial tumors.

13

CHAPTER 14

Endocrine Tumors

Thomas J. Kearney, David A. August

Introduction

Tumors of the endocrine system encompass a broad spectrum of clinical presentations, pathological findings, treatment options and outcomes. Even the most common endocrine tumors are relatively rare compared to frequently occurring epithelial tumors. Surgery is the primary treatment for most endocrine neoplasms although other modalities are used on occasion. Therefore, it is important for the surgeon to understand appropriate evaluation and management of endocrine tumors. This Chapter will cover the evaluation and treatment of the common tumors of the thyroid, parathyroid and adrenal glands as well as a discussion of endocrine tumors of the pancreas, carcinoid tumors and the multiple endocrine neoplasia (MEN) syndromes.

Thyroid

Scope of the Problem

Thyroid neoplasms are relatively common. Thyroid cancer represents approximately 90% of all endocrine malignancies. The estimated number of new cases in the U.S. in 1999 is 18,100.[1] Clinically detectable thyroid nodules occur in about 4-7% of the adult population. Nodules are more common in women. Most of these are benign follicular adenomas or a result of multinodular goiter. Only 5-10% of thyroid nodules are malignant in the general population. Thyroid nodules in a patient with a history of childhood neck irradiation are malignant in 35% of cases.

Among the nodules that represent thyroid carcinoma (Table 14.1), about 80-85% are classified as well-differentiated cancers, primarily papillary (75%) and follicular (10%), along with their variants. Medullary thyroid cancer (MTC), derived from parafollicular C-cells, represents about 10% of thyroid cancer. These cancers may be sporadic or associated with either a multiple endocrine neoplasia (MEN) syndrome or the familial medullary thyroid cancer (FMTC) syndrome. The remaining types of thyroid cancer include poorly differentiated and anaplastic thyroid cancers along with thyroid lymphoma.

Risk Factors

A history of radiation exposure in childhood is a significant risk factor for thyroid nodules and papillary thyroid cancer. This risk factor is present in about 10% of patients with thyroid cancer. Thyroid nodules are up to four times as common in women as in men. Thyroid nodularity increases with age with one study showing

---

Table 14.1. Overview of thyroid cancer histology

Type	Frequency	Important Features
Papillary	75%	Well-differentiated, good prognosis
Follicular	10%	Well-differentiated, FNA inaccurate
Medullary	About 10%	Derived from C-cells, secrete calcitonin, associated with MEN-2
Anaplastic	< 5%	Often lethal
Lymphoma	Rare, < 2%	Treated with radiation and chemotherapy

nodularity in 90% of women over the age of 70. Both sporadic and familial forms as well as an association with MEN-2 syndrome characterize medullary thyroid cancer. Thus, a family history of inherited cancers should alert the physician to the possibility of medullary thyroid cancer. Conversely, a new diagnosis of medullary thyroid cancer should prompt a re-examination of the family history.

Screening

Physical examination of the thyroid gland should be a part of all general physical examinations. In the setting of an MEN pedigree, screening for medullary thyroid cancer is crucial (see section on MEN syndromes).

Diagnosis

Most patients present with a thyroid nodule. A history of radiation exposure is important. Physical examination should be complete but should focus on the thyroid and surrounding nodes. Biochemical tests of thyroid function including TSH and T4 should be obtained. Diagnostic imaging may include ultrasound (US) which will differentiate cysts from solid lesions. However, US cannot reliably distinguish benign from malignant solid lesions. Isotope scanning with radioiodine (^{131}I or ^{125}I) or ^{99m}Tc-pertechnetate was previously popular. Hot nodules are almost always benign but most nodules are cold. Most cold nodules are benign too. Neither US nor isotope scanning can reliably exclude or confirm the presence of thyroid cancer.

Fine needle aspiration (FNA) with interpretation by an experienced cytologist is the procedure of choice for the initial evaluation of a thyroid nodule. Results should be characterized as benign, suspicious, malignant or inadequate. Patients with benign cytology can be safely observed with surgery reserved for nodules that grow. About 70% of all thyroid nodule FNA samples will be benign. False negative results occur in 1-5% of cases. Suppression of TSH secretion with thyroid replacement medication during observation is controversial. Patients with suspicious or malignant cytology will represent about 15% of all samples. These patients require surgery. The primary drawback to FNA cytology is the inability to differentiate follicular adenoma from follicular cancer, which requires histologic study. Therefore, patients with a diagnosis of follicular neoplasm on FNA cytology have a suspicious result and require surgical intervention.

Preoperative Evaluation

In addition to any preoperative testing based on concurrent medical problems, patients should have a chest x-ray to rule out pulmonary metastases. Serum calcium is measured to assess parathyroid function. Those patients with medullary thyroid cancer should be evaluated for MEN-2 associated abnormalities of the parathyroid glands and pheochromocytoma. Calcitonin measurements are usually only obtained when the history suggests medullary cancer or an MEN syndrome.

Staging

Due to the importance of age and histologic type in the prognosis of thyroid cancer, these factors are included in the American Joint Committee on Cancer (AJCC) staging system.[2] Tumors are divided into three size categories: 1 cm or less (T1), 1 to 4 cm (T2) and greater then 4 cm (T3). Tumors of any size with extracapsular extension are T4. Nodal metastases are either absent (N0) or present (N1). Nodal involvement is classified as ipsilateral (N1a) or bilateral, midline or contralateral (N1b). Distant metastases are either absent (M0) or present (M1). The TNM classification should be used to categorize the stage of the tumor (Tables 14.2 and 14.3).

Several other related classification systems (AMES, AGES, DAMES, MACIS and others) have been developed to identify patients at either low risk or high risk of distant disease. These systems are used by many endocrine surgeons as a basis for selecting the extent of thyroid resection that is required. A review of the AMES (Table 14.4) system demonstrates the close correspondence with the AJCC staging system. The other systems have similar components.

Treatment

The primary treatment of well-differentiated papillary and follicular thyroid cancer is surgical. There is ongoing debate about the extent of thyroid resection and lymph node removal that is required.[3,4] There are no prospective randomized trials to provide guidance. A number of retrospective reviews are available. At a minimum, lobectomy on the side of the tumor along with resection of the isthmus is necessary. For many patients with low risk disease, this is all that is required. Most patients are cured and any residual occult foci of cancer in the other lobe are likely to be clinically insignificant. The small but real risk of bilateral recurrent nerve injury or permanent hypoparathyroidism is avoided when lobectomy is performed. Those who advocate total thyroidectomy cite a low rate of complication by experienced surgeons. In addition, less then total thyroidectomy hampers the ability to treat residual disease with radioiodine ablation. One reasonable approach is to perform lobectomy and isthmusectomy for small (T1), well-differentiated papillary and follicular tumors in younger patients (age < 45). Patients with larger tumors, extracapsular extension, nodal metastases or older age should undergo total thyroidectomy. When nodal metastases are present, a modified neck dissection should be added. The dissection should remove the nodal metastases but spare the sternocleidomastoid muscle, jugular vein and spinal accessory nerve.

After total thyroidectomy, a radioactive thyroid scan may be used to identify any remaining occult thyroid tissue. An ablative dose of radioactive iodine should be used to eradicate all residual thyroid tissue. Following this, thyroid hormone is given to suppress TSH secretion. Cytotoxic chemotherapy is generally not used in the adjuvant setting for well-differentiated thyroid cancer.

Patients with medullary cancer should undergo a total thyroidectomy with paratracheal central neck dissection. Metastases to nodes are much more common

14

Table 14.2. TNM classification for papillary and follicular thyroid cancer

T	Tumor	N	Nodes	M	Metastases
T1	< 1 cm	N0	–nodes	M0	– metastases
T2	> 1 and < 4 cm	N1a	+ nodes: ipsilateral	M1	+ metastases
T3	> 4 cm	N1b	+ nodes: bilateral, midline or contralateral		
T4	Local extension				

Table 14.3. AJCC stage for papillary and follicular thyroid cancer

Stage	Age < 45	Age ≥ 45
I	Any T, any N, M0	T1, N0, M0
II	Any T, any N, M1	T2-3, N0, M0
III		T4, N0, M0
		Any T, N1, M0
IV		Any T, any N, M1

Table 14.4. AMES system

Characteristic	Low risk	High risk
Age	Males ≤ 40	Males > 40
	Women ≤ 50	Women > 50
Metastases	No distant disease	Distant disease
Extent	No extrathyroid extension	Extrathyroid extension
Size	< 5 cm	> 5 cm

than in papillary or follicular cancer. Adjuvant radioiodine has no role. Adjuvant chemotherapy does not affect eventual outcome. Patients with anaplastic tumors and poorly differentiated tumors often present with locally advanced disease. The median survival is five months with most patients dying from local disease and airway involvement. Treatment should consist of external beam radiation and doxorubicin chemotherapy. In occasional patients with resectable disease, a total thyroidectomy is done with postoperative radiation and doxorubicin chemotherapy.

Thyroid lymphoma is rare, representing less then 1% of all lymphomas. The diagnosis can be made by needle biopsy. Surgery is usually not indicated unless there is diagnostic uncertainty. The tumor is treated with external beam radiation and doxorubicin based chemotherapy.

Outcome

In general, the treatment of well-differentiated thyroid cancer is quite gratifying. Patients with Stage I disease have a 95% 10-year survival rate. This category includes all younger patients, even those with nodal metastases. Most thyroid cancer patients are in the low-risk category with 80-90% long-term cure rates. Patients with papillary cancer have slightly better survival (90-95%) then patients with follicular cancer (70-80%). About 10-20% of patients will recur and about 80% of the recurrences will be local, usually in lymph nodes. Most of these patients can be salvaged with additional surgery. Patients with distant metastatic recurrence and those who present with distant metastases usually succumb to their disease.

Medullary cancer is less favorable. Survival rates at 5 years vary between 50-80%. Given the lack of effectiveness of adjuvant therapies, the key to successful treatment rests on early diagnosis and total thyroidectomy. Patients with anaplastic thyroid cancer usually die quickly from locally advanced disease. Patients with thyroid lymphoma have 5-year survival rates of 70%.

Posttreatment Surveillance

Patients should be seen twice yearly postoperatively with decreasing frequency after several years have passed. Patients with well-differentiated thyroid tumors should have thyroglobulin levels measured. Rising thyroglobulin levels should occasion a radioiodine scan, searching for recurrent disease. Therapeutic radioiodine can then be given. Patients with medullary cancer of the thyroid are followed with calcitonin levels. Radioiodine scanning is not useful. Successfully resected patients with poorly differentiated tumors or thyroid lymphoma are followed with physical examination. There are no tumor markers or scans available.

Special Considerations

Medullary cancer of the thyroid can be sporadic, familial or associated with MEN. A detailed family history should be obtained and consideration given to screening family members. The genetic defect associated with MEN and FMTC has been identified and can be used for genetic screening of the patient and family members. This is discussed further in the section on MEN syndromes.

Parathyroid

Scope of the Problem

Parathyroid tumors are almost always benign. Adenomas account for roughly 80% of cases with hyperplasia accounting for almost all of the remainder. Cancer of the parathyroid gland is exceedingly rare and accounts for only 1-2% of cases of parathyroid tumors. Primary hyperparathyroidism is common, with an incidence of 40 cases per 100,000, accounting for about 100,000 new cases annually in 1999. Many of these patients will not require surgery. Surgery is usually performed on symptomatic patients, patients with high calcium levels (> 11-12 mg/dl) and patients under age 50. The total number of expected cases annually of parathyroid cancer in the U.S. is less then 100. The cancer cases are usually associated with severe hypercalcemia (> 14 mg/dl) and often a palpable mass.[5]

Risk Factors

There are no known risk factors for sporadic parathyroid adenoma, hyperplasia or carcinoma. Secondary hyperparathyroidism can occur in the setting of chronic renal failure. In this setting, hyperphosphatemia, decreased calcitriol levels, bone resistance to increasing PTH levels and other factors lead to increasing parathyroid function. If this hyperfunctioning becomes autonomous, the term tertiary hyperparathyroidism is used. A family history of hyperparathyroidism or other endocrine tumors should prompt consideration of an MEN syndrome.

Screening

Given the rarity of the problem, no formal screening is recommended for parathyroid tumors. However, the routine use of multi-test blood chemistry panels has increased the number of patients who carry the diagnosis of hypercalcemia. Thus, many more asymptomatic patients are identified today than in previous decades. The two most common causes of hypercalcemia are primary hyperparathyroidism and malignancy, accounting for 90% of all cases.

Diagnosis

The diagnosis of hyperparathyroidism is fairly straightforward. Patients with hypercalcemia should be tested for an increased blood level of intact PTH. Urinary calcium excretion is measured to exclude benign familial hypocalciuric hypercalcemia. Calcium levels above 14 mg/dl along with a neck mass should heighten the concern about parathyroid cancer.

Preoperative Evaluation

Renal function should be assessed with routine blood chemistries. Preoperative localization studies are not recommended in patients scheduled for an initial bilateral neck exploration by an experienced endocrine surgeon. For patients with persistent hyperparathyroidism after surgery, previous neck surgery or when cancer is a real concern, preoperative localization studies are indicated. Noninvasive options include ultrasound and conventional CT and MRI scans. The sensitivity of these three modalities can be above 80%. Nuclear imaging with thallium-201/technetium-99m pertechnetate subtraction scans has sensitivity up to 70%. Technitium-99m sestamibi scans have high sensitivity (almost 100%) with good specificity (90%). In addition patients can be scanned from multiple angles, thereby localizing glands. Increasingly, sestamibi scans are being used routinely for preoperative localization prior to minimally invasive parathyroid surgery. This technique involves removal of the single involved gland through a small incision without formal neck exploration.

When noninvasive tests are conflicting or equivocal, invasive studies can be performed. These include angiography and selective venous sampling for PTH. Sensitivities for these techniques can be up to 80%. These invasive tests are only used prior to re-exploration for recurrent or persistent hyperparathyroidism. Intraoperatively, ultrasound has been used to help localize parathyroid masses. In addition, intravenous methylene blue can be given. Parathyroid tissue selectively concentrates this dye, often aiding in identification.

Staging

Because of the rarity of parathyroid cancer, there is no staging system for parathyroid tumors.

Treatment

Classically, treatment consists of a bilateral neck exploration under general anesthesia through a collar incision. However, recently there has been an increased interest in minimally invasive approaches. In the classic operation, the strap muscles are retracted but not divided. The thyroid lobe is retracted medially. The dissection must be meticulous. Special care is taken to protect the recurrent laryngeal nerve. The standard locations of the parathyroid glands should be inspected. The surgeon must be familiar with the common hidden locations of "missing" parathyroid glands (Tables 14.5 and 14.6). Embryologically, the "lower" gland (derived from branchial pouch III) migrates from a superior position along with the thymus. The "upper" gland (derived from branchial pouch IV) migrates upward. Incomplete migration can lead to aberrant locations. The lower gland (parathyroid III) is more likely to occupy an aberrant position than the upper gland (parathyroid IV).[6,7]

If a single enlarged gland is identified, it should be removed after locating the other normal glands. The enlarged gland should be weighed and sent to pathology for confirmation of parathyroid tissue. If there is doubt about the nature of the remaining normal glands, one can be biopsied. If all three are biopsied, the risk of postoperative hypoparathyroidism is increased. The remaining normal glands should be marked with a clip or nonabsorbable suture in the event reoperation is required. If two or three enlarged glands are noted, they should be removed after the normal glands are identified.

In the event of parathyroid hyperplasia, a subtotal parathyroidectomy is indicated. The most normal looking gland should be biopsied and left intact. The other hyperplastic glands are removed once the viability of the biopsied gland is certain. Because a fifth gland will be present in the thymus in up to one quarter of patients with hyperplasia, some surgeons consider removing the thymus. When there is concern about the viability of remaining parathyroid tissue, hyperplastic tissue can be transplanted into a forearm muscle pocket or cryopreserved for delayed transplantation.

Some authors interested in the minimally invasive approach recommend preoperative localization studies combined with a unilateral approach. The presumed benefits are a lowered risk of hypoparathyroidism and a decreased risk of bilateral recurrent laryngeal nerve injury. Operative time is decreased. However, the surgeon must be prepared to explore the other side if a typical adenoma combined with a normal gland is not identified through the unilateral approach. With this minimally invasive approach, systemic methylene blue is often administered, as this is taken up by parathyroid (and pancreatic) tissue. The color can assist localization through a small incision.

The emerging technology of intraoperative radionuclide mapping may allow for the preoperative localization and intraoperative identification of parathyroid adenomas with a hand-held gamma probe.[8] In this approach, the surgeon uses the hand-held probe to identify a "hot-spot" in the neck within 2-2 1/2 hours of Tc99m-sestamibi injection. A small incision is made and the probe is used to locate the adenoma. Clearly, this technique is useful only for patients with a single adenoma. Further prospective studies will be needed to validate this approach. Regardless of the surgical approach used, a quick, intraoperative PTH assay is available to confirm the removal of a PTH secreting tumor.

14

Table 14.5. Usual and aberrant locations of parathyroid glands at initial surgery

Upper gland	%	Lower gland	%
Cricothyroid junction	77%	Lower pole of thyroid	42%
Upper pole of thyroid	22%	Thymic tongue	39%
Retropharyngeal	1%	Lateral to lower thyroid	15%
and retroesophageal		Carotid sheath	2%
		Miscellaneous	2%

Table 14.6. Aberrant locations of "missed" parathyroid glands at reoperation

Location	%
Superior posterior mediastinum	34%
Anterior mediastinum	21%
Upper pole of thyroid	19%
Lower pole of thyroid	10%
Thymic tongue	13%
Retroesophageal	3%

In the setting of suspected parathyroid cancer, an en-bloc resection of the tumor including any locally invaded tissue and the ipsilateral thyroid lobe should be performed. The tumor should not be biopsied in order to prevent seeding. Enlarged lymph nodes should be removed and an elective central node dissection (ipsilateral jugular nodes and suprathymic tissue) should be performed.

Outcome

Surgical therapy by an experienced surgeon will cure primary hyperparathyroidism in up to 95% of patients. The remaining 5% will have persistent or recurrent hyperparathyroidism. In this setting, single gland disease still represents the majority of the cases but the incidence of multiple gland disease doubles from 15-30%. An experienced team should evaluate persistent or recurrent disease. Localization studies should be liberally used in this circumstance.

For the rare patient with parathyroid cancer, five-year survival rates between 40-70% have been reported. Patients who suffer local recurrence after initial complete resection can occasionally be palliated with repeat surgery. Cure is rare in this situation. Radiation is not effective. A variety of chemotherapy regimens have been used but the rarity of this disease precludes any clinical trials with adequate sample size. Hypercalcemic crisis is often seen in patients with residual parathyroid cancer and is the leading cause of death. Biphosphanates are the most active agents for managing this condition.

Posttreatment Surveillance

Due to the 5% failure rate with surgery for primary hyperparathyroidism, it is reasonable to obtain serum calcium levels at 6 and 12 months after surgery. Follow-up beyond that time is probably unnecessary. Patients with successful resection of parathyroid cancer can have recurrences up to 10 years postoperatively. Therefore clinical follow-up for 10 years seems prudent.

Special Considerations

Most cases of primary hyperparathyroidism are sporadic. However, given the association with MEN-1 and MEN-2A, a thorough family history must be obtained from all patients.

Adrenal

Scope of the Problem

Adrenal tumors are similar in some respects to parathyroid tumors. Cancer of the adrenal gland is rare. The incidence is about 1 per 1,000,000 giving an annual rate of about 200 new cases per year in the U.S. There is a bimodal age distribution with the incidence peaking in childhood and again at about age 50. The primary surgical interest in adrenal tumors is focused on hypersecretion syndromes resulting from hyperfunctioning cortisol, aldosterone and sex hormone secreting adenomas. In addition, patients may have pheochromocytoma with the resultant signs and symptoms of catecholamine excess. The incidence of primary hyperfunctioning adrenal adenomas is only slightly higher then adrenal cancer. No more then 1000 cases of all types would be expected in the U.S. annually. Pheochromocytoma is more common than hyperfunctioning adrenal adenoma, with an incidence of 1 per 100,000, giving several thousand cases per year in the U.S. In addition, with the increasing use of diagnostic body CT scan, incidental findings in the adrenal gland have created the entity of the "adrenal incidentaloma". These asymptomatic, nonfunctioning adenomas do not require treatment or further evaluation if they are typical. Typical incidentalomas are characterized by size < 3 cm as well as CT or MRI appearance. Adrenal masses may be seen in over 1 of every 200 abdominal CT scans. About 35% of large incidentalomas (> 5 cm) are malignant. Metastases to the adrenal glands are fairly common in some cancers, particularly lung cancer, breast cancer or melanoma.

Risk Factors

There are no specific risk factors for functional adrenal adenomas. About 10% of pheochromocytoma patients have an affected family member. Screening may be appropriate in this setting or when the MEN-2 syndrome is suspected.

14

Screening

Screening for adrenal tumors is not indicated in the general population. It is reasonable to consider in certain high-risk groups or patients with specific signs and symptoms. Fewer then 1% of hypertensive patients have primary hyperaldosteronism. Persistent hypokalemia in a hypertensive patient should suggest the possibility of primary hyperaldosteronism. Patients with obesity, muscular weakness, striae, diabetes and hypertension may be exhibiting signs of glucocorticoid excess (Cushing's syndrome) and should be evaluated appropriately. Unusual virilization or feminization, particularly when combined with signs of hypercortisolism, should prompt a search for a sex hormone secreting tumor. Patients with paroxysmal headaches, hypertension, tachycardia and palpitations should be evaluated for pheochromocytoma.

Patients suspected of having MEN-2 syndromes or those diagnosed with medullary thyroid cancer should also be screened for pheochromocytoma.

Diagnosis

The diagnosis of adrenal tumors requires recognition of the syndrome and confirmation with biochemical testing (Table 14.7). The diagnosis of primary hyperaldosteronism (Conn's syndrome) is made by demonstrating hypokalemia and low levels of plasma renin combined with increased plasma aldosterone levels. About 85% of patients with primary hyperaldosteronism have an adrenal tumor. The remaining 15% have bilateral hyperplasia. CT scans will help differentiate between the two in many patients. In difficult cases, iodocholesterol nuclear scans and selective adrenal venous sampling can select patients for surgical intervention.

Cushing's syndrome can be ACTH (corticotropin) dependent or independent. The great majority of patients have ACTH dependent disease. The ACTH is secreted either from a pituitary adenoma (Cushing's disease, 70%) or an ectopic ACTH source (10%). About 20% of patients have ACTH independent disease, either from an adrenal adenoma (10%) or adrenal cancer (10%). Cushing's syndrome is initially evaluated with 24-hour urine cortisol measurement and an overnight low-dose dexamethasone suppression test. A small dose of dexamethasone (1 mg) is given at night. The serum cortisol is measured the next morning. Normally the level would be < 3-5 µg/dl. A normal level excludes Cushing's syndrome. A higher level is consistent with glucocorticoid excess although various other conditions can give a positive result (stress, pregnancy, estrogen or anticonvulsant use). Confirmation of Cushing's syndrome requires a high-dose dexamethasone suppression test. In the high dose test, 4 mg of dexamethasone is given over 48 hours and urine cortisol is measured during the last 24 hours. Failure to suppress urine cortisol confirms Cushing's syndrome. Differentiation between ACTH dependent and independent forms requires measuring levels of ACTH.

Once ACTH independent Cushing's syndrome is confirmed, CT scan is used to examine the adrenal glands for a tumor. Nuclear scans with iodocholestrol can also be useful in differentiating a unilateral adenoma from hyperplasia. If ACTH dependent hypercortisolism is found, brain MRI and venous sampling of the petrosal sinus can differentiate between pituitary adenoma and ectopic ACTH production.

Suspected sex hormone producing tumors should be evaluated with 24-hour urine collections for 17-ketosteroids, 17-hydroxysteroids and free cortisol. Serum measurements of testosterone and estrogen should be obtained.

The diagnosis of pheochromocytoma is based on measuring catecholamines and their metabolites, vanillylmandelic acid (VMA) and metanephrines. Direct plasma measurement of epinephrine and norepinephrine is available but it is unclear if they are better then the urinary tests. The most reliable screening test is a 24-hour urine collection for measurement of VMA and metanephrines. In equivocal or borderline cases, a clonidine suppression test is useful. Both CT and MRI can be used to localize tumors once the diagnosis is confirmed. Radioiodine labeled MIBG can be used for localization, particularly when ectopic locations are present. The "rule of 10" is useful when considering pheochromocytomas. Malignancy is seen in 10% of cases. Classically, 10% of cases are either bilateral or have an ectopic location. Children account for 10% of the cases.

Table 14.7. Adrenal tumor syndromes

Tumor syndrome	Substance	Patient characteristics	Test results
Conn's syndrome	aldosterone	Hypertension, hypokalemia	↑ aldosterone, ↓ renin
Cushing's syndrome (ACTH independent)	cortisol	Appearance, diabetes, cortisol, hypertension	↑ urinary + dexametha--sone suppression test, ↓ ACTH
Sex hormone excess	testosterone or estrogen	Inappropriate or ambiguous 2° sexual characteristics	↑ urine and serum testosterone and estrogen
Pheochromocytoma	catecholamines	Palpitations, hypertension, tachycardia, headaches	↑ metanephrines, ↑ VMA

Incidentally detected adrenal masses should be evaluated initially with a detailed history and physical exam. A personal history of cancer should prompt concern about metastases. The physician must decide if the adrenal mass is functional and if it is malignant. Patients with tumors larger then 5 cm and a low likelihood of adrenal metastases should be screened for pheochromocytoma and prepared for surgery. Patients with tumors < 3 cm and typical appearance on imaging can be followed with repeat imaging. Patients with tumors between 3-5 cm should be evaluated for functioning adrenal tumors by physical exam, blood pressure measurement and serum electrolyte determination. Abnormal findings require more extensive evaluation. This would include 24-hour urine collections for Cushing's syndrome, sex hormone tumors and pheochromocytoma. Primary hyperaldosteronism is unlikely in the absence of hypertension and hypokalemia. If the functional tests are positive, the tumor is managed surgically. If the functional tests are unremarkable, size and imaging characteristics guide management. MRI characteristics may be used to differentiate pheochromocytoma from adenomas based on the ratio of T2 signal intensity compared to normal spleen and liver tissue. Metastatic cancer and adrenal cancer have ratios intermediate between pheochromocytoma and adenoma. Nonfunctional, small tumors less than 3 cm are usually observed for stability at 3-6 month intervals. Tumors that grow or are larger then 5 cm should be removed. There is controversy about the management of masses between 3 and 5 cm in size. Small size cannot definitively rule out adrenal cancer and prognosis is directly related to size.

The possibility of metastatic disease to the adrenals in a patient already carrying a cancer diagnosis should be strongly considered. Metastases can be confirmed in some situations with FNA. Pheochromocytoma should always be excluded by appropriate evaluation prior to FNA of an adrenal mass. In general, resection of adrenal metastases is not performed. Occasionally, a patient with no other site of metastases will benefit from adrenalectomy.

14

Preoperative Evaluation

Due to the metabolic derangement produced by some adrenal tumors, particular attention must be paid to patient preoperative preparation. For patients with Cushing's syndrome and/or Conn's syndrome, hypokalemia, hyperglycemia and hypertension should be controlled. Patients with pheochromocytoma require α-blockade for blood pressure control several weeks preoperatively. These patients also tend to be dehydrated. Once blood pressure is controlled, patients who remain tachycardic should also be given a β-blocker. Anatomic considerations in adrenalectomy should be considered. Patients with large tumors involving the vena cava may require venovenous or cardiopulmonary bypass.

Staging

There is no staging system for benign adrenal tumors. The MacFarlane staging system for adrenal cancer (Tables 14.8 and 14.9) is based on tumor characteristics (size, the presence of invasion and the involvement of adjacent organs) as well as the presence or absence of lymphatic and distant metastases. The AJCC has not adopted a staging system.

Treatment

The treatment of benign adrenal tumors is surgical. The underlying metabolic abnormalities should be controlled preoperatively. A variety of surgical approaches are available. Traditional approaches include anterior midline approaches with a thoracoabdominal extension if needed, a flank approach and a posterior approach. Recently, laparoscopic approaches have been developed. The anterior open approach is valuable if the surgeon is concerned about local invasion and needs access to vascular structures. The anterior approach is required if cancer is suspected. In addition, when there are concerns about multiple sites (10% of pheochromocytomas are multicentric) the anterior approach provides greater flexibility. The flank and posterior approaches are useful for isolated tumors that are presumed to be benign. These two approaches are retroperitoneal and avoid some of the complications associated with abdominal surgery. Several authors have reported experiences with laparoscopic adrenalectomy and have compared their patient series with historical controls undergoing the posterior or flank approach.[9] The laparoscopic approach compares favorably, particularly with respect to postoperative pain. However, the operating time and hospital charges are higher for the laparoscopic approach. Over 10% of laparoscopic cases require conversion to an open approach. The laparoscopic technique will probably be developed fully only at specialty referral centers.

Outcome

The outcome from surgical removal of benign, functioning adenomas is good. Almost all patients with an aldosteronoma experience resolution of hypokalemia and 60 to 80% develop normal blood pressure. Patients with pheochromocytoma experience cure of hypertension in 85-90% of cases. Patients with Cushing's syndrome will have reversal of the stigmata of their disease but this often takes a year or more to stabilize.

The outcome for patients with adrenal cancer is related to the stage of the tumor. About 50% of patients present with Stage I or II disease. Patients who have a complete primary resection have 5-year survival rates of about 55%. Patients with

Table 14.8. Proposed TNM classification for adrenal cancer

T	Tumor	N	Nodes	M	Metastases
T1	≤ 5 cm	N0	–nodes	M0	–metastases
T2	> 5 cm	N1	+ nodes	M1	+ metastases
T3	Invasion of peri-adrenal fat				
T4	Invasion of adjacent organs				

Table 14.9. MacFarlane staging system for adrenal cancer

Stage	TNM
I	T1, N0, M0
II	T2, N0, M0
III	T3-4, N0, M0
	T1-2, N1, M0
IV	T3-4, N1, M0
	Any T, any N, M1

incomplete resection are not cured and have a median survival of 12 months following surgery. There is no proven role for adjuvant therapy. Patients with recurrent or metastatic adrenal cancer should be evaluated for a secondary complete resection. If this can be accomplished, 5-year survival is also around 50%.[10] Patients with unresectable recurrence or metastases may be treated with mitotane. Response rates of about 20% can be expected. Toxicity can be high and drug levels and dosages must be carefully controlled. Radiation can be used for the palliation of bone metastases.

At least 10% of patients with pheochromocytoma (and perhaps more) have malignant disease. In the setting of unresectable or metastatic disease, continued α-blockade is indicated for blood pressure control. Survival at 5 years varies between 30-60% in selected series. Several chemotherapy regimens are available including high dose streptozotocin and the combination of cyclophosphamide, vincristine and darcarbazine. Response rates of about 30-50% are reported but all of the series have small numbers of patients.

Posttreatment Surveillance

Patients with benign adenomas are usually cured with surgical resection. However, it may take a year or more for the hypothalamic-pituitary-adrenal axis to return to normal. Patients undergoing adrenalectomy for Cushing's syndrome may require glucocorticoid supplementation for up to two years. Thus, it is reasonable to continue follow-up until endocrine function is normal. Large adenomas (50-100 grams) that appear benign histologically have malignant potential. Patients should be followed for several years to exclude recurrence.

Patients with adrenal cancer are likely to suffer recurrence, even with complete resection. These patients should be seen regularly for at least five years. Traditionally, pheochromocytoma is malignant in 10% of cases although some series report malignancy rates of over 30%. Some experts therefore recommend long-term fol-

14

low-up of patients with measurement of urinary catecholamines and annual MIBG scans.

Special Considerations

Most patients with benign adrenal tumors can be cured of their disease. Proper diagnosis is essential. Due to the rarity of these tumors, there are no large randomized studies to guide the details of therapy. Referral to specialty centers is strongly encouraged.

Endocrine Tumors of the Pancreas

Scope of the Problem

Endocrine tumors of the pancreas are heterogeneous. They may be malignant or benign. About 75% are functional while the other 25% are nonfunctioning. Most of the tumors are within the pancreas although some reside in the duodenum. These tumors usually present as sporadic cases but can be linked to the multiple endocrine neoplasia (MEN) syndromes, which are discussed in a later section. Nonfunctioning endocrine tumors of the pancreas are reported in over 1% of patients in autopsy series. However, clinically relevant endocrine tumors are rare, with an incidence of about 4 per 1,000,000 or about 1000 cases per year in the United States. The majority of the patients (about 75%) present with symptoms of excess hormone production with the specific syndrome depending on the type of tumor.

The most common pancreatic endocrine tumor is the insulinoma, accounting for about a third of such tumors. These tumors arise from insulin secreting β cells in the pancreas. Insulinomas are benign in 90% of the cases and malignant in the other 10%. Malignancy is defined by the discovery of metastases rather then by histologic criteria. They usually are solitary. About 10% of insulinomas are associated with MEN and they are more likely to be multicentric. Rarely, nesidioblastosis (hyperplasia of islet cells) is present. Hypoglycemia in infancy is associated with nesidioblastosis.

Gastrinomas represent the next largest group. They probably arise from uncommitted precursor cells since the pancreatic islets do not normally contain gastrin-secreting cells. Gastrinomas are malignant in 50-90% of reported series. About 25% of gastrinomas are associated with MEN and gastrinoma is the most common pancreatic endocrine tumor seen in MEN. VIPomas (vasoactive intestinal peptide), glucagonomas, PPomas (pancreatic polypeptide) and somatostatinomas account for only a minority of pancreatic endocrine tumors but as a group represent about 20-25% of pancreatic endocrine tumors. More then half are malignant. Very few are associated with MEN. The remaining 20-25% of pancreatic endocrine tumors are nonfunctional. These nonfunctional tumors usually present with symptoms caused by local tumor growth such as obstructive jaundice or gastric outlet obstruction, similar to adenocarcinoma of the pancreas. The majority of nonfunctional tumors are malignant and they usually are not associated with MEN.

Risk Factors

There are no known risk factors for pancreatic endocrine tumors in the general population. Some of these tumors are familial as part of the MEN syndromes.

Screening

Screening for these tumors in the general population is not indicated given their rarity. In families with the MEN syndrome, radiological screening is appropriate. The role of genetic screening is addressed later in the MEN section.

Diagnosis

The diagnosis of a pancreatic endocrine tumor initially rests on recognition of the clinical syndrome (Table 14.10). Patients with gastrinoma often present with the Zollinger-Ellison syndrome characterized by severe recurrent peptic ulcers, gastric hypersecretion and a pancreatic mass. Upper endoscopy will document the severe ulcer disease, which often involves the duodenum and upper jejunum. An elevated fasting serum gastrin level (> 1000 pg/ml) confirms the diagnosis but is only present in 30% of patients. The majority of gastrinoma patients have fasting gastrin levels between 200-1000 (normal is 100-200). The secretin stimulation test may be used to confirm the diagnosis in these borderline cases.

Patients with insulinoma often present with Whipple's triad of symptomatic hypoglycemia, blood glucose levels below 50 mg/dl and relief of symptoms with glucose. Biochemical confirmation is obtained by measuring elevated insulin levels in conjunction with hypoglycemia. C-peptide and proinsulin levels are measured to rule out surreptitious self-administration of insulin.

Many of the less common pancreatic endocrine tumors are also associated with syndromes. VIPomas are associated with watery diarrhea, hypokalemia and achlorhydria (WDHA syndrome, Verner-Morrison syndrome, pancreatic cholera). An elevated serum VIP level is diagnostic. Glucagonomas can be associated with a severe rash known as necrolytic migratory erythema combined with mild diabetes. Elevated serum glucagon levels confirm the diagnosis. Somatostatinomas produce a syndrome of hyperglycemia, cholelithiasis and steatorrhea. Elevated somatostatin levels are confirmatory. Ppomas secrete pancreatic polypeptide (PP) which can be measured in serum but does not produce any recognizable syndrome. Nonfunctioning pancreatic endocrine tumors produce symptoms via mass effect, similar to pancreatic adenocarcinoma.

Preoperative Evaluation

Recommendations for localization of pancreatic endocrine tumors are essentially the same for the major tumor categories. Abdominal CT scan is the initial procedure of choice for preoperative localization of pancreatic endocrine tumors once the diagnosis has been established. Almost all pancreatic endocrine tumors other then insulinoma can be identified with CT scans. MRI has similar sensitivity and specificity as good quality CT. If localization is still in question after CT scanning, visceral angiography can occasionally provide further information. Upper endoscopy, particularly when combined with endoscopic ultrasound, can be useful in identifying duodenal tumors. Since many pancreatic endocrine tumors contain receptors for somatostatin, nuclear scanning with octreotide can also be helpful on occasion. Patients with insulinoma are often harder to localize given the small size of the tumors. If the tumor location is still unknown after CT scan and angiography, these patients can sometimes benefit from selective portal venous sampling with selective

14

Table 14.10. Pancreatic endocrine tumor syndromes

Tumor type	Secreted substance	Clinical syndrome
Insulinoma	Insulin	Hypoglycemia, Whipple's triad
Gastrinoma	Gastrin	Peptic ulcers, diarrhea
VIPoma	Vasoactive intestinal peptide	Watery diarrhea, hypokalemia, achlorhydria
Glucagonoma	Glucagon	Hyperglycemia, necrolytic migratory erythema
Somatostatinoma	Somatostatin	Hyperglycemia, gallstones, steatorrhea
PPoma	Pancreatic polypeptide	None described

intra-arterial calcium injection to localize the general area of an insulinoma. Intraoperatively, ultrasound can help identify small tumors, particularly insulinomas.

Staging

There is no AJCC accepted staging system for pancreatic endocrine tumors. Ellison has proposed a staging system for gastrinomas based primarily on size of the tumor (Tables 14.11 and 14.12).

Treatment

The treatment plan must address both the clinical syndrome and the tumor's potential for malignancy. Insulinomas are rarely malignant. Surgical intervention focuses on preoperative localization, full mobilization and exploration of the pancreas and the liberal use of intraoperative ultrasound. The majority of insulinomas can be enucleated. Tumors involving the pancreatic duct or which infiltrate the pancreas require either distal pancreatectomy or Whipple procedure, depending on the location. If no tumor can be localized or identified intraoperatively, a pancreatic biopsy is done to rule out β cell hyperplasia (nesidioblastosis).

Patients with gastrinoma are likely to have malignant disease with the potential for metastases. Treatment focuses on control of the ulcer syndrome, resection of the tumor and treatment of metastases. Prior to the introduction of H-2 blockers and proton pump inhibitors in the 1980s, patients routinely required total gastrectomy to eliminate the source of gastric acid. Now, effective acid blocking medications can control almost all cases of gastric hyperacidity. Given the cost of these medications, some authors suggest an acid reducing surgical procedure such as proximal gastric vagotomy or vagotomy and pyloroplasty although these suggestions are controversial. Most patients will be controlled with medication alone. Patients with sporadic gastrinoma should undergo surgery with extensive exploration of the "gastrinoma triangle". This area is defined as the anatomic region encompassed by the cystic duct superiorly, the third portion of the duodenum inferiorly and the junction of the neck and body of the pancreas medially. About 25% of gastrinoma patients have MEN syndrome with multifocal disease. Several authors feel that these patients cannot be cured and do not recommend surgery although agreement is not universal. Intraoperative ultrasound should be liberally used. Intraoperative endoscopy helps

14

Table 14.11. Proposed TNM classification for gastrinoma

T	Tumor	N	Nodes	M	Metastases
T1	≤ 1 cm	N0	–nodes	M0	–metastases
T2	1.1–2 cm	N1	+ nodes	M1	+ metastases
T3	2.1–2.9 cm				
T4	≥ 3 cm				

Table 14.12. Ellison gastrinoma staging system

Stage	TNM
I	T1-2, N0-1, M0
II	T3-4, N0-1, M0
III	Any T, any N, M1

identify tumors in the duodenum by transillumination. Duodenotomy should be performed if needed. Enucleation is performed if possible with Whipple procedure and distal pancreatectomy reserved for patients with larger tumors.[11,12] Patients with resectable liver metastases should undergo formal resection or wedge resection.

The techniques applied to treatment of insulinoma and gastrinoma are generally applicable to the rare tumors as well. The rare pancreatic endocrine tumors are generally larger and resection is usually required rather than enucleation. Patients often have local invasion. Patients with metastatic disease to the liver may have five-year survival rates exceeding 50%. Surgical debulking is a reasonable option in good risk patients and may be associated with increased survival. Interstitial hepatic ablation techniques using cryotherapy and radiofrequency ablation are being studied.

Chemotherapy with streptozocin and 5-FU along with doxorubicin has response rates of 50% in some series of patients with metastatic gastrinoma. Alpha interferon has been used successfully in some patients. Somatostatin analogues can be used to control symptoms but usually do not cause any significant tumor response. In metastatic insulinoma, diazoxide can control symptoms in 50% of patients.

Outcome
The outcome depends primarily on the malignant potential of the tumor. Almost all patients with insulinoma are cured of their syndrome. Among the small minority of patients with malignant insulinoma, about 65% will have recurrences. Some of these patients can have repeat resection. Median survival is about 5 years for patients with malignant insulinoma who undergo resection.

Survival from sporadic gastrinoma is related to the size of the tumor and the presence of liver metastases. Patients with tumors smaller than 3 cm who undergo complete resection enjoy 10-year survival exceeding 90%. Patients with unresectable liver metastases have five-year survival rates of 20-50% in selected series. Complete resection of liver metastases appears to increase the five-year survival and should be attempted in good risk patients. Overall, the five-year survival rate for all patients with gastrinoma is about 60%. Patients with the more rare pancreatic endocrine tumors (VIPoma, glucagonoma, and somatostatinoma) benefit from complete surgical resection when possible. One small series of VIPoma patients reported a 30% cure rate.

Posttreatment Surveillance

There are no prospective studies to guide follow-up of patients with pancreatic endocrine tumors. Patients with benign insulinomas are usually cured and probably do not need intense follow-up. Patients with gastrinoma and other tumor types often recur. In one large study only 35% of patients had normal gastrin levels at 5 years. Patients can often be palliated by additional surgery. Thus, regular follow-up seems prudent. Levels of the appropriate hormone can be followed. A secretin stimulation test can often "unmask" patients with recurrent disease that still have normal gastrin levels. Regular CT follow-up appears prudent.

Special Considerations

Patients with pancreatic endocrine tumors may have an MEN syndrome. The family history should be reviewed. Patients with gastrinoma in the setting of MEN-1 usually have multifocal disease and several experienced authors advise against surgery directed at the gastrinoma. These controversies are discussed further in the section on MEN syndromes.

Carcinoid Tumors

Scope of the Problem

Carcinoid tumors represent one of the four tumor types that account for almost all primary malignancies of the small bowel. The others are adenocarcinoma, lymphoma and sarcoma. Carcinoids represent about 25-30% of primary small bowel cancers, accounting for about 1200 to 1500 of the estimated 4800 primary small bowel cancers seen annually. However, the small bowel is the site for only about 35% of carcinoids with roughly another third being seen in the appendix, 15% in the rectum, 10% in the lung and the remainder scattered about the GI tract (Table 14.13). One report suggests that rectal carcinoids, because they often can be asymptomatic, may be more prevalent than classically described. Thus there are about 4000-4500 cases of carcinoid tumor annually in the US. In the appendix, carcinoid is the most common primary malignancy, representing about 75% of all primary appendiceal cancers.

Carcinoids of the midgut (small bowel and appendix) usually secrete serotonin (5-hydroxytryptamine) along with a variety of other hormones and related substances. Carcinoids of the foregut (bronchus, stomach, duodenum and pancreas) tend to secrete 5-hydroxytryptophan, a precursor of serotonin. Hindgut carcinoids (colon, rectum) usually do not secrete these substances. The metastatic potential of carcinoid tumors is directly related to their size.

Risk Factors

Carcinoid tumors are associated with pernicious anemia, use of omeprazole and the Zollinger-Ellison syndrome. This suggests that hypergastrinemia leading to enterochromaffin-like (ECL) cell hyperplasia may be an initial step in the development of at least some carcinoid tumors.

Table 14.13. Frequency and prognosis of carcinoid tumors, all stages combined

Location	Frequency	Survival (5-year)
Small bowel	35%	40-70%
Appendix	30-35%	90-99%
Rectum	15%	75-100%
Lung	10%	85%
Colon, stomach	< 5%	50%

Screening

There is no recommended screening for carcinoid tumors.

Diagnosis

Only about 10-20% of patients with GI carcinoid tumor present with elements of the carcinoid syndrome (hepatomegaly, diarrhea, flushing and right heart valve disease). Most patients present with signs and symptoms of small bowel obstruction, abdominal pain and weight loss, which can represent a number of entities. A diagnosis of small bowel tumor can often be suggested by preoperative imaging studies but determination of the exact nature of the tumor requires surgical exploration. Patients with carcinoid syndrome can be identified preoperatively if the syndrome is recognized. Serum should be tested for serotonin and urine for excretion of the metabolite 5-hydroxyindole acetic acid (5-HIAA). About 50% of all carcinoid patients, even if asymptomatic, will have elevated 5-HIAA levels. Nuclear scanning with MIBG can identify carcinoid tumor in about 50% of cases.

Preoperative Evaluation

Since most patients with GI carcinoid will present with nonspecific findings of small bowel obstruction or will be recognized at the time of appendectomy, there will rarely be an opportunity to perform a preoperative evaluation specific for carcinoid. If the syndrome is recognized and confirmed with serum and urine studies, imaging for metastases is important. Nuclear imaging with MIBG or somatostatin analogues can detect metastatic disease. CT scan is helpful in evaluating patients for liver resection for palliation of the carcinoid syndrome. There are rare case reports of cardiopulmonary collapse from anesthesia in patients with the carcinoid syndrome. Therefore, patients with the carcinoid syndrome should be treated preoperatively with octreotide.

Staging

The AJCC has published staging systems for both small bowel and large bowel including the appendix. However, carcinoid is excluded from both of these staging systems.

Treatment

Surgical therapy for carcinoid tumors found unexpectedly at laparotomy depends upon the size and extent of disease. Patients with small bowel carcinoids regardless of size should undergo a wide small bowel resection including mesentery and lymph nodes. Appendiceal carcinoids are treated based on their size. Metastases are rare

14

with appendiceal carcinoids < 1 cm. Appendectomy is adequate. If the base of the appendix is involved, cecectomy is recommended. Appendiceal carcinoids > 2 cm or those with clinically involved lymph nodes require formal right colectomy. There are conflicting suggestions for tumors between 1 and 2 cm. The treatment of rectal carcinoids is also based on size and the presence of liver metastases. Tumors smaller then 1 cm can be treated with fulguration or wide local excision. Rectal carcinoids larger then 2 cm are often associated with lymphatic spread. A formal rectal resection (low anterior or abdomino-perineal) is recommended. As is seen with appendiceal carcinoids, there are conflicting recommendations for rectal carcinoids 1-2 cm in size. Individual patient characteristics should guide the decision. It is reasonable to attempt wide local excision of these intermediate size tumors.

Almost all symptomatic patients and many with initially localized disease will have liver metastases. Octreotide can control symptoms well. Long acting depot preparations are available. In some patients, octreotide is associated with tumor shrinkage. Patients with resectable metastases should be evaluated for surgical resection. Chemotherapy responses have been poor. Interferon has been used to control symptoms but tumor response is poor. Ablation of liver metastases with cryotherapy, radiofrequency ablation and hepatic artery embolization has been used with some success in highly selected patients.

Outcome

Outcome is related to the presence of metastases. Patients with complete resection of local disease are almost all cured. Patients with liver metastases who are resected have five-year survival rates of about 60%. Patients who cannot be resected can still have five-year survival rates over 30% with optimal control of symptoms.

Posttreatment Surveillance

Small carcinoids rarely metastasize. Metastases can often be treated surgically so regular follow-up after resection seems prudent. Levels of 5-HIAA can be measured. CT scan is also available to screen for liver metastases.

Special Considerations

Carcinoid tumors are rare. The diagnosis is often a surprise. Given the relatively good survival, even with metastatic disease, patients with liver metastases should be evaluated for surgical resection.

Multiple Endocrine Neoplasia (MEN)

Scope of the Problem

Multiple endocrine neoplasia (MEN) is a rare disorder that has interested endocrine and oncologic surgeons for much of the last 50 years. There is much overlap between this syndrome and many of the endocrine tumors discussed in previous sections of this Chapter. Most patients with endocrine tumors do not have the MEN syndrome. It is difficult to assess the incidence of MEN. Given that the endocrine tumors involved are rare and MEN represents the minority of patients with these tumors, the number of cases per year in the US cannot exceed several hundred. Currently there are three recognized MEN syndromes: MEN-1, MEN-2A and MEN-2B. MEN-1 (Wermer's syndrome) involves parathyroid hyperplasia, pituitary tumors and pancreatic endocrine tumors (the three P's). In addition, patients can also have foregut carcinoids, angiofibromas, thyroid adenomas and other unusual tumors. MEN-2 patients all have medullary cancer of the thyroid. Pheochromocytoma is often seen in both subtypes. MEN-2A (Sipple's syndrome) may also involve parathyroid hyperplasia and colonic congenital aganglionosis (Hirschsprung's disease). MEN-2B (Gorlin's syndrome) involves multiple facial neuromas as well as GI tract ganglioneuromas. Patients with MEN-2B often have a "marfanoid" appearance due to skeletal abnormalities.

These syndromes are caused by autosomal dominant genetic mutations (Table 14.14). The gene involved in the MEN-1 syndrome, *MEN-1*, is a tumor suppressor gene located at 11q13.[13] The gene product, menin, is a 610 amino acid nuclear protein. Numerous mutations in multiple exons of the gene have been identified leading to loss of function secondary to protein truncation or missense mutation. The gene involved in the MEN-2 syndrome is the *RET* proto-oncogene on chromosome 10 which codes for a receptor tyrosine kinase.[14] This mutation is also implicated in familial medullary thyroid cancer (FMTC). The location of the mutation in exons 10 and 11 (the extracellular domain) is associated with MEN-2A and FMTC. Mutation in the intracellular domain (specifically the tyrosine kinase catalytic core at position 918) is associated with MEN-2B. It is interesting to note that about 30% of sporadic medullary thyroid cancer patients who are tested have this same mutation at amino acid 918.

Risk Factors
The primary risk factor for MEN is belonging to a family with the syndrome.

Screening
Screening of the general population is not indicated. However, screening of relatives in known MEN families is important. Given the recent increase in knowledge about the genetic basis of MEN, genetic screening and early detection is now possible under some circumstances. Because hyperparathyroidism is seen in 90% of cases of MEN-1 and 20-60% of cases of MEN-2A, it is reasonable to consider the possibility of MEN in all patients with hyperparathyroidism, particularly those patients presenting at a young age. A new diagnosis of pancreatic endocrine tumor should lead to consideration of screening of family members. Newly diagnosed patients with medullary cancer of the thyroid or pheochromocytoma should be screened as well given the association with MEN-2. About 80% of cases of medullary thyroid cancer are sporadic. The remaining 20% are either familial or associated with MEN-2.

The initial screening test in suspected MEN syndrome should be a detailed family history. Until recently, suspected MEN patients would be subjected to a variety of biochemical tests and imaging techniques. Biochemical testing is readily available to screen for hyperparathyroidism and pheochromocytoma. Patients with suspected medullary thyroid cancer can be screened with calcitonin measurement but this often needs to be supplemented with a pentagastrin challenge. Imaging or biochemical testing for secreted tumor substances is required to screen for pancreatic endocrine tumors. All of these tests rely upon the patient developing the tumor in question and identifying the tumor in a preclinical stage. Screening tests therefore need to be repeated at regular intervals if negative.

14

Table 14.14. MEN and related syndromes

Syndrome	Gene	Protein	Tumor	%
MEN-1	*MEN-1*	Menin nuclear protein	Parathyroid	90
			Pancreas	80
			Pituitary	50
MEN-2A	*RET*	Receptor tyrosine kinase, extracellular domain	Medullary thyroid	100
			Pheochromocytoma	20-40
			Parathyroid	20-60
FMTC	*RET*	Receptor tyrosine kinase, extracellular domain	Medullary thyroid	100
MEN-2B	*RET*	Receptor tyrosine kinase, intracellular domain- amino acid 918	Medullary thyroid	100
			Pheochromocytoma	20-40
			Neuromas and skeletal deformities	varies

The development of genetic testing alters the entire approach to screening patients for some of the MEN syndromes. The ability to identify *RET* mutations with a variety of molecular techniques allows testing at an early age for MEN-2 and FMTC. Sensitivity exceeds 90% and specificity is 100%. A positive result allows for prophylactic total thyroidectomy, even in early childhood, prior to the development of the malignancy. A negative test obviates any further testing. Genetic testing for *MEN-1* mutation is in earlier stages of development than for *RET* mutations. This is due in part to the multiplicity of genetic mutations that can lead to the syndrome. In addition, it is currently unclear which interventions beyond parathyroid surgery would be offered to MEN-1 patients identified in the preclinical stage. In a research setting, however, molecular techniques have been used to identify patients and confirm the presence of MEN-1 syndrome.

Diagnosis

The evaluation of the individual tumors of the MEN syndromes is similar to evaluation for sporadic forms of the same tumors. Diagnosis of the syndrome depends upon recognizing and confirming the presence of multiple endocrine tumors. Increasingly, genetic testing can be performed. Among patients with MEN-1, about 90% will have hyperparathyroidism, usually from hyperplasia rather than an adenoma. Patients will have pancreatic endocrine tumors in 50-80% of cases and pituitary tumors in 50% of the cases. Adrenal and thyroid tumors can also be present. Among patients with MEN-2 syndromes, 100% will have medullary cancer of the thyroid. Pheochromocytoma is present in 20-40%. Therefore, all patients suspected of having an MEN-2 syndrome must undergo urine screening for pheochromocytoma prior to thyroid surgery or invasive testing. Among patients with MEN-2A, parathyroid involvement is seen in up to 60% of cases.

Preoperative Evaluation

The same preoperative evaluations are done as has been recommended for the sporadic forms of the same tumors.

Staging

There is no staging system for the MEN syndromes. Some of the individual tumors have accepted or proposed staging systems (thyroid and gastrinoma).

Treatment

Patients with MEN-1 usually present with hyperparathyroidism. The disease is usually due to hyperplasia. These patients should undergo either resection of 3 1/2 glands or total parathyroidectomy with forearm autotransplantation. The subtotal procedure can result in recurrence in up to 40% of patients. Total parathyroidectomy with autotransplantation produces recurrence rates of about 5% and is preferred over subtotal resection. The treatment of the pancreatic tumors seen in MEN-1 is controversial. The tumors are multifocal and often cannot be completely resected. Many authors recommend controlling the syndromes with medication and avoiding operation. Other authors feel that surgery is indicated in some cases because many of the small tumors are not clinically relevant and the clinical syndrome is often caused by the largest, most easily identified tumor. Given the rarity of the syndrome, there will be no prospective studies to resolve this dispute. The pituitary tumors are almost always prolactinomas and can usually be controlled with bromocriptine. Trans-sphenoidal hypophysectomy is used for patients failing medical therapy or for those with nonprolactin secreting tumors.

Patients with MEN-2 require total thyroidectomy for treatment of the medullary thyroid cancer. Family members from an MEN-2 kindred can be genetically screened in childhood for the presence of the syndrome. Identification of these children permits prophylactic total thyroidectomy prior to the appearance of malignancy. Those patients with MEN-2A and hyperparathyroidism should have this condition addressed at the time of total thyroidectomy. Total parathyroidectomy with forearm autotransplantation is the procedure of choice. Those patients with pheochromocytoma require adrenalectomy. The adrenalectomy should be performed first to avoid hypertensive crisis during the thyroid surgery. Pheochromocytoma is often bilateral in MEN-2. A large time span can separate development of the second adrenal tumor. Controversy exists concerning the extent of adrenalectomy required. Some surgeons recommend bilateral adrenalectomy initially. Others prefer to perform unilateral adrenalectomy and then follow the patient with imaging. No prospective studies are available to offer guidance. The use of laparoscopic adrenalectomy in pheochromocytoma is not yet defined. Most of the laparoscopic experience has been with adrenal adenomas. The primary concern with laparoscopic adrenalectomy for pheochromocytoma is excessive tumor manipulation which may provoke a hypertensive crisis.

Outcome

The outcome from the MEN syndromes is primarily dependent upon the stage of the pancreatic endocrine tumors in MEN-1 and the medullary thyroid cancer in MEN-2. Due to the multifocal nature of the pancreatic tumors, MEN-1 patients are usually not cured. However, some patients with MEN-1 do not have pancreatic tumors. In MEN-2, cure rates can approach 90%, particularly with early diagnosis and surgery.

Posttreatment Surveillance

The follow-up of patients with MEN syndromes is indefinite, given the possibility of pancreatic tumor malignancy in MEN-1 and the malignant nature of medullary thyroid cancer combined with the late development of pheochromocytoma in MEN-2. Patients with medullary thyroid cancer will require long-term calcitonin screening for recurrence.

Special Considerations

The MEN syndromes are rare diseases. A high level of suspicion with a detailed family history is often the key to recognition of the syndrome. Genetic testing is becoming more available and should be offered to family members.

Selected Readings

1. Landis SH, Murray T, Bolden S et al. Cancer statistics, 1999. CA A Canc J Clin 1999; 49:8-31.
    A standard source for cancer statistics. This is updated annually.
2. American Joint Commission on Cancer. Cancer Staging Manual. New York, Philadelphia: Lippincott-Raven 1997.
    The standard staging manual, now in its fifth edition.
3. Cady B. Papillary carcinoma of the thyroid gland: Treatment based on risk group definitions. Surg Oncol Clin N Am 1998; 7:633-644.
    This article presents the argument favoring minimal surgery for well-differentiated thyroid cancer.
4. Chen H, Udelsman R. Papillary thyroid carcinoma: Justification for total thyroidectomy and management of lymph node metastases. Surg Oncol Clin N Am 1998; 7:645-664.
    This article presents the counterpoint to Cady's article.
5. Shane E, Bilezikian JP. Parathyroid carcinoma: A review of 62 patients. Endoc Rev 1982; 3:218-226.
    This older review provides an excellent description of the clinical syndrome.
6. Wang C. The anatomic basis of parathyroid surgery. Ann Surg 1976; 183:271-275.
7. Wang C. Parathyroid re-exploration. Ann Sur 1977; 186:140-145.
    These two articles were written several decades ago but provide an excellent discussion of parathyroid embryology and anatomy.
8. Norman J, Chheda H. Minimally invasive parathyroidectomy facilitated by intraoperative mapping. Surgery 1997; 122:998-1004.
    A description of an emerging technology by one of the innovators.
9. Thompson GB, Grant CS. Laparoscopic versus open posterior adrenalectomy: A case-control study of 100 patients. Surgery 1997; 122:1132-1136.
    This is one of several recent descriptions of a new technique. This case series is from the Mayo Clinic.
10. Schulick RD, Brennan MF. Long-term survival after complete resection and repeat resection in patients with adrenocortical carcinoma. Ann Surg Oncol 1999; 6:719-726.
    This paper reviews the extensive Memorial Sloan-Kettering experience with this rare cancer.

14

11. Ellison EC. Forty year appraisal of gastrinoma. Ann Surg 1995; 222:511-524.
This article describes the forty-year Ohio State University experience with this tumor. The proposed staging system is explained and correlated with prognosis.

12. Norton JA, Fraker DL. Surgery to cure the Zollinger-Ellison syndrome. NEJM 1999; 341:635-644.
This paper describes the NCI clinical experience with this rare tumor. It includes patients with sporadic gastrinoma and MENI.

13. Marx S et al. Multiple endocrine neoplasia type 1: Clinical and genetic topics. Ann Int Med 1998; 129:484-494.
This is the report of a symposium convened several months after the identification of the MEN-1 gene. It contains a good discussion of the problems associated with genetic testing for MEN-1.

14. Goodfellow PJ, Wells SA. RET gene and its implications for cancer. J Nat Canc Inst 1995; 87:1515-1523.
This is a review of the basic research findings that led to the discovery of the genetic basis for MEN 2 and FMTC. The rationale and results with genetic testing and prophylactic thyroidectomy are discussed.

14

CHAPTER 15

Principles of Chemotherapy

Dennis Sanders

The use of chemotherapy for patients with neoplastic diseases has evolved over the last three decades. Initially, chemotherapy was used as a last ditch salvage attempt for patients with end-stage, metastatic cancer. During these early days, the efficacy of the chemotherapeutic drugs for palliation tumor response was marginal, and treatments would result in significant side effects for patients. This often brought up the question, is the treatment worse than the disease? As our understanding of the biochemical and molecular processes underlying malignant diseases improved, so have the chemotherapeutic agents in terms of efficacy and specificity. These improvements have allowed clinicians to utilize chemotherapeutic agents in earlier stages of malignancy, which has altered the natural history of the majority of malignant illnesses, resulting in improved patient survival. So-called adjuvant chemotherapy for cancer of the breast, colon, ovary, testes, and lung are considered the standard of care. In addition, the use of cytoreductive chemotherapy either alone, or in combination with radiation therapy, has provided surgeons an opportunity to excise previously unresectable tumors such as cancers of the esophagus, breast, lung and rectum. More importantly, so-called neoadjuvant chemotherapy has also allowed surgeons to use organ-sparing techniques to preserve rectal sphincter function in colorectal cancers, or preserve cosmetic appearances in cancers of the breast. The latter therapeutic approaches can be applied without compromising patient survival.

As you learn how to manage and treat patients with malignant diseases, many of the clinical challenges involve dealing with potential side effects of chemo-therapeutic drugs. These clinical problems can be solved more readily if there is an understanding of the mechanisms of action of the various chemotherapeutic agents. This knowledge will provide the ability to predict when such side effects may present, and hopefully to prevent them.

Many practitioners have used the analogy of "controlled poisoning" in discussions regarding chemotherapy. This concept is not too far from the mark, in that many of the targets for chemotherapy that trigger cellular death and apoptosis in malignant cells, are shared in nonmalignant cells. This can lead to a narrow therapeutic window where the balance of cell kill in normal tissues and organs results in bone marrow suppression, neurologic impairment, alimentary tract disorders, renal dysfunction and other imbalances which provide clinical challenges.

This Chapter is designed to provide a basic understanding of how chemotherapy causes cytotoxicity to both malignant and benign cells. By understanding the different classes of chemotherapeutic agents, you will understand how these drugs are used in certain combinations and schedules. Through this knowledge, it will become

Surgical Oncology, edited by David N. Krag. ©2000 Landes Bioscience.

apparent why certain clinical side effects are observed, and how clinicians can prevent many adverse side effects with newer supportive care drugs such as antiemetics, bone marrow growth factors, cytoprotective agents, and antidiarrheal medications.

Malignant Transformation and Molecular Targets for Chemotherapy

Malignant cells have acquired multiple mutations to key genes and are under very little restraint to progress through mitosis. This allows some tumors to double in size every 6-12 weeks. The cell cycle has many control points for each of the four phases of mitosis, and through mutations of oncogenes and tumor suppressor genes, these control points are lost. In general, oncogenes code for proteins that function as signal transduction proteins, and they function to direct a cell to progress through mitosis and proliferate. For example, the cellular membrane protein RAS initiates a sequence of biochemical steps through the cytoplasm into the nucleus and directs the cell to gear up for mitosis. For the majority of solid tumors, overproduction of this mutated RAS membrane protein is associated with rapidly growing malignant cells. Tumor suppressor genes, on the other hand, in general code for proteins that inhibit progression of the cellular machinery through mitosis. For example, the tumor suppressor gene p53, codes for a protein that binds to other DNA-binding proteins which allows mitosis to continue unchecked. Therefore, the combination of unchecked signal transduction for mitosis from oncogenes and the loss of "brakes" in the form of mutations in tumor suppressor genes results in malignant transformation (Fig. 15.1).

Defining the cellular kinetics of malignant cells has provided information about the cell cycle, and discrete phases in the mitotic process have been described. The mitotic phase defines the interval when a cell divides into two daughter cells with identical sets of chromosomes. Following mitosis, the proliferating cell enters a growth phase, called G1. It is in this phase that cells produce proteins and enzymes which, in turn, produce the basic molecular building blocks for DNA synthesis. It is in the G1 phase where a strong "ying-yang" effect from mutated oncogenes and tumor suppressor genes can influence whether a cell progresses forward through the cell cycle or shifts into a nonproliferative mode for cellular differentiation of normal tissues. The G1 phase is also a significant interval where a cell will go into cell death and apoptosis if the replication machinery is damaged. If the net signal to the cellular nucleus is for proliferation, the cell will progress into the phase of DNA synthesis, so-called S phase. During S phase the molecular machinery for DNA strand replication is engaged to double the amount of genetic material in the cell for eventual mitosis. After completion of the S phase, the proliferating cell enters a second growth phase, called G2. It is during this phase that the molecular machinery for chromosomal separation is synthesized and assembled to allow the cell to progress into M phase for cellular division (Fig. 15.2).

Without reviewing in detail the in vitro and in vivo experiments which have characterized the cellular kinetics of tumor growth rates, some general conclusions can be made. Tumors that are highly responsive to chemotherapeutic agents (e.g., testicular cancer, lymphomas) tend to have a very rapid doubling time compared to tumors that are less responsive to chemotherapy (e.g., pancreatic and prostate cancers).

15

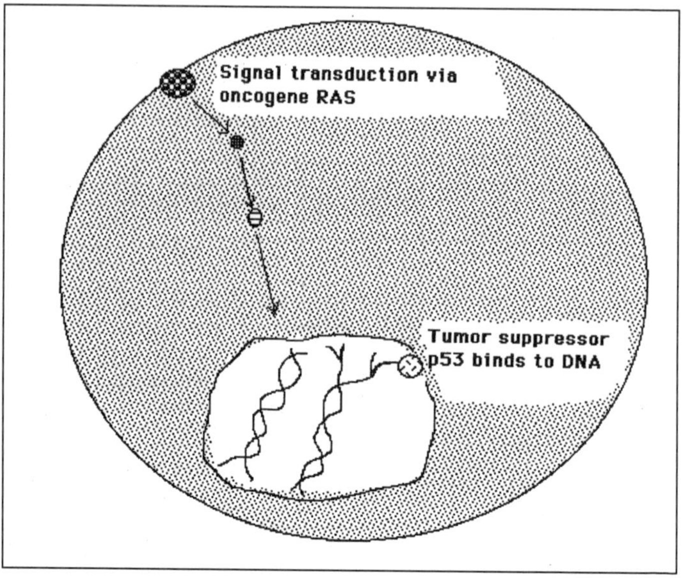

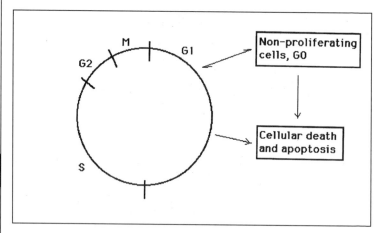

Fig. 15.1. Malignant trasformation is the net result of mutations to oncogenes and tumor suppressor genes, which control how a cell progresses through the cell cycle.

Fig. 15.2. The cell cycle is divided into four phases of mitosis. G1, S, G2 and M. Chemotherapeutic drugs and radiation often exert their maximum lethal activity during specific phases of the cell cycle.

The tumor doubling time of metastases from any primary tumor tend to be more rapid than the cells found in the primary site.

These general conclusions as well as our improved understanding of the phases of cellular proliferation have allowed development of chemotherapeutic agents to take advantage of poisoning certain critical pathways in each of the phases of mitosis, resulting in cell death and apoptosis. Unfortunately, some of these same pathways can develop mutations in the genes that code for the protein targets, and allow a cell to escape the poisoning from a given drug. For example, many drugs act to bind or alter key nucleotides in DNA synthesis. However, the DNA repair enzymes utilized during S phase can be over-produced through mutation to correct any chromosomal damage and prevent cellular death. Another example is the membrane protein pumps that are responsible for removing toxic metabolites and drugs from cells. Chronic exposure of malignant cells to certain drugs causes upregulation of the these so-called multidrug resistance membrane p-glycoproteins. This results in rapid expulsion of drug from a cell before it can poison the cellular proliferative process.

The molecular targets of chemotherapy are varied and affect all levels in the cellular replicative process. The targets include nucleoside analogues that interrupt or inhibit DNA strand synthesis, antimetabolites that inhibit formation of nucleosides for DNA synthesis, vinca alkaloids that inhibit microtubule function and prevent mitosis, alkylating agents that cause direct damage to DNA strands which result in miscoding of genes for proliferation resulting in cellular apoptosis, and drugs that inhibit DNA unwinding enzymes called topoisomerases. Some of the drugs we will review are highly effective in cells that are actively dividing, whereas others do their damage at any phase of the cell cycle.

Not covered in this Chapter are endocrine therapies against malignancies and biologic response modifiers which include immune modulating agents such as interferons, interleukins, and monoclonal antibodies.

Classes of Chemotherapeutic Agents

Antimetabolites

Antifolate drugs were some of the first used clinically against childhood leukemias. These compounds interrupt the synthesis of purine nucleosides, thus preventing DNA synthesis for cellular proliferation(Table 15.1). In the metabolic pathway of purine synthesis, cycling of reduced and oxidized folate compounds provides the substrate for purine ring synthesis. Antifolate drugs such as aminopterin and methotrexate inhibit this cycle by binding to the enzyme dihydrofolate reductase, thus preventing formation of tetrahydrofolate. This leads to accumulation of inactive dihydrofolates and depletion of reduced folates. In addition, metabolites of methotrexate are thought to also bind to other folate-requiring enzymes in the synthesis of the pyrimidine, thymidylate. In this way, two pathways for nucleoside synthesis are inhibited, thus preventing the formation of molecular building blocks for DNA synthesis.

Antifolates have their effects predominantly upon actively dividing cells, and the observed normal tissue toxicities reflect this. The primary toxic effects observed are myelosuppression and gastrointestinal mucositis. Since these two organ systems have

15

Table 15.1. Antimetabolite agents

Methotrexate
AraC
Gemcitabine
5-Flourouracil
Capecitabine
Hydroxyurea
Pentostatin
Fludarabine
2'-Chlorodoxyadenine

---

the highest relative cellular turnover in normal tissues, they are particularly susceptible to cell cycle specific chemotherapy. Methotrexate is thought to have a prolonged
half-life in the epithelium of the GI tract, and symptoms of mucositis and diarrhea
predominate. These symptoms are usually observed 3-7 days after administration of
methotrexate.

Because of the more rapid doubling time of tumor cells relative to normal tissue
cells, the strategy of antifolate rescue has been used to circumvent severe bone marrow and GI tract toxicities. With the use of high dose methotrexate, plasma levels
that will cause myelosuppression and mucositis can be reduced 24 hours after chemotherapy administration by using the reduced folate leucovorin. This allows purine
and pyrimidine synthesis to resume in normal tissues, while the more rapidly growing tumor cells die from lack of nucleoside synthesis in the previous 24 hours.

Other toxicities observed from methotrexate include transient hepatoxicity at
high doses, and portal cirrhosis observed with chronic dosing in patients with rheu-
matoid arthritis. Rare cases of methotrexate pneumonitis are observed and resolve
when the drug is discontinued. When methotrexate is used intrathecally for carci-
nomatous meningitis or for treatment of CNS lymphomas, symptoms of extremity
motor paralysis, cranial palsy and seizures may be observed after 3 or more weeks of
therapy. These toxicities appear to worsen if a patient is receiving concurrent CNS
radiation.

Cytidine Analogues

Cytidine analogues are a unique class of antimetabolites in that they are incorporated into DNA strands during the S phase of mitosis. This results in chain termination and eventually cellular death. Some of the analogues are also able to inhibit
enzymes that produce deoxynucleotides which deplete actively growing cells of the
nucleotides for DNA synthesis.

The prototypical cytidine analogue is arabinosylctosine, also known as AraC.
AraC, when used with antracyclines, is commonly used to treat leukemias and lymphoid malignancies. AraC has very little activity in solid tumors. However, further
modifications have been made on the basic arabinose nucleoside molecule, and related cytidine analogues have demonstrated significant activity against cancer of the
pancreas, lung, breast and bladder.

The mechanism of action of AraC is well defined. AraC undergoes phosphoryla-
tion to form the nucleoside analogue araCTP. AraCTP will bind to DNA polymerase
and be incorporated into the growing DNA strand. However, after incorporation,

DNA polymerase will terminate the replication strand and halt further extension. Enzymes that are programmed to remove damaged nucleosides cannot recognize AraCTP, thus making the gene code unreadable for protein production. This leads to cellular death by apoptosis.

One of the more important developments in treatments for solid tumors from the pancreas, lung, bladder and breast is the cytidine analogue gemcitabine. Like AraC, gemcitabine is phosphorylated in a stepwise fashion to form a triphosphate nucleotide which is then incorporated into DNA via the enzyme DNA polymerase. This results in chain termination of DNA strands. Also, gemcitabine will deplete nucleoside pools by inhibiting the enzyme ribonucleoside reductase. One of the key features to overcoming cellular resistance to cytadine analogues is the ability of gemcitabine to inhibit the enzyme dCMP deaminase, thus prolonging the half-life of gemcitabine in the cell.

Both drugs can cause significant bone marrow suppression which occurs 10-14 days after treatment. Gemcitabine can cause prolonged thrombocytopenia, and both drugs can cause nausea and diarrhea. At high doses used for leukemia protocols, AraC can cause significant neurologic toxicities such as cerebellar ataxia, slurred speech and dementia. These side effects usually resolve when the drug is discontinued.

Flourinated Pyrimidines

Fluorinated pyrimidines are a unique group of antimetabolites that were synthesized to block a specific pathway. Early in vitro observations suggested that certain tumor types utilize the pyrimidine precursor uracil more rapidly than do nonmalignant cells. This observation led to the synthesis of several compounds designed to behave like uracil, thus blocking further steps in pyrimidine synthesis. The prototypical drug used in the clinic is 5-flourouracil (5-FU). 5-FU is a prodrug and is converted to 5-flourodeoxyuridine (5FDUMP) in the cytoplasm. 5FDUMP then binds to the enzyme thymidine nucleosides. In addition, 5-FU is also incorporated into the synthesis of flourouridine triphosphate (FUTP). FUTP is incorporated into all classes of RNA, which leads to mistakes in translation of proteins and cell death.

Early on, it was observed that the inhibitory effect of FDUMP was transient, and could be overcome with accumulation of the natural deoxyuracilmonophosphates that results from inhibition of thymidylate synthetase. It was observed that the addition of reduced folates, such as leukovorin, formed more stable tertiary complexes with FDUMP/TS. This combination enhanced cellular toxicity since pyrimidine synthesis could be inhibited. As you remember from the previous section, leukovorin acts to rescue the toxicity induced by the antifolate methotrexate. 5-FU and FDUMP exert cytotoxicity upon actively dividing cells. Similar to methotrexate, the most susceptible normal organs for toxicity include the bone marrow and gastrointestinal system. 5-FU can induce significant stomatitis and diarrhea in patients, and these side effects are observed more frequently when the drug is administered by continuous intravenous infusion. Bone marrow suppression is more frequently observed in patients receiving the drug by bolus intravenous infusion. For most patients, the addition of leucovorin to the treatment regimen increases the likelihood of mucositis and diarrhea.

15

Other unique toxicities from 5-FU include hand-foot syndrome manifest as severe exfoliation of the palms and soles. 5-FU may also induce neurotoxicities such as cerebellar ataxia, somnolence and upper motor neuron signs. Lastly, cardiotoxicity is observed at high doses given by continuous intravenous infusion, and chest pain with EKG changes have been reported.

One of the short comings of 5-FU is its short half-life in the plasma secondary to rapid hepatic clearance. Although this can be overcome by delivery of the drug by 24-hour continuous infusion, this can result in central venous line infections or thrombosis of catheterized veins. Oral fluorinated pyrimidines have been developed and are in common use in Japan and Europe. Current trials in the United States have confirmed the activity and safety of these compounds in tumor types such as colorectal cancers, and its use will become more widespread.

Doxifluidine, known as capecitabine, is an oral 5-FU prodrug that is preferentially converted to 5-FU at the tumor. Given either alone, or with leucovorin, antitumor activity in breast and colon cancer are comparable to IV 5-FU. Another oral agent soon to be utilized in the United States is UFT, which combines the 5-FU prodrug tegafur, with uracil. Uracil acts to block the degradation enzyme dihydropyrimidine dehydrogenase (DPD), which leads to higher sustained levels of 5-FU. Another DPD inhibitor, eniluracil, has been tested with oral 5-FU with similar improvement in circulating 5-FU levels.

Other Antimetabolites

Hydroxyurea is one of the oldest chemotherapeutic molecules. It was first synthesized in 1869, but its clinical utility was not revealed until the 1940s, when experimental animals developed leukopenia and anemia when exposed to the drug. This suggested an ability to inhibit cells which are rapidly dividing. Although hydroxyurea has not demonstrated much direct activity against solid tumors, it has been the mainstay for many myeloproliferative disorders such as chronic myelogenous leukemia and essential thrombocytopenia. Hydroxyurea has also been used in conjunction with radiation therapy and is used as a radiosensitizing agent in tumors of the brain, head and neck, and ovary.

Hydroxyurea inhibits the ribonucleotide reductase enzyme system, which is responsible for conversion of ribonucleotides to deoxynucleotides. The net result is depletion of nucleotides which slows DNA synthesis and DNA repair. Cells escape the effects of hydroxyurea by upregulating production of the enzyme ribonucleotide reductase. Hydroxyurea can cause mild symptoms of nausea and will cause bone marrow suppression with high or chronic dosing.

Another group of antimetabolite drugs are those which inhibit the enzyme adenosine deaminase. By inhibition of this enzyme, accumulation of adenine nucleotide derivatives such as dATP occurs. These nucleotide derivatives in turn inhibit ribonucleotide reductase and greatly reduce nucleic acid synthesis. There are currently three compounds of this type in clinical treatments for lymphoid and leukemic malignancies. They are 2'-chlorodeoxyadenine (2cDA). All three compounds are able to cure patients with hairy cell leukemia, and they revolutionized treatment for patients with indolent grade non-Hodgkins lymphomas.

Microtubule Inhibitors

Tubulin is the structural protein that polymerizes to form microtubules. These microtubules play a key role in functions of nerve conduction, neurotransmission and mitosis. Microtubules form the mitotic spindles which allow for daughter chromosomal separation in cellular division. There are several drugs that bind to microtubules to interrupt assembly, or conversely, inhibit disassembly of microtubules. The net result is prevention of cellular division.

The prototypical microtubular inhibitor is colchicine, which is used to treat gout. Colchicine binds on the protein heterodimer and prevents assembly of the tubule.

The classic microtubule inhibitors are derived from vinca alkaloids and have widespread antitumor activity against solid tumors and myeloproliferative diseases. The periwinkle is the source of vincristine and vinblastine. These two complex molecules differ in structure by a single carbon substitution, but this greatly influences two types of toxicities observed with each compound.

Vincristine and vinblastine both bind to the heterodimer of tubulin and inhibit microtubule formation. This results in cells arrested in the mitosis phase of cell division. Resistance to the cytotoxic effects of vinca alkaloids is thought to occur by two mechanisms. The first is through mutations in the binding site of tubulin which decreases the affinity of the vinca compounds for tubulin. The other mechanism of resistance is thought secondary to the effects of cellular membrane proteins that pump natural product drugs out of the cell before they can exert their toxic effects. These proteins are p-glycoproteins in the class of multidrug resistance molecules (MDR). MDR membrane proteins are thought to also play a role in chemotherapy resistance for other drugs such as anthracyclines, epipodophyllotoxins, and actinomycin D which will be discussed later.

The toxicity of vincristine is predominantly neurologic in nature and is usually based upon cumulative doses. The most common complaints are distal sensory and motor neuropathies, with loss of deep tendon reflexes in the lower extremities. This is followed by dysesthesias in the fingers and toes, and eventually by loss of strength of dorsiflexors of the lower extremities, the hand and the wrist. At high cumulative doses, central nervous effects are seen on cranial nerves, as well as hyponatremia secondary to inducing the syndrome of inappropriate antidiuretic hormone (SIADH).

Vinblastine usually causes bone marrow suppression and mucositis. Vinblastine may cause myalgias, but in general does not cause peripheral neuropathies seen with vincristine.

Navelbine is a semisynthetic derivative of the vinca alkaloids, with a similar mechanism of action against microtubule formation. Navelbine has significant activity against tumors from the lung and breast.

Another relatively new class of microtubule inhibitors are the taxanes, which are alkaloid esters derived from the yew tree. Taxol is the prototypical clinical drug, with a newer semisynthetic derivative taxotere now in use. Unlike the vinca alkaloids, the taxanes markedly enhance all aspects of tubulin polymerization, and the microtubules that are formed are more tightly packed. It is thought that the taxanes prevent the ability to partially or fully depolymerize the mitotic spindle during mitosis. The taxanes have been used extensively as adjuvant drugs in tumors of the ovary and breast resulting in improved survival in higher stage patients. The taxanes have

---

Table 15.2. Microtubule inhibitors

Vincristine
Vinblastine
Navelbine
Paclitaxel (Taxol)
Docetaxel (Taxotere)

---

significant activity against tumors of the lung, head and neck, bladder and prostate. Taxanes are also used as radiosensitizing agents with concurrent radiation therapy.

Because the taxanes have limited ability to solubilize in water, they must be mixed in Cremophor and ethanol. This can cause anaphylactic allergic reactions in patients unless they are premedicated with steroids, benadryl and H2-blocking medications. The most commonly observed toxicities of the taxanes include neuropathies, neuralgias and bone marrow suppression.

Alkylating Agents

Alkylating drugs are a class of molecules that form covalent bonds to cytoplasmic and nuclear macromolecules, which results in dysfunction of these molecules in cellular replication. The best defined mechanism of cytotoxicity is interstrand DNA cross-linking, which results in inactivation of the DNA template, inhibition of DNA synthesis and cell death. It is thought that alkylation takes place preferentially in sites of DNA transcription. This interrupts accurate gene expression and leads to apoptosis.

Victims of sulfur mustard gas in World War I experienced lymphoid aplasia, pneumonitis of the lungs and mucositis of the mouth. The prototypical drug, nitrogen mustard, was developed based on these observations and was first used in patients with lymphoid malignancies. Promising activity led to further drug synthesis of alkylating agents, and this class of drugs is currently one of the major building blocks of modern chemotherapy for all tumor types.

The most commonly used oral alkylating agents are melphalan and chorambucil. These drugs are used for patients with myeloproliferative disorders such as multiple myeloma, lymphoma and chronic lymphocytic leukemias. The dosing is usually given for four days at intervals of 28 days to allow bone marrow recovery.

The other commonly used alkylating agents are cyclophosphamide and ifosfamide. These are prodrugs in that they have no inherent alkylating activity, but once injected into the blood, they are converted to active metabolites. These drugs are commonly used in the adjuvant treatment of breast cancer, as well as first-line therapy against high grade lymphomas.

Nitrosoureas are another type of alkylating agent which are lipophilic and can cross the blood-brain barrier. As such, they commonly used for treatment of primary malignancies of the brain. However, nitrosoureas also have significant activity against lymphomas and multiple myeloma.

The dose limiting toxicity of alkylating agents is myelosuppression, with all cell lines affected. Typically hematologic nadirs occur 10-14 days after drug administration. The exception is for nitrosoureas, where nadir effects on marrow function peak at three to four weeks after dosing. Unlike other chemotherapeutic drugs, alkylating

15

Table 15.3. Alkylating agents

Nitrogen mustard
Melphalan
Chlorambucil
Cyclophosphamide
Ifosfamide
Nitrosoureas
Procarbazine
Dacarbazine

agents can cause significant damage to hematopoietic stem cells. There have been reports of increased incidences of leukemias in patients four to ten years after a "cure" for a prior malignancy (e.g., lymphomas, breast cancer) treated with alkylating agents.

Other toxicities that can be observed if precautions in hydration are not taken include renal and bladder toxicity. Hemorrhagic cystitis can be observed with cyclophosphamide or ifosfamide. The uroprotective agent MESNA is usually infused along with ifosfamide to prevent these toxicities. Interstitial pneumonitis can be seen with cyclophosphamide, and chronic use of busulfan can cause pulmonary fibrosis. In addition, high doses of cyclophosphamide used in bone marrow transplant therapies is associated with an increased incidence of cardiac toxicity, including congestive heart failure which can be severe.

Other drugs with alkylating activity against macromolecules that are not derived from the nitrogen mustard compounds include procarbazine, dacarbazine, and hexamethylmelamine. All of the agents are prodrugs, and must be activated intracellularly for cytotoxic effects. Procarbazine is useful in the treatment of lymphomas and brain tumors. Dacarbazine is active in melanomas, sarcomas, and Hodgkins lymphoma. All of these drugs cause myelosuppression and flu-like symptoms with nausea and myalgias.

Antitumor Metabolites (Table 15.4)

This class of chemotherapeutic drugs includes agents with various mechanisms to induce cytotoxicity. They were discovered from byproducts of microorganism fermentation. Some of these compounds have already been described under other sections such as nucleoside analogues. In this section we will learn about the clinically significant compounds actinomycin D, mitomycin and bleomycin.

Actinomycin D (DACT) is derived from the yeast species Streptomyces. DACT is known to bind strongly to DNA. It is thought that DACT interrupts transcription of RNA from DNA templates and disrupts synthesis of essential proteins for cellular division. Cells tend to develop resistance to DACT by pumping the drug out via cellular membrane proteins in the MDR class of proteins, so-called p-glycoproteins.

DACT has been a mainstay in therapy for various tumors in pediatric oncology including Wilm's tumor, rhabdomyosarcoma and Ewing sarcoma. The drug can cause myelosuppression 8-14 days after infusion. Acute toxicities include nausea, diarrhea and hair loss.

15

Mitomycin C is another antitumor metabolite and is isolated from the yeast species *Streptomyces caespitosus*. Mitomycin C behaves as an alkylating agent, inducing DNA cross-strand linking, chromosomal breakage, and inhibition of DNA synthesis. Like other alkylating agents, mitomycin C must be activated intracellularly. Cytotoxic resistance is thought to occur through a combination of drug efflux and increased DNA repair enzyme activity.

Mitomycin C is very active in tumors from the lung, breast and gastrointestinal system. However, mitomycin C has several toxicities that limit the amount of drug that can be used. Aside from mild leukopenias, cumulative doses of mitomycin C cause profound thrombopenia. At cumulative doses above 30-50 mg/m^2, an increased incidence of the hemolytic uremic syndrome and pulmonary interstitial pneumonitis is observed.

Another yeast species, *Streptomyces verticillus,* produces a mixture of peptides known as bleomycin. Unlike most chemotherapy, bleomycin lacks significant toxicity of bone marrow function, which allows it to be combined with a wide variety of classic chemotherapeutic agents. Bleomycins are glycoproteins that are taken up by tumor cells and will bind with iron molecules to form a bleomycin-Fe (II) complex. This complex binds to DNA and results in the generation of oxygen free radicals which lead to DNA strand cleavage. Cytotoxic resistance is mediated by increased activity of DNA repair enzymes.

Bleomycin is used in the treatment of Hodgkins lymphoma and testicular cancer. Bleomycin is also active in squamous cell cancers of the head and neck and esophagus. The most important toxic actions of bleomycin are on the lungs and skin. Bleomycin can cause interstitial pneumonitis, which is complicated in later stages by progressive interstitial fibrosis and hypoxia. Therefore, patients with a history of tobacco use or underlying emphysema are at particular risk when receiving bleomycin. For patients with prior bleomycin exposure there is a risk of inducing interstitial pulmonary disease if exposed to inspired oxygen concentrations greater than 70%. This is important to document in patients going to surgery with general anesthesia. Roughly half of patients treated with bleomycin will experience skin erythema while a smaller proportion will develop peeling of the skin.

Topoisomerase Inhibitors

Topoisomerases function as nuclear enzymes that control the shape and form of DNA as it goes through replication and RNA transcription. DNA is usually in the form of a very tight coil which is twisted and curled back upon itself, called supercoiling. Topoisomerase I and II mediate ligation and religation reactions for 'unknotting' DNA strands. If these enzymes are inhibited, DNA strands break which results in cell death.

The epopodophyllotoxins are derived from the mayapple plant which has been used in folk remedies for generations. These basic podophyllotoxic compounds bind to tubulin structures like the vinca alkaloids, and block mitosis. However, if these compounds are glycosylated, they form epopodophyllotoxins, and bind to topoisomerase II and DNA strands, thus disrupting DNA strand uncoiling, which blocks progression through the cell cycle. Etoposide (VP-16) and teniposide (VM-26) are examples of topoisomerase inhibitors. Etoposide is most widely used in lymphomas, germ cell tumors, and lung cancer. Cellular resistance is via the

---

Table 15.4. Antitumor metabolites

Actinomycin D
Mitomycin C
Bleomycin

---

membrane p-glycoprotein MDR efflux, and via increased DNA repair enzyme activity. Etoposide can be administered by oral or intravenous routes. It can cause myelosuppression, nausea, diarrhea, and hair loss.

The other topoisomerase inhibitors fall into the class of drugs called camptothecins. These are plant alkaloids first isolated in the 1970s. Semisynthetic derivatives, irinotecan and topotecan are currently in clinical use. Camptothecins bind to topoisomerase I and form a complex with DNA. This results in single strand breaks and eventual cell death. Irinotecan has been used in metastatic colon cancer and is currently being tested in clinical trials in combination with 5-FU for adjuvant therapy of colorectal cancers. Topotecan has a wider distribution of antitumor activity and is being used in patients with cancers of the ovary, lung, head and neck. A newer camptothecin 9-aminocamptothecin has demonstrated strong activity in lymphomas and is being tested in clinical trials.

The dose-limiting toxicity of the camptothecins is myelosuppression. Both irinotecan and topotecan can induce significant secretory diarrhea. This may require the use of atropine to prevent abdominal cramping at the time of infusion.

Anthracyclines

Like alkylating agents, anthracyclines have broad clinical activity against lymphomas, leukemias and a wide range of solid malignancies. These drugs, which were discovered from byproducts of microbial organism fermentation, include daunomycin, doxorubicin, and the newer related compounds mitoxantrone, idarubicin, and epirubicin.

The development of platinum analogues represents the wonderful serendipitous nature of many of the great discoveries in the world of medicine. In 1965, B. Rosenberg and colleagues were studying the effects of electric current on growing bacteria. They noted that after being exposed to alternating current from platinum electrodes, or when exposed to growth media that was exposed to alternating current through platinum electrodes, bacterial cultures would stop growing. These scientists were able to find that under the mechanism of electrolysis, platinum was released as hexachloroplatinate. When combined with ammonium salts and light, the complex called cisplatin was formed. This agent, and subsequent related compounds revolutionized the treatment of germ cell tumors and greatly improved the response rates in tumors from the ovary, lung, head and neck and esophagus.

Platinum compounds have reactive ring complexes that bind covalently to DNA which alters the structure and function of the DNA strand. This mechanism is similar to that observed with alkylating agents. Platinum compounds can cause interstrand cross-links in DNA which impair DNA replication. Cellular resistance to platinum compounds occurs by reaction with sulhydryl groups on proteins and also by increased DNA repair enzyme activity. Many of the observed toxic side-effects

15

Table 15.5. Topoisomerase Inhibitors

Etoposide
Teniposide
Irinotecan
Topotecan
9-Aminocaptothecin

from cisplatin can be attenuated when given with sulfur-containing compounds such as amifostine.

Cisplatin is potentially a very toxic drug and must be given under conditions of prolonged intravenous hydration to prevent nephrotoxicity. In addition, cisplatin can also induce significant acute myelosuppression and nausea, with cumulative effects on hearing and peripheral neuropathies. The sulfur containing compound amifostine, which was developed to protect bone marrow toxicity of radiation exposure from nuclear weapons, has been shown to reduce the degree of neuro- and nephrotoxicity of cisplatin. However, the analogue carboplatin is also available for use clinically, and it spares the patient adverse effects on the renal and neurologic systems. For many tumor types including those arising from the ovary, lung, and head and neck, carboplatin can be substituted for cisplatin.

A new diaminocyclohexane platinum derivative, called oxaliplatin, has been tested in Europe. Oxaliplatin has demonstrated impressive antitumor activity against 5-FU-resistant colon cancers, and is currently being tested in the United States. Like cisplatin, oxaliplatin can cause significant peripheral neurological toxicities and bone marrow suppression.

Future Drugs

Chemotherapeutic drugs have been the mainstay in treatment of malignancies both in the adjuvant and metastatic setting. The predominant mechanism of action has been to induce cytotoxicity. However, other approaches are being investigated to take advantage of our knowledge of how cells are transformed stepwise from premalignant changes to invasive disease and metastasis. Tumor progression is dependent upon cell signaling to the nucleus. Various pathways have been elucidated for these signals of proliferation, thus creating new targets to inhibit cell growth. In addition, the formation of tumors involves a complex interaction with surrounding tissues, requiring blood supply for oxygen and nutrients to the growing malignant cells. Compounds have been developed to impair tumor blood vessels and possibly slow or prevent tumor development. One can envision using these compounds in patients who have been treated to remission with chemotherapy, and then use agents that block proliferation signals, or prevent blood vessel formation as additional adjuvant treatment against malignancies.

Signal Transduction Inhibitors

Oncogenes can produce cellular proteins that induce signals to the nucleus to push a quiescent cell to begin proliferating. Many oncogene products are cellular membrane proteins that are commonly found in most tumor types. Some of these proteins are derivatives of growth factor receptors that become activated by mutation

Table 15.6. Anthracyclines

Daunomycin
Doxorubicin
Idarubicin
Epirubicin
Mitoxantrone

Table 15.7. Platinum analogues

Cisplatin
Carboplatin
Oxaliplatin

and produce inappropriated signals for cell growth. The HER2 membrane receptor is an oncogene that is overexpressed in tumors of the breast, prostate and lung. Current studies have examined the roles of monoclonal antibodies that bind to these receptors and make the receptors nonfunctional. Trastuzumab is a monoclonal antibody that has demonstrated activity alone and in combination with chemotherapy in metastatic breast cancers that over-express the HER2 receptor. Ligands are being developed against other oncogene membrane growth factor receptors and are being tested in the clinics.

The other way of attacking the overexpression of growth factor receptors is to inhibit the downstream effects on biochemical signal transduction. For example, the normal ras protein is attached to the cellular membrane by a carbon moiety called farnesyl. A new compound called R115777 inhibits the enzyme farnesyl protein transferase, resulting in the ras protein not being farnesylated. This prevents the molecule from attaching to the cellular membrane and prevents cellular signal transduction.

Cyclin dependent kinases (CDK) regulate the function of proteins called cyclins. Cyclins in turn regulate various control points in the cell cycle. Flavopiridol is a flavone that inhibits CDKs and induces growth arrest in the G1 phase of the cell cycle. This drug is currently being tested in the clinic.

Antiangiogenesis Drugs

Angiogenesis refers to the process of new blood vessel development. Angiogenesis and vascular growth factors play a prominent role in formation of new blood vessels to a developing tumor. The tumor interstitium produces small molecules called integrins that signal vascular endothelial migration and rearrangement which helps to provide the growing tumor with blood vessels. Drugs such as TNP-470 and thalidomide are being tested in patients after demonstrating antiangiogenesis activity in preclinical models.

Endogenous angiogenesis inhibitors are being developed after it was observed that we have protein fragments in our tissues that inhibit endothelial cell growth. Endostatin is a fragment from collagen XVIIIk that is thought to compete with endogenous endothelial growth factors for blood vessel formation. Angiostatin is a

15

fragment of plasminogen and has been shown to cause regression of tumors in animals. Both of these compounds are currently in clinical trials.

Selected Readings

1.   Chabner BA. Clinical strategies for cancer treatment. In: Chabner BA, Collins JM eds. Cancer Chemotherapy: Practices and Principles. 1st ed. Grand Rapids: JB Lippincott Co., 1990d:1-15.
     This Chapter and others in the book provide good technical overview and review of cancer chemotherapy and pharmacokinetics.
2.   Mayer RJ. Chemotherapy for metastatic colorectal cancer. Cancer 1992; 70:1414-1424.
     This article reviews the use of antimetabolite chemotherapy in colorectal cancers.
3.   Michael M, Moore M. Clinical experience with gemcitabine in pancreatic carcinoma. Oncology 1997; 11:1615-1621.
     This article reviews the activity of the cytidine antimetabolites.
4.   Takamoto CH, Arubck SG. Clinical status and optimal use of topotecan. Oncology 1997; 635-1639.
     This article provides a good review of the topoisomerase I inhibitors, and discusses new combinations for solid tumors.
5.   Fidler IJ, Ellis LM. The implications of angiogenesis for the biology and therapy of cancer metastasis. Cell 1994; 79: 185-188.
     A good review on using angiogenesis inhibitors to interrupt growth of malignant tumors.
6.   Milross CG, Mason KA, Hunter NR et al. Relationship of mitotic arrest and Apoptosis to antitumor effect of paclitaxel. J NCI 1996; 88:1308-1314.
     A good review of the mechanisms of action in cytotoxicity induced by taxanes.

15

CHAPTER 16

Major Principles of Radiation Oncology

Thomas A. Roland

Historical Perspective

Konrad Roentgen discovered x-rays in 1895 and soon thereafter the Curies published their discovery of Radium in 1898. The biological effects of ionizing radiation were quickly recognized, and the birth of radiation therapy ushered in the new century. During the first two decades of the twentieth century, radiobiology was poorly understood and technologic advances began to drive the interest in developing radiation therapy as a true medical discipline.

The International Congress of Oncology was held in Paris in 1922 and at that time the first real scientific evidence was presented that advanced head and neck cancers could be treated effectively for cure with radiation. During this period of time and through the thirties, the concept of fractionation developed as a basis for radiation treatments. By 1936 published reports indicated that protracting the radiation over several weeks with fractionated schedules would improve the therapeutic ratio, increasing dose to tumor and allowing more repair to normal tissues. While higher energy x-rays were being produced by new technology, at the same time the art of brachytherapy was developing as well. By the early 1900s Radium needles were being placed within malignant tumors in many anatomic locations with long-term control of tumors.

During the 1940s and 50s the development of higher energy generators and Cobalt 60 ushered in the megavoltage era which again improved the armamentarium for curing cancer. The last three decades have seen considerable advances in radiation treatment by improved knowledge of radiation physics, radiation biology, and treatment planning with the use of sophisticated computers. In conjunction with these developments the imaging improvements in Diagnostic Radiology with CT and MRI have added another dimension for accurate placement of radiation fields.

More recently the field of Oncology has stressed the importance of combined modality approach in treatment, emphasizing a much greater interaction among Radiation Oncologists, Medical Oncologists, and Surgical Oncologists. In addition, improved irradiation techniques as well as improved effectiveness of cytotoxic drugs and advanced surgical techniques have all increased the overall cure rate.

It is now realistic to expect a cure in over 50% of newly diagnosed cancer patients.

Clinical Perspective

The clinical application of radiation involves a complex combination of processes including staging, a knowledge of the particular pathology, and defining the therapeutic goals such as cure or palliation. Determination of the optimal dose of

Surgical Oncology, by David N. Krag. ©2000 Landes Bioscience.

radiation will depend on the volume of tissue to be treated and the overall general condition of the patient including other underlying major medical problems. This process involves the Radiation Oncologist working closely with the Physicist, the Dosimetrist operating the treatment planning computers, and the Therapists who actually set up the patient each day and operate the treatment equipment.

Treatment planning is the hallmark of modern radiation therapy and involves a process of defining the target volume to be treated. The utilization of various imaging modalities such as MRI, CT and nuclear scans all help to define the target to be irradiated and the normal tissue to be spared. Figure 16.1 is a representation of the volumes in radiation therapy defining the shrinking field technique. The goal is to treat only the tissue which may harbor the gross or microscopic cells and to decrease the amount of radiation to the normal tissues as possible.

Many malignancies can be cured with radiation as the sole modality, and in general the higher the T stage the less likelihood of long-term control with single modality treatment. With the addition of sensitizing chemotherapy such as carboplatin and taxol or cisplatin, advanced T-stage or bulky tumors such as head and neck malignancy or gynecologic malignancy can now be controlled without major surgery. Also, the use of novel delivery techniques, such as targeted high dose intraarterial cisplatin with high dose radiation, has produced excellent results and organ sparing in advanced T4 head and neck malignancies. This modality may be utilized in the future in other bulky sites of disease in the body.

Table 16.1 lists a group of anatomic sites which are potentially curable with organ sparing.

Radiation Therapy—Physical Basis

Radiation therapy uses electromagnetic radiation and particle radiation, primarily electrons. Neutrons and protons are also used in clinical radiation oncology but less commonly (Fig. 16.2). These forms of ionizing radiation have varying energies and are produced by various types. External beam units are in use throughout the world but all have similar characteristics in producing a focused beam of energy directed towards the target volume of tumor (Fig. 16.3).

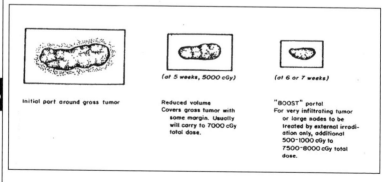

Initial port around gross tumor

(at 5 weeks, 5000 cGy)

Reduced volume
Covers gross tumor with some margin. Usually will carry to 7000 cGy total dose.

(at 6 or 7 weeks)

"BOOST" portal
For very infiltrating tumor or large nodes to be treated by external irradiation only, additional 500-1000 cGy to 7500-8000 cGy total dose.

Fig. 16.1. Example of shrinking field technique. Rectangles show borders of irradiated field.

16

Table 16.1. Curable cancer—anatomic site

Head and neck	Hodgkin's lymphoma
Skin	Non-Hodgkin's lymphoma—early stage
Cervix	Esophagus
Endometrium	Breast
Lung	Ewing's Tumor
Bladder	Penis
Prostate	

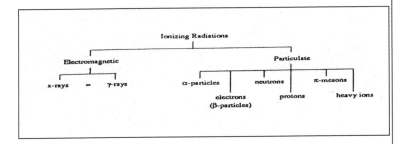

Fig. 16.2. Categories of ionizing radiation.

Gamma and x-rays are both ionizing radiation with similar biologic effect but are produced differently. X-rays are obtained from extranuclear processes when an accelerated electron strikes a target such as gold or tungsten. Gamma rays are emitted from the nucleus during radioactive decay. A Cobalt 60 Unit is a classic example of a gamma ray emitter. The half-life of a radioisotope is the time required for that isotope to decay to one-half of its initial activity. Thus, all radioisotopes used for therapeutic purposes must be periodically replaced according to their half-life.

In addition to x-rays and gamma rays, particulate radiation is also commonly used in the Radiation oncology clinic. Electrons, because of their charge and low mass, can be accelerated to high energies through a linear accelerator or a betatron. The electrons have a limited range, and thus are often used to treat more superficial lesions of the body such as chest wall tumors, skin malignancies or lymph nodes. Gamma rays and x-rays are absorbed exponentially in matter and are more suited to deeper lesions as in prostate, lung or pelvis (Fig. 16.4).

The absorbed dose is the unit of measurement for radiation. It is the amount of energy absorbed from a beam of radiation per unit mass of absorbing material. The rad (100 ergs absorbed per gram) is still a commonly used unit but is rapidly being replaced by the current term of Gray (Gy), which is defined as one joule absorbed per kilogram (1 Gray equals 100 rad and 1 centigray [cGy] equals 1 rad).

The majority of patients receiving radiation therapy are treated with external beam units which deliver the radiation from a source external to the patient. Commonly used equipment in a modern Radiation Therapy Clinic includes linear accelerators, Cobalt Units, and occasionally Betatrons. These units produce high energy x-rays which have a characteristic penetration dependent upon their particular energy.

16

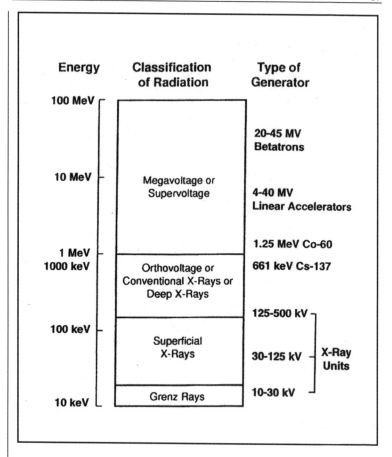

Fig. 16.3. Classification of radiation used clinically.

The Cobalt 60 Unit, developed after the second World War, is one of the first classic high energy megavoltage units and has a radiation energy averaging 1.25 MeV. Linear accelerators have average energy ranges from 4-18 MV and are often utilized for large volumes or deeper malignancies. Betatrons range from 25-35 MV and have even greater penetrating depth doses compared to accelerators (Table 16.2).

Brachytherapy is also commonly used in gynecologic and urologic malignancies. This type of therapy uses radioactive sources which are placed inside the tumor (interstitial therapy) or close to the tumor as in the vaginal canal for cervix cancer (intracavitary therapy). Each radioactive source in brachytherapy has a characteristic depth dose and sources are selected according to individual treatment situations. The use of high dose rate brachytherapy (HDR) is now becoming more popular in the treatment of gynecologic malignancy because it allows the patient to be treated as an outpatient.

16

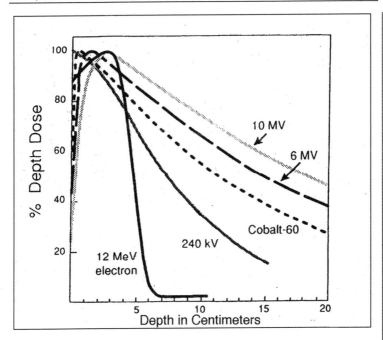

Fig. 16.4. Depth dose characteristics for several selected radiation beams.

Table 16.3 gives the characteristics of frequently used radionuclides.

The techniques used to consistently reproduce a daily treatment require proper immobilization of the patient and the placement of field marks on the patient or on the immobilization device. Specially devised immobilization masks or plastic devices conforming to the body contours are used. Often small skin tattoos aid in lining up the treatment field with the use of lasers in the treatment room and in the simulator room.

With the use of proper immobilization devices, information obtained from CT scans or MRI scans, and the use of treatment planning computers, conformal treatments can be devised which shape or "conform" the treatment to the volume of tumor while eliminating as much irradiation to normal tissue as possible.

Stereotactic radiosurgery, another form of highly focused radiation in a conformal pattern, utilizes many small radiation beams directed at a smaller target, usually less than 4 cm in size and often within the brain. The use of this localized, highly intense radiation destroys the local tissue. AV malformations, meningiomas, astrocytomas, glioblastomas, or brain metastases may be treated with stereotactic radiation.

Radiation Therapy—Biological Basis

Ionizing radiation produces a biologic response in tumor tissue as well as normal tissue. The observed clinical effect on these tissues is a result of a chain of chemical reactions brought about through the ionization process. Many studies are now under-

16

Table 16.2. Skin sparing and depth at 50% dose for commonly used XRT beams

E_{max}	d_{max} (cm)	DD_{50} (cm)
250 kV (x-ray unit)	Surface	7
1.25 MeV (60Co)	0.5	11
Linear accelerators		
4 MV	1.0	14
6 MV	1.5	16
10 MV	2.0	18
15 MV	3.0	20
18 MV	3.1	21
Betatrons		
25 MV	4.0	23
35 MV	5.0	26

E_{max}: Maximum photon energy
d_{max}: Depth of maximum dose.
DD_{50}: Depth of 50% depth dose in a 10 x 10 cm2 field size.
A disadvantage of cobalt-60 unit is that the source is several cm in diameter and the edge of the treatment field is not sharply defined due to the large penumbra. Also, the depth of penetration is not as good with a cobalt unit as with a linear accelerator or betatron.

Table 16.3. Characteristics of frequently used radionuclides

Radionuclides	Half-life	Radiation avg. E(MeV)	HVL (cm lead)
Cobalt-60	5.26 yrs	1.25	1.2
Cesium-137	30 yrs	0.661	0.6
Iridium-192	74 days	0.350	0.4
Gold-198	2.7 days	0.412	0.3
Radon-222	3.8 days	0.047-2.4	1.66
Radium-226	1622 yrs	0.047-2.4	1.66
Iodine-125	60 days	0.028	0.003
Palladium-103	17 days	0.021	0.001

way to determine the appropriate and optimal dose fractionation schedules for treatment. Radiation damage can be enhanced in tumor tissue with the addition of sensitizing chemotherapy but with enhancement of side effects in normal tissue as well.

Linear energy transfer (LET) describes the density of the ionization energy along the pathway of the radiation beam. This term refers to the amount of energy deposited in tissue per unit tracked length. High LET, which is densely ionizing radiation, occurs in neutron and alpha particles and is more efficient than gamma or x-ray radiation in producing biologic damage. Most radiation used in the standard radiation clinic is low LET since high LET energies, especially neutron beam, are used in only a few facilities throughout the United States.

Relative biological effectiveness (RBE) refers to the efficiency of radiation in producing a biologic response. Standard megavoltage machines in the clinic have a RBE which is somewhat less than lower energy x-ray machines while neutrons and alpha particles have both high LETs and high RBEs (Table 16.4).

Most of the damage at the cellular level is due to an indirect effect through the formation of reactive free radicals. Most DNA damage is through this indirect effect, although damage to DNA may occur by direct ionization of the molecule as well. A major factor in radiation induced cell killing comes from damage to the chromosomes.

The radiosensitivity of a tumor represents the relationship between the volume of cell kill with the dose of radiation delivered. A marked difference in sensitivity to radiation exists between various cell lines in the human body. As high as a sevenfold difference in sensitivity exists between the most radiosensitive and the least radiosensitive cells. Tumors arising from the hematopoietic system, seminoma, and dysgerminoma are quite radiosensitive with relatively low doses in the range of 2,000-4,000 cGy needed to eradicate them. In the intermediate group, requiring between 5,000-6,000 cGy, would be tumors of organ parenchyma such as colon, breast, lung, skin, cervix, and uterus. The least radiosensitive tumors would include melanomas, glioblastomas, soft tissue sarcoma, and bone tumors. These tumors require greater than 6,000 cGy for eradication. In addition to inherent sensitivity, the size of the tumor itself plays a role in terms of overall control. In Figure 16.5 one sees the radiosensitivity as a relationship of escalating dose and tumor control also as a factor of overall dose.

The complete eradication of a tumor requires total annihilation of all tumor cells within that mass. This eradication depends not only on the radiosensitivity of the tumor but also on the ability of the surrounding normal tissue to repair itself. The therapeutic ratio (TR) is a measure of radio-curability and is described by the following relationship:

$$TR = \frac{\text{Normal tissue tolerance dose}}{\text{Tumor lethal dose}}$$

Determination of the appropriate dose in modern radiation therapy has come about as a result of many clinical studies. The challenge in each new case presented to the Radiation Oncologist is to determine that dose which will produce tumor control while at the same time keeping that dose limited in order to reduce the risk of complications (Fig. 16.6).

Table 16.4. Approximate LETs and RBEs of several types of radiation

Radiation type	LET (ke V/µm)	RBE
Linac (6-15 MeV)	0.3	-0.8
Beta particle (1 MeV)	0.3	0.9
Cobalt-60	0.3	0.8-0.9
250 kVp X-rays	2	1.0
Protons	up to 100	~1.0
Helium ions	up to 200	~1.0
Neutrons (19 MeV)	0.5-15	1-2
Alpha particles	30-100	5-10

16

In addition to the total amount of radiation given over a treatment course and the volume of tissue to be treated, the size of fraction delivered at each treatment is also an important factor. In general, smaller fraction sizes in the range of 1.8-2.5 Gray are given on a daily basis 5 days per week. However, new fractionation schedules with increased daily fraction size or multiple fractions on each day are being tested to try to improve the therapeutic ratio.

Interdisciplinary Management—Surgery, Radiation, and Medical Oncology as a Team

Close communication between Surgery and all disciplines of Oncology has become necessary for optimal management of cancer. By working to enhance communication at all levels through multidisciplinary tumor boards, pre-review sessions prior to planned surgery, and simple telephone discussion, many cases can be optimized prior to surgery; and the ultimate outcome in cases may include improved function, better cosmesis and ultimately better tumor control.

The following are examples of improvement in outcome as a result of better communication:

1. Reduction in tissue removal with improved function in soft tissue sarcoma.
2. Incision site placement and orientation prior to breast irradiation resulting in improved cosmesis and less scarring.
3. Selection of field arrangements in preoperative chemo/radiation planned for colorectal malignancies reducing unnecessary radiation dose to perianal tissues.
4. Radio-opaque marker placement to help in treatment planning to improve reduction of normal tissue radiation dose as in lumpectomy sites, nodal sites, and sarcoma bed.
5. Endobronchial mapping prior to high dose brachytherapy to reduce overall dose to endobronchial tissues.
6. Decision about axillary lymph node incision site to reduce scarring and optimize in radiation planning for tangential breast radiation.
7. Surgical and radiation pre-planning for stereotactic radiosurgery.
8. Surgical and radiation planning for permanent seed prostate brachytherapy.
9. Intraoperative consultation between Surgeons and Radiation Oncologists to visualize, palpate, and clip tumor prior to closure.

While the rationale for good communication among specialists is obvious, this is one of the most neglected aspects of good cancer management in modern medicine today. A quick phone call, e-mail message, or use of the fax machine can often improve the outcome of cancer management by setting forth the overall treatment plan prior to the institution of treatment.

Curative Radiation

Higher doses of radiation may produce better tumor control, and numerous dose-response curves have been produced. However, by the use of multi-modality therapy such as surgical reduction in bulk or reduction in size of tumor with pre-radiation chemotherapy, the curative radiation dose may be reduced. Many tumors, however, can be cured with radiation even in bulky stages after a simple

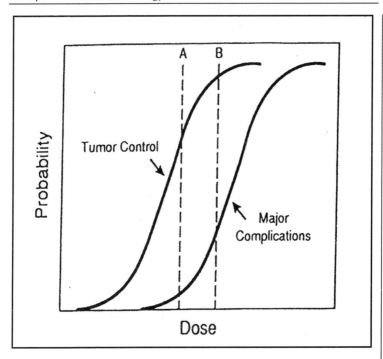

Fig. 16.5. Tumor control and complications as a function of dose delivered.

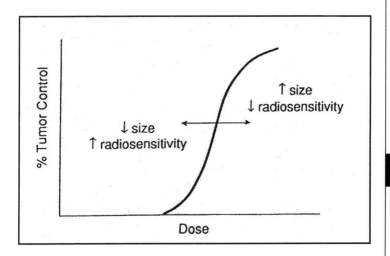

Fig. 16.6. Sigmoid curve of tumor control as a function of dose. Differences in tumor size and radiosensitivity may shift the curve to the right or left as illustrated.

16

biopsy. The tumors most commonly cured with radiation following histologic confirmation include Hodgkin's Disease, some non-Hodgkin's lymphoma, skin cancer (basal and squamous cell), carcinomas of the head and neck region, cervical and vaginal cancer, prostate cancer, and non-small cell lung cancers (less than 3 cm). With the increased use of combined chemotherapy and radiation, the interaction of anticancer drugs with radiation is better understood. While this combined treatment often results in better local control and sometimes survival, increased toxicity to various organ sites may be seen (Table 16.5).

For clinically palpable tumors in the T1 and T2 category, 60-70 Gray at 2 Gray per day, 5 fractions weekly are necessary to eradicate tumor. With increasing size of tumor and increasing T stage the doses necessary to eradicate tumor increase to 75-80 Gray. Above that level it is difficult to sterilize tumor without incurring major complications to normal tissues. With combined modality therapy, including preoperative chemotherapy or chemo/radiation, tumors can often be controlled with lower doses of radiation when the bulk of the tumor has been reduced. In general 45-50 Gray often serves as a good subclinical disease dose which can eradicate microscopic disease while at the same time being less damaging to normal tissues. It is now expected that with combined modality therapy in early stages, the following tumors should be cured in a high percentage of cases with relatively moderate morbidity to normal tissues: seminoma, dysgerminoma, Wilm's tumor, neuroblastoma (early stage), non-Hodgkin's lymphoma (low grade), skin cancer (basal and squamous cell), breast cancer, medulloblastoma, retinoblastoma, Ewing's tumor, head and neck malignancies, cervix and endometrial cancer, and prostate cancer. Table 16.6 lists the potentially curative dose ranges for various types of tumors.

Adjuvant Radiation Therapy

Radiation therapy, both preoperatively and postoperatively, can be used in conjunction with chemotherapy and surgery to improve the overall local control of disease. Doses in the range of 45-50 Gray will sterilize subclinical or microscopic disease in greater than 90% of cases. Some of the uses for radiation in an adjuvant setting would include the following:

1. Preoperative rectal treatment to a dose of 45-54 Gray.
2. Treatment of the breast post-lumpectomy (50 Gray).
3. Treatment of the pelvis following hysterectomy (45 Gray).
4. Postlaryngectomy or other head and neck malignancy (50-60 Gray).
5. Postorchiectomy for seminoma (25 Gray).
6. Postresection for skin cancer, positive margin, (45-60 Gray).
7. Postoperative lung cancer for positive margin (50-60 Gray).
8. Postoperative high grade brain tumor (60 Gray).
9. Postoperative endometrial and cervical carcinomas (45 Gray).
10. Postoperative soft tissue sarcoma (55-65 Gray).

Palliative Radiation Therapy

Radiation therapy has always played a major role in the palliation of symptoms produced by cancer. Nearly 50% of all cancer cases come to radiation therapy at some point, and many of these patients require palliative management. It is important to select the proper dose and time schedule for palliative management of patients

16

Table 16.5. Interaction of anticancer drugs with radiation

Drug	In combination with XRT	Toxicity
5FU	Increased cell kill with increased concentration or exposure time	Lower GI, mucous membranes, skin, heart, marrow
Cisplatin	Cell kill enhancement occurs to greater extent in cells sensitive to cisplatin. Concentration dependent enhancement Postradiation enhancement also occurs and may be most important clinically	Skin, mucous membranes, CNS
Mitomycin	Scheduling with XRT unimportant. May work best early on when fraction of hypoxic cells is greatest	Lung, marrow
Adriamycin or Actinomycin	Post-XRT usage may result in recall of radiation reaction	Heart, lung, kidney, skin, esophagus, mucous memb., lower GI
Hydroxyurea	May synchronize cells by inhibiting entry into S from G1	Marrow
Bleomycin	Tumor dependent variation in optimum treatment schedule	Lung
Nitrosourea	Can penetrate CNS	Marrow
Cytoxan	Hemorrhagic cystitis can occur with or without XRT	Marrow, bladder
Vincristine	Peripheral neuropathy may occur with or without XRT	Peripheral nerves
Methotrexate	Often excluded when treating breast cancer patients to avoid excessive skin reactions	CNS, skin

who often have a major difference in performance status, extent of their disease, and life expectancy. One must also consider that many of these patients have had previous treatments including surgery and chemotherapy. While many patients treated curatively will undergo radiation therapy for 6-8 weeks, the palliative care patient, often due to poor performance status and significant pain, must be treated with as short a course as possible to produce symptom relief. In general the vast majority of patients receiving radiation therapy do achieve palliative benefit from this modality. In particular, pain from bone metastases, intracranial pressure from metastases, hemorrhage, and visceral blockage are usually well palliated by radiation.

Table 16.7 lists palliative situations and the suggested treatment with consequent side effects.

16

Table 16.6. Curative doses of radiation for different tumor types

20-30 Gy
   Seminoma
   Dysgerminoma
   Acute lymphoctyic leukemia
30-40 Gy
   Seminoma (bulky)
   Wilms' tumor
   Neuroblastoma
40-50 Gy
   Hodgkin's disease
   Lymphosarcoma
   Seminoma
   Histiocytic cell sarcoma
   Skin cancer (basal and squamous cell)
50-60 Gy
   Lymph nodes, metastatic (N0, N1)
   Squamous cell carcinoma, cervical cancer, and head and neck cancer
   Embryonal cancer
   Breast cancer, ovarian cancer
   Medulloblastoma
   Retinoblastoma
   Ewing's tumor
   Dysgerminomas
60-65 Gy
   Larynx (< 1 cm)
   Breast cancer, lumpectomy
70-75 Gy
   Oral cavity (< 2 cm, 2-4 cm)
   Oro-naso-laryngo-pharyngeal cancers
   Bladder cancers
   Cervix cancer
   Uterine fundal cancer
   Ovarian cancer
   Lymph nodes, metastatic (1-3 cm)
   Lung cancer (< 3 cm)
80 Gy
   Head and neck cancer (> 4 cm)
   Breast cancer (> 5 cm)
   Glioblastomas (gliomas)
   Osteogenic sarcomas (bone sarcomas)
   Melanomas
   Soft tissue sarcomas (> 5 cm)
   Thyroid cancer
   Lymph nodes, metastatic (> 6 cm)

From: Rubin P, McDonald S, Qazi R, eds: Clinical Oncology: A Multidisciplinary Approach, ed. 7. Philadelphia: WB Saunders, 1993:72.

16

Emergencies Requiring Radiation Therapy

The need for rapid treatment with radiation is relatively uncommon and is determined by the clinical situation, its severity and the rate of progression of symptoms. The primary radiotherapy emergencies are superior vena caval syndrome and spinal cord compression. Intracranial pressure and extensive brain metastases may require fairly prompt attention but are usually not considered an emergency since steroid therapy can often alleviate the symptoms relatively rapidly.

Obstruction of the superior vena cava from lung cancer or lymphoma is considered a medical emergency, and prompt use of radiation therapy can alleviate symptoms. Again, the use of corticosteroids often plays a role in the treatment of this problem. Approximately 80% of cases of superior vena caval syndrome are due to malignancy, and lung cancer accounts for approximately 70% of the cases. Small cell carcinoma is the most common histologic type with squamous cell carcinoma as the second most common. Non-Hodgkin's lymphoma and Hodgkin's disease may also produce

Table 16.7. Palliative radiation

Condition	Work-Up	Dose Schedule	Acute Side Effects
Bronchial Obstruction (Lung Collapse)	Bronchoscopy CT Imaging	30 Gy/10 Fractions 20 Gy/5 Fractions	Cough Mild Skin Changes Esophagitis
Brain Metastases	CT/MRI Brain	35 Gy/15 Fractions 30 Gy/10 Fractions	Alopecia Fatigue Skin Erythema
Limited Skeletal Metastases	Bone Scan Skeletal Survey	30 Gy/l0 Fractions	Mild Skin Changes Neutropenia
Extensive Skeletal Metastases	Bone Scan Skeletal Survey	Systemic Radiotherapy with Strontium-89 /Sumarium-97	Neutropenia Thombocytopenia Fatigue
Hepatic Metastases	LFT's Abdominal CT /Ultrasound	21-24 Gy/7-8 Fractions	Mild Nausea
Extensive Nodal Enlargement	Physical Exam CT/MRI Imaging	20-30 Gy/10-15 Fractions	Mild Skin Changes
Intraabdominal Carcinomatosis (Ascites)	CT Imaging Ultrasound	30 Gy/15-20 Fractions	Nausea Fatigue
Esophageal Obstruction	Barium Swallow CT Imaging	30-40 Gy/10-15 Fractions	Esophagitis
Obstructive Jaundice (G.B./Pancreas)	CT Imaging Ultrasound	45 Gy/25 Fractions 30 Gy/10 Fractions	Fatigue Nausea

16

this syndrome. Rarer situations of mediastinal fibrosis, goiter, aneurysm, and thrombosis may produce symptoms of SVC as well. The histologic confirmation of malignancy is important but not always essential to the institution of radiation therapy. If the patient is deteriorating rapidly, a single fraction of 300-400 cGy may begin rapid resolution of compression and allow the surgeon to obtain tissue following the treatment. Lymphomas will require approximately 30-40 Gray total while carcinomas require a higher dose in the range of 40-50 Gray.

Spinal cord compression is truly a medical emergency. The longer the cord is compressed the more likelihood that it will be devascularized and permanent loss of function will ensue. For this reason immediate attention at any time of day or night is necessary for this condition and will necessitate radiation therapy even in the middle of the night.

Pain is the most common symptom of spinal cord compression and often precedes weakness or sensory deficit by many weeks. Work-up includes a careful history and physical examination with neurologic assessment and immediate radiographic work-up including plain films of the spine and usually CT or MRI. When the diagnosis has been made clinically, corticosteroids should be administered in high doses usually in the range of 16-32 mg per day with a starting IV dose of 10 mg dexamethasone.

Surgical intervention may be necessary in the presentation of spinal cord compression and depends on the stability of the spine, the rate of progression of neurologic dysfunction, and the overall condition of the patient. Following surgical decompression the patient will require postoperative radiation.

Patients presenting with slowly progressive neurologic symptoms, rapid resolution of symptoms with steroids, and overall good performance can usually be treated with radiation immediately and followed closely for any neurologic progression. In general the literature suggests that radiation therapy is preferable to surgical intervention as long as the patient is neurologically stable.

The total dose of radiation varies according to the individual case, the histology, and the progression of disease. Lymphoma usually receives 30-40 Gray in 15-20 fractions of 2 Gy while carcinoma would be treated with 30-40 Gray in 10-15 fractions of 2.5-3 Gy.

Selected Readings

1. Principles and Practice of Radiation Oncology, Third Edition: Perez, Carlos and Brady, Luther. Lippincott-Raven 1988.
    This comprehensive textbook covers all anatomic sites as well as specific discussion of benign disease and pain management.
2. Introduction to Clinical Radiation Oncology, Second Edition: Cora, Lawrence R. and Maylan, David J. Medical Physics Publishing 1994.
    This short, readable text is useful for quick reference and directed toward medical students and Radiation Oncology residents.
3. Moss'Radiation Oncology, Rationale, Technique, Results, Seventh Edition: Moss, William T, Cox, James D. Mosby 1994.
    A less comprehensive textbook but one which covers all anatomic sites and is an excellent quick reference on techniques for Radiation Oncology treatments

16

4. Textbook of Radiotherapy, ed 3, Fletcher GF: Philadelphia, Lea & Febiger, 1983.
    A classic small textbook of Radiation Oncology written by Gilbert Fletcher who has direct lineage with Roentgen himself.

5. Chemoradiation: An Integrated Approach to Cancer Treatment, Madhu, J. John, et al, Philadelphia, Lea & Febiger, 1993.
    An excellent textbook covering the biology of chemoradiation, clinical applications, toxicity, and future trends in mulit-modality treatments.

6. Radiation Therapy for Head and Neck Neoplasms, Third Edition, Wang, C.C., Wiley-Liss, 1997.
    An excellent small textbook written by an international expert in Radiation Oncology of head and neck cancer. This textbook covers most anatomic sites of the head and neck and also discusses important aspects of dental care and complications of radiation therapy.

7. Pediatric Radiation Oncology, 3rd Edition, Halperin, Edward C et al, Lippincott Williams & Wilkins, 1999.
    A comprehensive but small textbook covering pediatric cancers and discusses their natural history, techniques of combined chemo/radiation and surgery, as well as the outcomes of treatment approaches. This text has excellent graphics and photographs of treatment techniques.

16

CHAPTER 17

Ultrasound Imaging in Surgical Oncology

Jay K. Harness

Introduction

Working knowledge of ultrasound technology and its clinical applications has emerged in the last decade as an essential skill for practicing and in-training general surgeons. Ultrasound imaging has moved far beyond radiology departments into arenas not thought of in the past. Surgeons apply this technology as "clinical sonographers". Real-time ultrasound enhances the physical examination of an area under study (i.e., breast, abdomen, rectum, etc.); can be used to guide biopsies of palpable and nonpalpable lesions (i.e., FNA, core); localize occult lesions (i.e., liver metastasis); guide resections (i.e., of liver and pancreatic lesions); and guide ablative procedures (i.e., radiofrequency ablation, cryoablation, or laser ablation).

The practice of surgical oncology will utilize more image-guided procedures in the 21st century. Radiofrequency, cryo-, or laser ablation treatments of breast cancers will become the "norm", particularly for nonpalpable, early staged cancers. These ablative techniques will be expanded beyond their current limited applications (i.e., liver metastasis, prostate cancer) to an ever-increasing number of sites. Ultrasound will be the primary technology used to guide these treatment procedures.

This Chapter will provide surgeons with a limited and condensed overview of what has become a very large field of study. Greater detail and expanded overviews can be found in textbooks and formal articles on ultrasound that have been produced primarily for surgeons (see Suggested Readings).

Basic Principles

It is beyond the scope of this Chapter to provide an in-depth discussion of ultrasound physics, or the mechanics of ultrasound machines and transducers (probes). The most basic concepts and terminology will be described in order to give an appropriate starting-point for a student new to ultrasound.

Sound is measured in units called hertz (Hz). One million cycles per second equals 1 megahertz (MHz). Medical ultrasound operates within a range of 1 MHz to 30 MHz. Sound travels through human tissue at an average speed of 1540 m/second (slowest in fat, fastest in bone). The number of interfaces of different tissues and degree of difference in speed of sound in these tissues (acoustic mismatch) determines how much sound is reflected at any given interface and how much sound is transmitted through to the next tissue. The greater the difference in tissue densities (i.e., bone vs. fat) between tissue types the greater reflection or the larger (brighter) the echoes.

Surgical Oncology, edited by David N. Krag. ©2000 Landes Bioscience.

Ultrasound waves are generated and received in the ultrasound machine's transducer (probe). Transducers may have as few as 1 crystal to well over 100 crystals. The generation and receiving of sound by these crystals is based on the piezoelectric effect (the conversion from mechanical to electrical energy and vice versa). As ultrasound waves interact with tissue, energy is lost by reflection, scattering, absorption, and refraction. Reflection and scattering are important properties because they return ultrasound signals to the transducer, which in turn provides diagnostic information in the form of an image. Absorption and refraction have the opposite effect by diverting sound energy away from the receiving transducer. Acoustic energy is constantly being lost from an ultrasound beam. This phenomenon is called attenuation. The sounds that travel the furthest distance will have the weakest echoes (darker).

Ultrasound echoes or reflections are displayed on the ultrasound machine's monitor as various shades of gray (a total of 256 shades). The gray scale is based on the intensity of the echoes received from acoustic interfaces within tissue. A large amount of acoustic mismatch (strong echo) is displayed on the monitor as a light shade of gray or even white. A small acoustic mismatch (weak echo) may be displayed as a darker shade of gray or even black. Weak echoes can also result from attenuation. In order to display a uniform image on the monitor, time-gain compensation (TGC) is used to compensate for attenuation. If TGC were not used, every image would be bright white closest to the transducer (near-field) and would become progressively darker as the distance from the transducer increases (far-field). By using TGC, near-field echoes can be suppressed and far-field echoes enhanced, which creates a uniformly gray image.

There are multiple types of transducers (probes) available. They vary by frequency, shape, and size. The frequency settings for transducers range from 2.25 MHz–15 MHz. The higher frequency of a transducer the greater the possible resolution, but the poorer the depth of penetration. The opposite is true for lower frequency transducers (greater depth of penetration, but poorer resolution). Surgeons should choose the highest frequency transducer possible for the specific application and depth (Table 17.1).

Two important settings on any ultrasound machine are the gain control and focal point settings. The gain control works like the volume control on a radio. Too much or too little gain impact image quality. Proper adjustments of the gain and the TGC controls are critical to insure an even balance of gray scale throughout the image. Somewhat like a camera, clear and focused images are best obtained with proper use of electronic focusing found with modern ultrasound machines. Focal points can be single or multiple and moved from the near-field to the far-field and vice versa. A focal point is a narrowing of the ultrasound beam, and provides the clearest image of a targeted area or structure.

It is important that surgeons know and properly use the conventional vocabulary of ultrasound. The following terms are fundamental in describing ultrasound images:

Hyperechoic—refers to a shade of gray that is bright white in appearance or brighter than surrounding structures (Fig. 17.1).

Hypoechoic—refers to a dark shade of gray or less bright than surrounding structures (Figs. 17.1, 17.2).

17

Table 17.1. Typical frequencies for various ultrasound applications

Transducer	Application
2.25 MHz	Obese patients and deep abdominal
3.5 MHz	General abdominal, obstetrics and gynecological
5.0 MHz	Neonates, pediatric patients, peripheral vascular, endoscopic ultrasound
7.5 MHz	Cerebrovascular, breasts, testicles, thyroid, parathyroid, endoscopic ultrasound
10 MHz	Breast, ocular imaging, vein mapping, superficial soft tissue, thyroid, parathyroid.
12-15 MHz	Breast, breast ductal system, thyroid, parathyroid, superficial soft tissue, endoscopic ultrasound (12 MHz).

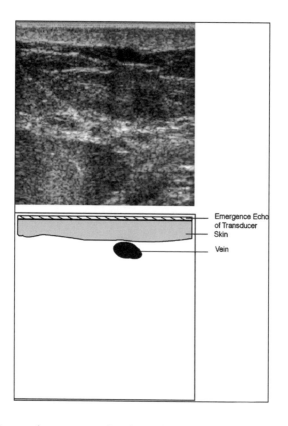

Emergence Echo of Transducer

Skin

Vein

Fig. 17.1. Image of emergence echo of transducer, hyperechoic skin, anechoic (black) subcutaneous vein, and hypoechoic subcutaneous fat. Reprinted with permission from: Harness JK, Wisher DB, eds. Ultrasound in Surgical Practice: Basic Principles and Clinical Applications 2000; 1-524. © 2000 John Wiley & Sons.

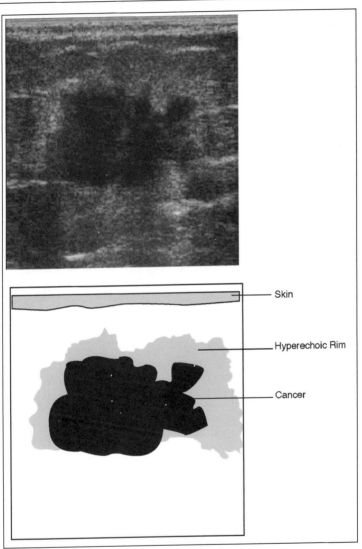

Skin

Hyperechoic Rim

Cancer

17

Fig. 17.2. Image of hypoechoic (nearly anechoic) breast carcinoma. The hyperechoic rim is from the desmoplastic response around the tumor. Reprinted with permission from: Harness JK, Wisher DB, eds. Ultrasound in Surgical Practice: Basic Principles and Clinical Applications 2000; 1-524. © 2000 John Wiley & Sons.

Isoechoic – refers to a shade of gray that is the same as surrounding structures.

Anechoic – refers to a structure without internal echoes (i.e., a simple cyst) which appears black on ultrasound (Figs. 17.1, 17.3).

Echogenic – refers to a structure that is bright white in appearance because of its sharp reflection of ultrasound echoes (Fig. 17.4).

Homogeneous – is a term used to describe the uniform appearance of shades of gray present throughout a structure or organ (i.e., the normal liver is homogeneous in appearance) (Fig. 17.5).

Heterogeneous – the shades of gray throughout an organ or structure are not uniform (opposite of homogeneous). Areas have a mixed grayscale appearance containing both hyperechoic and hypoechoic shades of gray (Fig. 17.6).

Posterior shadow – represents an artifact (a black or anechoic shadow) resulting from significant absorption or deflection of the ultrasound beam from a structure or substance (i.e., calculi, ribs, bowel gas, air, etc.) (Fig. 17.7).

Posterior enhancement – represents an artifact of a hyperechoic area behind a structure (flashlight effect) because the ultrasound beam is less attenuated as it passes through the structure (i.e., simple cyst) than the tissue surrounding the structure (Fig. 17.8).

Near-field – is the first one-half of the image displayed on the monitor (transducer to mid-field).

Far-field – is the bottom one-half of the image displayed on the monitor (mid-field to the farthest end of the image).

Breast Ultrasound

The diagnosis and treatment of patients with breast cancer is a central focus of the field of surgical oncology. Mammographic screening has increased the number of patients with nonpalpable or indeterminate lesions. High-resolution, real-time modern ultrasound equipment and transducers make breast ultrasound an important diagnostic tool. Using ultrasound, the surgeon becomes a "sonographic clinician", synthesizing the patient's history, physical findings, and mammographic and ultrasound findings into a probable diagnosis and/or a further diagnostic work-up. Surgeons, performing their own real-time ultrasound, provide patients with more complete and efficient care.

Breast ultrasound equipment must provide excellent spatial and contrast resolution. Breast sonography should be conducted with high-frequency, real-time linear array transducers ranging from 7.5 MHz to 15 MHz. The 7.5 MHz transducer is used for routine scanning, while the higher frequency transducers (10-15 MHz) are particularly helpful for imaging the ductal system.

When scanning, the patient is placed in the same position utilized for physical examination – supine with the ipsilateral hand behind the head. This position thins the breast to the greatest extent and allows the tissue planes of the breast to be pulled parallel to the skin. Imaging the nipple-areola complex, and its underlying ducts, requires special maneuvering of the transducer (angling the beam under the complex at the edge of the areola) because of the dense connective tissue in this area.

Surgeons generally perform targeted breast ultrasound examinations for a specific clinical and/or mammographic finding. Lesions of the breast should always be

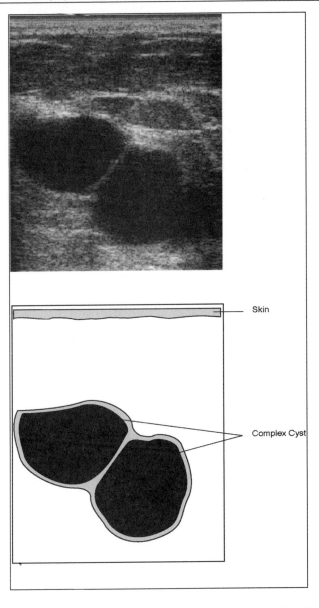

Fig. 17.3. Image of a complex breast cyst, with internal septation. The left side of the cyst is anechoic and the right side is nearly anechoic. Reprinted with permission from: Harness JK, Wisher DB, eds. Ultrasound in Surgical Practice: Basic Principles and Clinical Applications 2000; 1-524. © 2000 John Wiley & Sons.

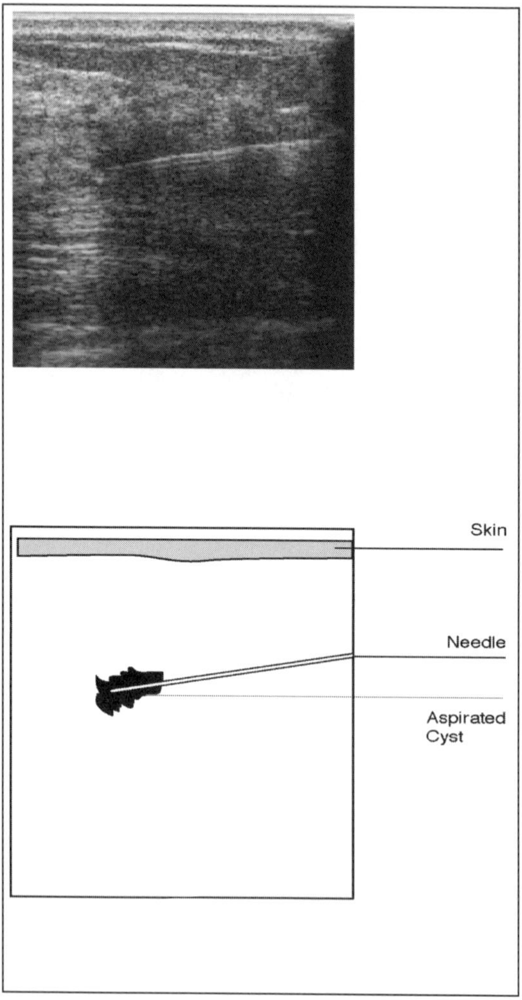

Fig. 17.4. Image of a postaspiration breast cyst. The needle is very echogenic and bright white in appearance. Reprinted with permission from: Harness JK, Wisher DB, eds. Ultrasound in Surgical Practice: Basic Principles and Clinical Applications 2000; 1-524. © 2000 John Wiley & Sons.

scanned in two planes, radial (like the hands of a clock) which runs parallel with the ductal system and antiradial (at the right angle to the radial direction) (Fig. 17.9).

The normal anatomic components of the breast and surrounding structures have characteristic sonographic features. These include: skin, subcutaneous fat, Cooper's ligaments, superficial fascia, parenchyma, nipple region, deep fascia, retromammary space, pectoralis muscle, ribs, pleura, and intramammary and axillary lymph nodes.

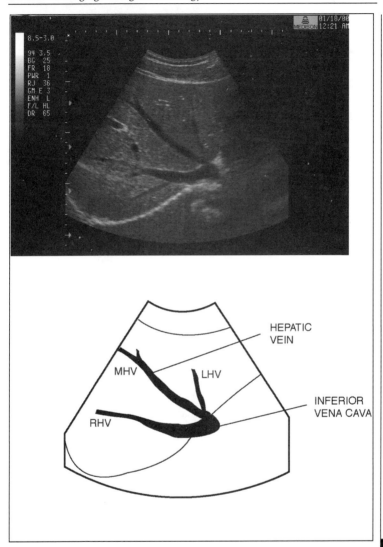

Fig. 17.5. Transverse view of liver (homogeneous) at the level of the take-off of the hepatic veins (anechoic) from the inferior vena cava (anechoic). Reprinted with permission from: Harness JK, Wisher DB, eds. Ultrasound in Surgical Practice: Basic Principles and Clinical Applications 2000; 1-524. © 2000 John Wiley & Sons.

17

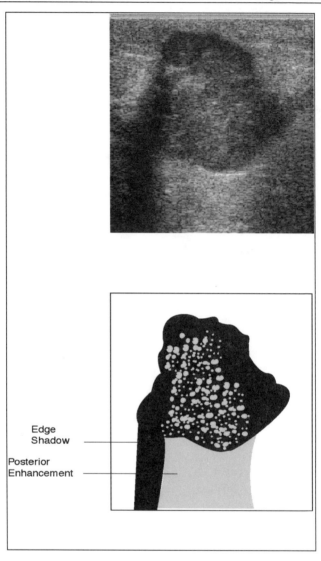

Edge
Shadow

Posterior
Enhancement

17

Fig. 17.6. Breast carcinoma demonstrating heterogeneous internal echo pattern, a single edge shadow and some posterior enhancement. Reprinted with permission from: Harness JK, Wisher DB, eds. Ultrasound in Surgical Practice: Basic Principles and Clinical Applications 2000; 1-524. © 2000 John Wiley & Sons.

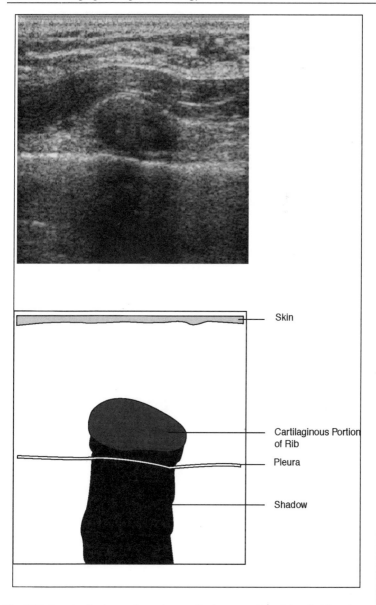

Fig. 17.7. Longitudinal scan through a cartilaginous portion of the rib. The pleura (hyperechoic) and a posterior shadow are also seen. Reprinted with permission from: Harness JK, Wisher DB, eds. Ultrasound in Surgical Practice: Basic Principles and Clinical Applications 2000; 1-524. © 2000 John Wiley & Sons.

17

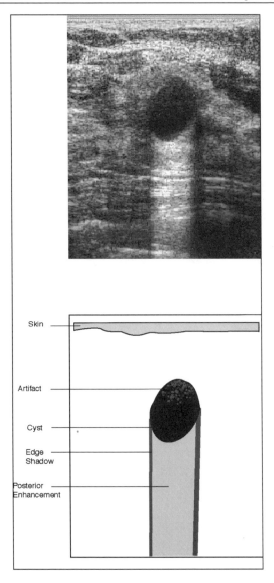

Fig. 17.8. Classic posterior enhancement (flashlight effect) behind a simple breast cyst. Also seen are bilateral edge shadows and reverberation artifact. Reprinted with permission from: Harness JK, Wisher DB, eds. Ultrasound in Surgical Practice: Basic Principles and Clinical Applications 2000; 1-524. © 2000 John Wiley & Sons.

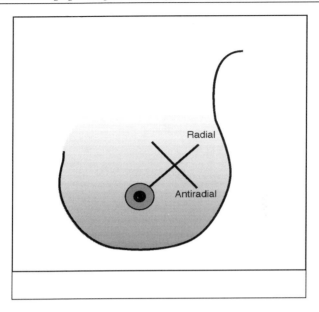

Fig. 17.9. Artist's view of transducer placement for radial and antiradial scanning. Reprinted with permission from: Harness JK, Wisher DB, eds. Ultrasound in Surgical Practice: Basic Principles and Clinical Applications 2000; 1-524. © 2000 John Wiley & Sons.

Fig. 17.10. Ultrasound of normal breast structures: hyperechoic Cooper's ligaments; hypoechoic subcutaneous fat; and hyperechoic breast parenchyma. Pectoralis muscle is posterior to the breast. Reprinted with permission from: Harness JK, Wisher DB, eds. Ultrasound in Surgical Practice: Basic Principles and Clinical Applications 2000; 1-524. © 2000 John Wiley & Sons.

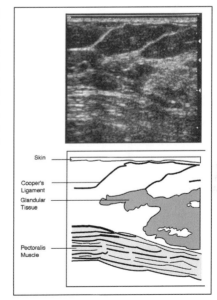

17

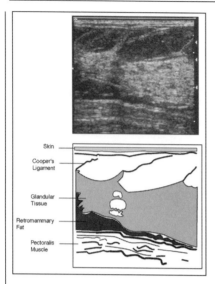

Fig. 17.11. Ultrasound of breast which demonstrates retromammary fat. Reprinted with permission from: Harness JK, Wisher DB, eds. Ultrasound in Surgical Practice: Basic Principles and Clinical Applications 2000; 1-524. © 2000 John Wiley & Sons.

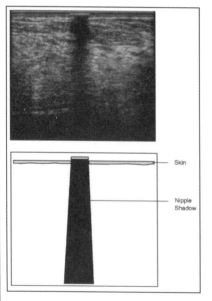

Fig. 17.12. Deep posterior shadow from a transducer place over a breast nipple. Reprinted with permission from: Harness JK, Wisher DB, eds. Ultrasound in Surgical Practice: Basic Principles and Clinical Applications 2000; 1-524. © 2000 John Wiley & Sons.

It is important that surgeons be able to identify all of these structures (Figs. 17.7, 17.10-17.13).

Benign and malignant breast lesions have a wide range of sonographic appearances. Figure 17.14 outlines the major areas of analytic focus. Combinations of features are better predictors of malignancy than any single feature. No characteristic is 100% specific. In addition, lesions that are "taller-than-wide" (greater in the

Fig. 17.13. Normal oval lymph node with fatty (hyperechoic) hilus. Reprinted with permission from: Harness JK, Wisher DB, eds. Ultrasound in Surgical Practice: Basic Principles and Clinical Applications 2000; 1-524. © 2000 John Wiley & Sons.

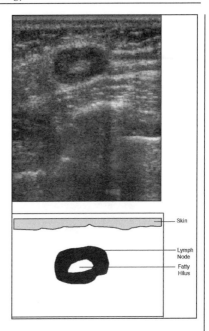

Skin

Lymph Node

Fatty Hilus

anterior-posterior direction) are more likely to be malignant because they are growing across normal tissue planes. Benign lesions tend to grow within tissue planes and are therefore "wider-than-tall" or ellipsoid.

Ultrasound is not a good modality for viewing microcalcifications; the best modality is mammography. However, microcalcifications can often be imaged and are seen as bright echogenic spots against the relatively hypoechoic backgrounds of both benign and malignant lesions (Fig. 17.15).

Invasive breast carcinomas have different sonographic appearances depending on their size, location, and degree of desmoplastic reactions. Typical invasive carcinomas are irregularly marginated; taller-than-wide; have a dense posterior shadow(s); have a thick echogenic rim (from desmoplastic reaction and compression of parenchyma); destroy adjacent fascial planes; and obstruct adjacent ducts (Figs. 17.16-17.18). Most invasive breast carcinomas image as hypoechoic to nearly anechoic lesions because of the desmoplastic reaction (which attenuates the ultrasound beam) and the fact that the tumor may have replaced all other normal tissue (Figs. 17.19, 17.20).

The most common solid benign lesion imaged with breast ultrasound is a fibroadenoma. Most fibroadenomas are oval, wider-than-tall, sharply marginated, homogeneous, and have posterior enhancement with bilateral edge shadows (Fig. 17.21).

Interventional breast ultrasound procedures include: cyst aspirations; catheter drainage of breast abscesses; FNA biopsy; spring-fired core biopsy; and vacuum-assisted core biopsy. Most of these procedures are performed using a "free-hand" technique. The advantage of ultrasound guidance for these procedures is the

17

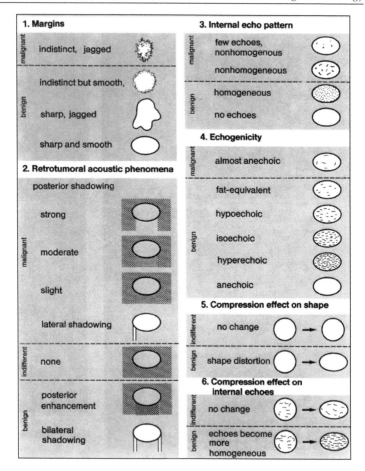

Fig. 17.14. Schematic representation of analytic criteria for the interpretation of focal breast sonographic lesions. (Reprinted with permission from: Leucht D, Madjar H, eds. Teaching Atlas of Breast Ultrasound. 1996; page 23. © 1995 George Thieme Uerlag).

documentation and assurance that the needles have been accurately placed (Figs. 17.22, 17.23). Ultrasound-guided breast procedures have been demonstrated to be accurate, facile, and reproducible. Many indeterminate lesions may be found to be benign and surgeons can avoid unnecessary open excisional biopsy of such lesions. If a malignant lesion is identified, treatment planning is facilitated and only one definitive operative procedure will generally be needed.

Additional use of breast sonography includes: guiding FNA-biopsy of suspicious axillary lymph nodes; placing localizing wires into suspicious masses prior to biopsy or formal lumpectomy, guiding real-time dissections of the ductal system; localizing

Fig. 17.15. Microcalcifications seen in an isoechoic to hypoechoic breast carcinoma. The internal echo pattern is heterogeneous. Reprinted with permission from: Harness JK, Wisher DB, eds. Ultrasound in Surgical Practice: Basic Principles and Clinical Applications 2000; 1-524. © 2000 John Wiley & Sons.

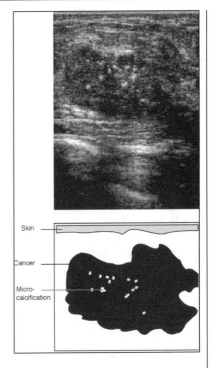

nonpalpable lesions; determining the thickness of skin flaps during skin-sparing mastectomy.

The 21st century treatment of early breast carcinomas will be more and more image-guided. The advent of stereotactic and ultrasound guided breast biopsy techniques means that no patient should have to go to an operating room for a diagnostic biopsy procedure. Core biopsy techniques should provide enough tissue for complete diagnostic testing and determination of markers. The application of radiofrequency, laser, and cryoablation techniques will be studied on a larger scale to determine if they can completely destroy smaller breast carcinomas. These procedures will primarily be ultrasound guided. Breast MRI may then be used to determine complete kill of cancers. The effectiveness of percutaneous ablative techniques will need to be determined in prospective, randomized clinical trials.

Breast ultrasound should be seen by surgeons as an important adjunct to their clinical decision-making process and as an increasingly important tool to guide diagnostic and treatment procedures.

Thyroid Carcinoma and Neck Metastases

Carcinoma of the thyroid gland accounts for only about 1% of all cancers. However, the incidence approaches 30% in single, solitary thyroid nodules. The risk of a cancerous nodule in a multinodular goiter is less than 1 percent.

The superficial location of the thyroid gland makes this organ easy to study with high-resolution, real-time ultrasound (both grayscale and color Doppler). The same

17

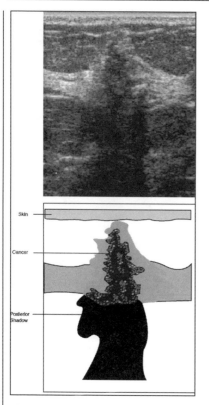

Fig. 17.16. Taller-than-wide breast carcinoma with irregular margins, heterogeneous internal echoes, and a wide posterior shadow. Reprinted with permission from: Harness JK, Wisher DB, eds. Ultrasound in Surgical Practice: Basic Principles and Clinical Applications 2000; 1-524. © 2000 John Wiley & Sons.

ultrasound equipment and range of transducers (7.5-15 MHz) used in breast sonography is used in thyroid ultrasound examinations. Normal thyroid anatomy and pathological conditions can be imaged with remarkable clarity (Fig. 17.24). The same is true of the remaining structures and locations of the neck that are routinely evaluated by surgeons (i.e., jugular chain, supraclavicular fossa, posterior cervical triangle, etc.).

For the thyroid ultrasound examination, the patient is placed in the same position used in the operating room for surgical exploration (i.e., supine, neck extended, with a pad placed under the shoulders). The entire gland is examined, including the isthmus. The thyroid should be visualized (both transversely and longitudinally) from the superior poles through the most inferior portion of the lower poles. The examination should be extended laterally (both on the right and left sides) to include the internal jugular veins, the internal jugular lymph nodes, and the carotid arteries. The internal jugular lymph nodes from the submandibular region to the supraclavicular fossa should be imaged, especially if thyroid or another head and neck carcinoma is suspected. Lymph nodes in the posterior cervical triangle should also be examined with the patient sitting up. Normal lymph nodes are often oval and have an echogenic (bright) fatty hilus (Fig. 17.13).

Fig. 17.17. Irregularly marginated in-filtrating ductal carcinoma with a hyperechoic (desmoplastic) rim and a posterior shadow. Reprinted with permission from: Harness JK, Wisher DB, eds. Ultrasound in Surgical Practice: Basic Principles and Clinical Applications 2000; 1-524. © 2000 John Wiley & Sons.

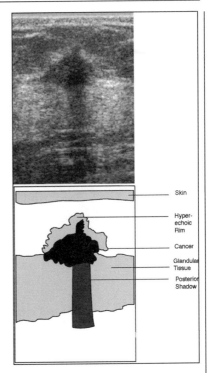

Skin

Hyper-echoic Rim

Cancer

Glandular Tissue

Posterior Shadow

Similar to breast sonography, benign and malignant lesions of the thyroid gland have certain sonographic appearances. The normal thyroid parenchyma has a homogeneous, medium-to-high level echogenicity (Fig. 17.24). This appearance makes the detection of focal hypoechoic or cystic lesions easy in most cases. If the thyroid capsule is imaged, it should appear as a thin hyperechoic line.

As with breast carcinoma, no single sonographic feature defines thyroid carcinoma. Well-differentiated papillary or papillary-follicular cancers account for 75-90% of all cases diagnosed in the United States and Canada. Medullary, pure follicular, Hürthle cell, and anaplastic carcinomas combine for the remaining 10-25% of thyroid carcinomas.

The sonographic appearance of thyroid carcinoma varies, but in general these lesions typically display irregular margins and are hypoechoic relative to thyroid parenchyma. The hypoechogenicity of thyroid carcinomas is often due to the minimal colloid substance and tightly packed cell content. Papillary carcinoma often have microcalcifications (psoamomma bodies) appearing as tiny, punctate, irregular hyperechoic foci (Fig. 17.25). Papillary cervical lymph node metastases may also be cystic as a result of extensive degeneration (Fig. 17.26).

The sonographic appearance of pure follicular carcinoma depends on the type encountered. The more advanced form (with capsular and vascular invasion) demonstrates irregular tumor margins and a thick, irregular halo. There are no unique

17

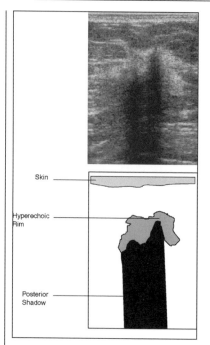

Fig. 17.18. Hypoechoic, infiltrating ductal carcinoma with a hyperechoic rim and posterior shadow. Reprinted with permission from: Harness JK, Wisher DB, eds. Ultrasound in Surgical Practice: Basic Principles and Clinical Applications 2000; 1-524. © 2000 John Wiley & Sons.

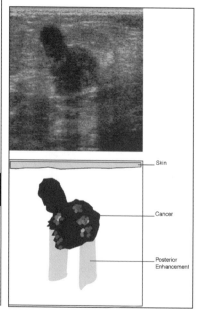

Fig. 17.19. Taller-than-wide, nearly anechoic breast carcinoma with some posterior enhancement. Reprinted with permission from: Harness JK, Wisher DB, eds. Ultrasound in Surgical Practice: Basic Principles and Clinical Applications 2000; 1-524. © 2000 John Wiley & Sons.

Fig. 17.20. Radial scan which demonstrates ductal extension (white arrow) of an infiltrating carcinoma. Reprinted with permission from: Harness JK, Wisher DB, eds. Ultrasound in Surgical Practice: Basic Principles and Clinical Applications 2000; 1-524. © 2000 John Wiley & Sons.

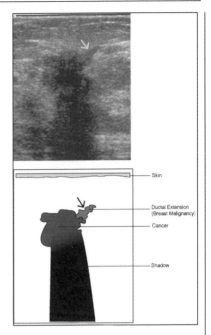

Skin

Ductal Extension (Breast Malignancy)

Cancer

Shadow

sonographic features that allow differentiation of minimally invasive follicular carcinoma from a benign follicular adenoma.

Medullary carcinoma has a similar sonographic appearance to that of papillary carcinoma. The amyloid deposits seen in medullary carcinoma appear as bright echogenic foci. Microcalcifications can also be seen with this carcinoma. Medullary lymph node and hepatic metastases may have bright echogenic foci from microcalcifications and/or amyloid.

The clinical applications of high-resolution ultrasound in the evaluation of thyroid nodules and cervical masses are several:

1. To characterize the palpable nodule or neck mass (i.e., solid, cystic, mixed, thyroid gland, lymph node, vascular mass, benign or malignant features).
2. To determine the exact location of palpable or nonpalpable neck masses.
3. To detect recurrent or metastatic carcinoma (Fig. 17.27).
4. To guide FNA biopsies of thyroid nodules or cervical lymph nodes.
5. To characterize the noninvolved portion of the thyroid gland (i.e., the lobe on the opposite side of the palpable thyroid nodule).

Ultrasound examination of the neck provides the surgeon with a wealth of practical information. It differentiates a neck mass as thyroid versus nonthyroid. It describes the characteristics of a neck lesion (i.e., solid, cystic, mixed, etc.). It provides important information about the anatomy and status of other structures (i.e., lymph nodes, the opposite thyroid lobe, etc.). It can be used to guide FNA biopsies (assuring accurate placement of the needle tip) of thyroid nodules and neck masses. This

17

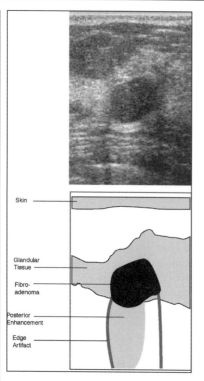

Fig. 17.21. Oval, sharply marginated, homogeneous fibroadenoma with bilateral edge shadows (artifact) and posterior enhancement. Reprinted with permission from: Harness JK, Wisher DB, eds. Ultrasound in Surgical Practice: Basic Principles and Clinical Applications 2000; 1-524. © 2000 John Wiley & Sons.

Skin

Glandular
Tissue

Fibro-
adenoma

Posterior
Enhancement

Edge
Artifact

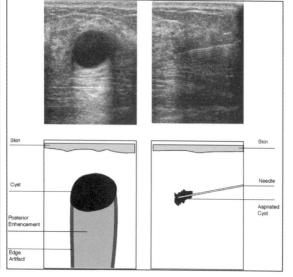

Fig. 17.22. Side-by-side images of a cyst (left image) and the postaspiration view (right image) with a needle still in place. Reprinted with permission from: Harness JK, Wisher DB, eds. Ultrasound in Surgical Practice: Basic Principles and Clinical Applications 2000; 1-524. © 2000 John Wiley & Sons.

Skin

Cyst

Posterior
Enhancement

Edge
Artifact

Skin

Needle

Aspirated
Cyst

17

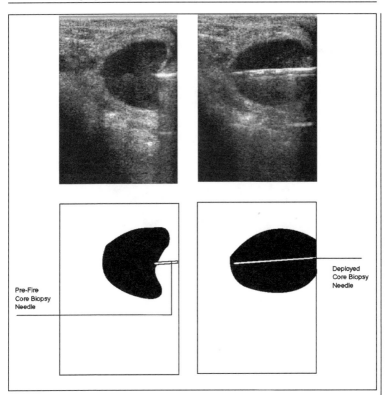

Fig. 17.23. Side-by-side images showing position of spring-fired core biopsy needle prior to firing (left image) and after firing (right image). Reprinted with permission from: Harness JK, Wisher DB, eds. Ultrasound in Surgical Practice: Basic Principles and Clinical Applications 2000; 1-524. © 2000 John Wiley & Sons.

wealth of information provides the surgeon with important preoperative data, which can result in a more focused and effective surgical exploration.

Ultrasound Imaging in Cancer Staging

Accurate preoperative and/or pretreatment staging of cancer has assumed a greater role with an interdisciplinary approach to cancer treatment. An increasing number of cancers are initially treated, after accurate staging of both tumor size and nodal status, with induction (neoadjuvant) chemotherapy and/or radiation therapy. Imaging modalities (i.e., CT, MRI, PET scanning, and ultrasound) can provide invaluable information for pretreatment staging. When an imaging modality can also be used for FNA and/or core biopsies, its value increases significantly. Ultrasound guided biopsies of axillary and cervical lymph nodes for staging have already been briefly described. In recent years, ultrasound has emerged as an important modality for staging esophageal, gastric, pancreatic, rectal, and gynecological carcinomas.

17

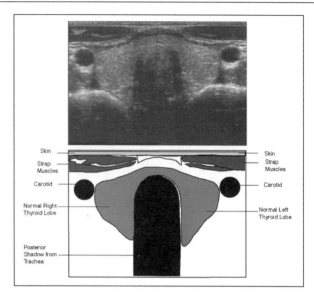

Fig. 17.24. Combined split screen image of the transverse view of a normal thyroid gland and carotid arteries. Reprinted with permission from: Harness JK, Wisher DB, eds. Ultrasound in Surgical Practice: Basic Principles and Clinical Applications 2000; 1-524. © 2000 John Wiley & Sons.

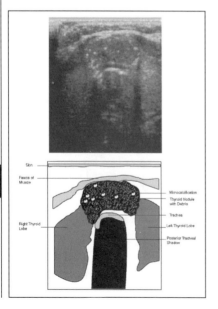

Fig. 17.25. Transverse view of a papillary carcinoma of the thyroid isthmus in a 45-year-old woman. Note the prominent micro-calcifications (psammoma bodies) seen. Reprinted with permission from: Harness JK, Wisher DB, eds. Ultrasound in Surgical Practice: Basic Principles and Clinical Applications 2000; 1-524. © 2000 John Wiley & Sons.

17

Fig. 17.26. Transverse view of an enlarged metastatic left internal jugular lymph node. Note the cystic degeneration and microcalcifications (psammoma bodies). Reprinted with permission from: Harness JK, Wisher DB, eds. Ultrasound in Surgical Practice: Basic Principles and Clinical Applications 2000; 1-524. © 2000 John Wiley & Sons.

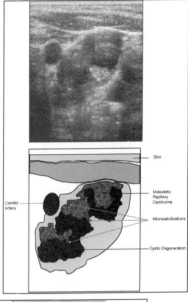

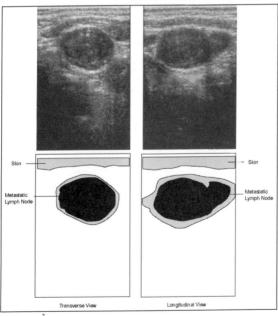

Fig. 17.27. Transverse (left image) and longitudinal (right image) views of a metastatic cervical lymph node. Reprinted with permission from: Harness JK, Wisher DB, eds. Ultrasound in Surgical Practice: Basic Principles and Clinical Applications 2000; 1-524. © 2000 John Wiley & Sons.

17

Endoscopic ultrasound (endoscopic endosonography) offers detailed images of the walls of the esophagus, stomach, and rectum and also allows imaging of adjacent structures (including regional lymph nodes). Endoscopic ultrasound permits the clinician to accurately determine the depth of tumor invasion (T stage) and the presence of metastatic lymph nodes without performing a surgical biopsy procedure. Table 17.2 summarizes the tumor invasion staging system utilized for gastrointestinal malignancies.

Two equipment designs dominate the endoscopic ultrasound market. First is a segmental scanning probe, which produces a wedge shaped, 100°, noncircumferential image (Fig. 17.28). The second is a radial probe, which produces a 360° "sweep" image. The entire circumference may be visualized at one time (Fig. 17.29). The 100° echoendoscope scans parallel to the shaft of the scope, scans at 7.5 MHz and 12 MHz, and may be capable (depending on the scope model) of color flow and Doppler imaging of vascular structures. The 360° echoendoscope scans perpendicular to the shaft of the endoscope and has the option for scanning at 7.5 MHz or 12 MHz. Both of these echoendoscopes have side viewing, fiberoptic endoscopic views. Clinicians using these devices should be familiar with using side-viewing endoscopes. In endorectal ultrasound, the same type of ultrasound equipment is used (100° and a 360° probe), without side-viewing endoscopic capability. The end of the echoendoscope transducer (as well as the nonviewing probe models) is covered with a water-filled balloon to allow close contact with the transducer to the lesion being evaluated. With gastric ultrasound, the stomach is filled with 200-500 ml of water to allow better images of the wall structures without any pressure artifact (Fig. 17.30).

The 100° sector scanning echoendoscope is the device currently used for ultrasound directed biopsies (FNAs) of a pancreatic mass or lymph nodes adjacent to the esophagus, stomach, pancreas or rectum. The real-time imaging of the needle position can be determined by the sector scanner. New instruments are under development, which will allow FNA biopsies with a 360° echoendoscope.

Endoscopic ultrasound images of the esophagus, stomach, and rectum delineate the several layers of the walls of these structures. Understanding the principles of ultrasonography makes the interpretation of the images viewed with endosonography possible. Tissue high in water content, such as muscle, appears hypoechoic (darker). Tissue with lower water content and more collagen (such as mucosa, submucosa, serosa, and fat) appear hyperechoic (brighter). From inside to out, mucosa is hyperechoic, the muscularis mucosa is hypoechoic, the submucosa is hyperechoic, the muscularis propria is hypoechoic, and the serosa is hyperechoic (Fig. 17.31).

Endoscopic and endoluminal ultrasound may be performed by different specialists, including general, thoracic, and colorectal surgeons as well as gastroenterologists. Individual specialists performing these studies will not only have to be very accomplished endoscopists, but they will also need additional training in endoscopic and endoluminal scanning techniques and interpretation of the images obtained. Additional experience will also be needed in performing endoscopic and endoluminal FNA biopsies.

The following limited examples of endoscopic and endoluminal ultrasound images are presented as illustrations of the potential uses of this technology. Figure 17.32 is a large squamous cell carcinoma of the distal esophagus in a 70-year-old male. The endoscopic ultrasound demonstrates invasion of the tumor through the muscularis

Table 17.2. Tumor invasion (T Stage) for gastrointestinal malignancies

T_1 – Tumor infiltration of the mucosal and submucosal layers; the muscularis propria remains uninvolved and intact.
T_2 – Tumor infiltration of the muscularis propria and subserosa.
T_3 – Tumor penetration through the serosa without invasion of adjacent organs.
T_4 – Tumor invasion of adjacent organs or structures.

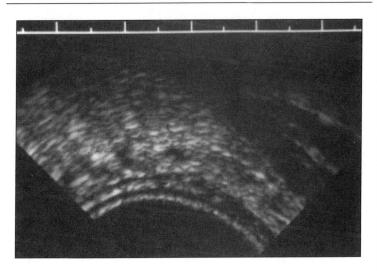

Fig. 17.28. A 100° sector scan of a rectal wall. Reprinted with permission from: Harness JK, Wisher DB, eds. Ultrasound in Surgical Practice: Basic Principles and Clinical Applications 2000; 1-524. © 2000 John Wiley & Sons.

propria and a large lymph node measuring 1.3 x 1.4 cm, which was proven to be metastatic at the time of surgical resection.

Figure 17.33 is an example of a gastric carcinoma, which infiltrates through the muscularis propria. The transducer is within a water-filled balloon and the tumor is against lower pole of the transducer.

Figure 17.34 is an example of a pancreatic mass imaged with a 100° sector echoendoscope. The hyperechoic, diagonal line at the top one-third of the mass is a FNA needle.

Figure 17.35 is an example of an adenocarcinoma of the rectum imaged with a 100° sector transducer. This lesion does not penetrate through the muscularis propria (the outer hypoechoic layer) but does penetrate through the submucosa. The perirectal fat is hyperechoic. The imaged lymph node is round in shape and hypoechoic, which suggests that it is metastatic.

Conclusion

The current and future practice of surgical oncology requires skills and understanding of basic ultrasound physics, operating an ultrasound machine, performing

17

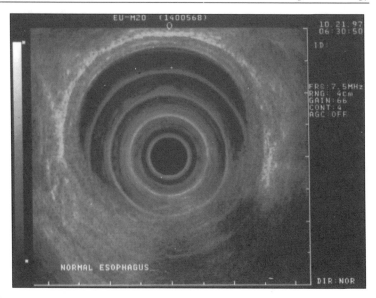

Fig. 17.29. A 360° "sweep" image of normal esophagus. Reprinted with permission from: Harness JK, Wisher DB, eds. Ultrasound in Surgical Practice: Basic Principles and Clinical Applications 2000; 1-524. © 2000 John Wiley & Sons.

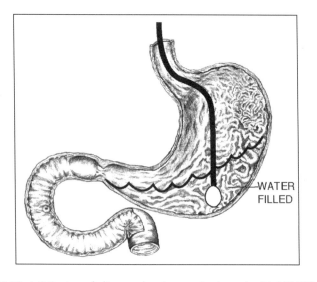

17

Fig. 17.30. Artist's view of ultrasound endoscope in stomach with 200-500 ml of water. Reprinted with permission from: Harness JK, Wisher DB, eds. Ultrasound in Surgical Practice: Basic Principles and Clinical Applications 2000; 1-524. © 2000 John Wiley & Sons.

Fig. 17.31. Layers of normal rectal wall imaged with a sector probe. Reprinted with permission from: Harness JK, Wisher DB, eds. Ultrasound in Surgical Practice: Basic Principles and Clinical Applications 2000; 1-524. © 2000 John Wiley & Sons.

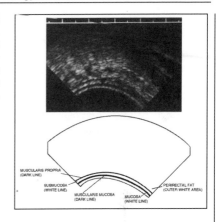

Fig. 17.32. Carcinoma of the distal esophagus. A metastatic lymph node and aorta are also imaged. Reprinted with permission from: Harness JK, Wisher DB, eds. Ultrasound in Surgical Practice: Basic Principles and Clinical Applications 2000; 1-524. © 2000 John Wiley & Sons.

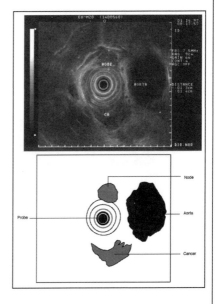

and interpreting ultrasound scans, and performing ultrasound-guided biopsy procedures. Future treatment of some carcinomas (i.e., breast) will involve ultrasound-guided ablative techniques. The in-training general surgery resident is now required by the American Board of Surgery to have experience in most aspects of ultrasound by the conclusion of their 5 year clinical general surgery residency program. The required skills and experience can be obtained by a combination of didactic training sessions (in their residency program or at regional/national ultrasound skills courses) and hands-on experience.

General surgeons will continue to push ultrasound imaging beyond just being used for diagnostic studies. Ultrasound technology will continue to evolve in the

17

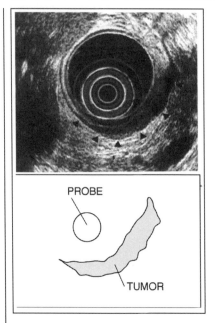

Fig. 17.33. Gastric carcinoma which infiltrates through the muscularis propria. The serosa appears intact (outlined by black triangles). Reprinted with permission from: Harness JK, Wisher DB, eds. Ultrasound in Surgical Practice: Basic Principles and Clinical Applications 2000; 1-524. © 2000 John Wiley & Sons.

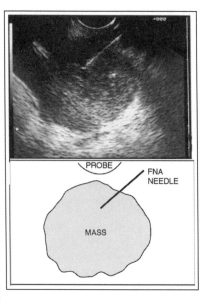

Fig. 17.34. Sector scan of a pancreatic mass with FNA biopsy needle seen. Reprinted with permission from: Harness JK, Wisher DB, eds. Ultrasound in Surgical Practice: Basic Principles and Clinical Applications 2000; 1-524. © 2000 John Wiley & Sons.

17

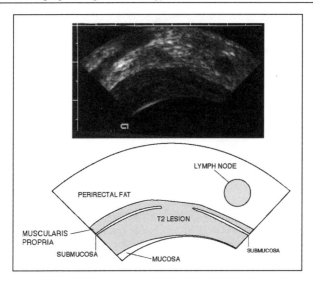

Fig. 17.35. Sector scan of a rectal carcinoma with round, hypoechoic metastatic lymph node. Reprinted with permission from: Harness JK, Wisher DB, eds. Ultrasound in Surgical Practice: Basic Principles and Clinical Applications 2000; 1-524. © 2000 John Wiley & Sons.

hands of general surgeons as a fundamental aspect of surgical practice similar to the evolution of laparoscopic technology and limited-access surgical techniques. Simply stated, no one specialty "owns" ultrasound technology. Surgeons should assume their appropriate leadership position in this aspect of medical technology and lead the advances of ultrasound capabilities and applications in the 21st century.

Selected Readings

1. Machi J, Sigel B eds. Ultrasound for Surgeons. Tokyo: Igaku-Shoin Ltd., 1997:1-368.
2. Staren ED, Arregui ME eds. Ultrasound for the Surgeon. Philadelphia: Lippincott-Raven, 1997:1-292.
3. Rozycki GS ed. Surgeon-Performed Ultrasound—Its Use in Clinical Practice. The Surgical Clinics of North America. Philadelphia: W.B. Saunders Co., 1998; 78(2):1-365.
4. McCaffrey J ed. World Progress in Surgery: Role of Ultrasound in Surgery. World J Surg 2000; 24(2):1-248.
5. Harness JK, Wisher DB eds. Ultrasound in Surgical Practice: Basic Principles and Clinical Applications. New York: John Wiley and Sons, 2000:1-524.

17

CHAPTER 18

Sentinel Lymph Node Biopsy

Frederick L. Moffat

Sentinel lymph nodes (SLNs) are defined as the first set of lymph nodes which receive lymph flow from a primary tumor, and are therefore the initial nodes encountered by tumor cells metastasizing through lymphatic channels. In principle, the histological status of the SLNs should accurately reflect regional nodal stage, thus permitting avoidance of formal lymphadenectomy in pathological node-negative (pN-) cancer patients and improving selection of patients for adjuvant therapy.[1]

Sentinel lymph node biopsy (SLNB), first described in penile carcinoma in 1976, has come into its own in the 1990s as a highly sensitive method of discriminating pathological node-positive (pN+) from pN- melanoma patients.[2] Initial trials employed vital blue dye as the lymph node labeling agent. However, extensive dissection was often required to find the blue-stained lymphatics leading to the SLNs, resulting in wound infections and skin flap necrosis in some patients. More recently, radiocolloids have gained favor as nodal labeling agents. Preoperative lymphoscintigraphy (LS) and a surgical gamma detection probe (GDP) are used alone or with blue dye to identify and resect SLNs. Radiocolloid methods are learned more quickly by novice surgeons than the blue dye technique, localize SLNs in a higher proportion of patients and have a lower complication rate. GDP-guided identification of cutaneous hot spots by the GDP permits placement of the SLNB incision directly over the radiolabeled sentinel node(s). The GDP guides dissection directly to the SLNs by the shortest route, avoiding disturbance of surrounding tissues. SLNB in clinical node-negative (cN-) melanoma using the GDP and radiocolloid with or without blue dye results in retrieval of sentinel nodes in 96-100% of cases, with an acceptable false-negative (FN) rate.[3,4]

Gamma Detection Probes

Surgical gamma detection probes (GDPs) are exquisitely sensitive, highly directional radiation detection devices.[1,2] The radiation detectors in these probes are either scintillation crystals (sodium or cesium iodide) or semiconductors (cadmium telluride). These detectors are shielded, collimated, and linked to appropriate electronics which convert the signal generated by incoming photons into variable-pitch audio feedback and a digital readout.

For SLNB, ^{99m}Tc is the radioisotope of choice. The incident photons of this gamma emitter have energies which are tightly distributed around a 140 keV energy "photopeak". The directionality of GDPs is dependent on shielding, collimation and discrimination of incident photons (in the energy photopeak) from scatter (lower energy) photons by the instruments' electronics.

Surgical Oncology, edited by David N. Krag. ©2000 Landes Bioscience.

Incident photons are those which travel from their point of emission to the GDP without colliding with matter (water or constitutive tissue molecules). Scatter photons have collided with matter before reaching the GDP, losing energy and changing direction or "scattering" in the process (the Compton effect). Scatter photons interfere with the directional accuracy of GDPs as they are no longer travelling in a straight line from their point of emission (the SLNs). The surgeon's ability to follow a clearly defined "line of sight" to radiolabeled SLNs would be compromised if GDPs were not designed to screen out scatter photons.

Surgical GDPs have an adjustable energy "threshold", usually set at or near the lower limit of the radioisotope's photopeak; photons with energies below the threshold are ignored by the GDP's electronics while those above are counted. There is also an adjustable energy "window" the lower limit of which is set on the threshold, the window usually being centered on the photopeak. Thus most scatter photons are screened out and most incident photons enumerated. In the University of Vermont prospective trials,[3,5] the energy threshold was set at 130 keV, eliminating almost all scatter photons, and the window widened to 40 keV to include as much of the upper end of the ^{99m}Tc photopeak as possible in the GDP count.

Lymphatic Migration of Radiocolloids

In LS and SLNB, radiocolloid particle size correlates inversely with percent migration from the injection site to the lymph nodes, and directly with the degree to which radioactivity is confined to the sentinel node(s) without time-dependent radiolabeling of more distal (nonsentinel) lymph nodes in a regional basin.[1,6]

Radiocolloid migration rate from the site of injection is dependent on density of lymphatic capillaries in interstitial tissues and on interstitial tissue pressure. Increased interstitial pressure induced by the volume of radiocolloid injectate opens the patent junctions between lymphatic endothelial cells, dramatically increasing the rate of radiocolloid ingress into the lymphatic lumen. Thus, in melanoma patients in whom radiocolloid is injected intradermally, relatively low activities (0.3-0.5 mCi) and small injection volumes (0.1-0.5 ml) are sufficient as the dermis has a rich lymphatic network. Moreover, as the dermis is comprised largely of densely packed bundles of collagen, acute increases in interstitial pressure are achieved with very small volumes of injectate.

When radiocolloid is injected into peritumoral breast parenchyma in breast cancer patients, radiocolloid activity should be greater (0.5-1.0 mCi) and volume much larger (at least 4 ml, and possibly 8-16 ml)[6] as breast lymphatics are more sparse and the consistency of breast parenchyma much looser than that of the dermis. The extra volume of fluid is required for optimal lymphatic ingress and migration of radiocolloid.[1,2,6]

Deposition of radiocolloid around the primary tumor site is an important consideration in SLNB, as demonstrated in Figure 18.1. If there is discontinuity in distribution of radiocolloid between two or more points of injection around a primary tumor, the radiocolloid may fail to enter one or more of the lymphatic ducts leading from that tumor, putting the patient at risk for an incomplete and/or FN SLNB. In situations in which the primary tumor or biopsy site is large, a larger volume and more sites of peritumoral radiocolloid administration may be necessary.[2] The recently opened National Surgical Adjuvant Breast and Bowel Project (NSABP) B-32

18

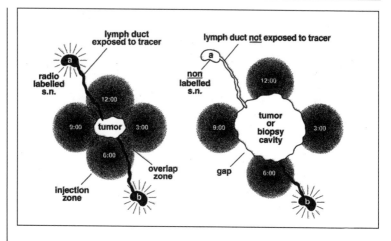

Fig. 18.1. Radiocolloid injection in four equal aliquots around a small and a large primary tumor or biopsy site. It is apparent that for the larger tumor, parts of the adjacent tissues may be missed, and therefore one or more clinically significant lymphatic trunks may not take up radiocolloid, leading to an incomplete and potentially false-negative SLNB. Reproduced with permission from Krag DN. Curr Probl Surg 1998; 35(11):951-1018.

prospective randomized trial of SLNB versus SLNB plus axillary node dissection (AND) in breast cancer patients has taken this into consideration. For primary tumors or biopsy sites of 2 cm diameter or greater, the volume of radiocolloid injectate is increased from 8 ml in four injections to 16 ml in eight injections, the radiocolloid activity remaining unchanged.

Whereas smaller radiocolloids such as antimony sulfide colloid, microsulfide colloid or noncolloidal human serum albumin are excellent for dynamic LS to determine the order in which lymph nodes in a given nodal basin receive lymph flow, rapid transit through the nodal basin and extensive labeling of nodes would render GDP-guided SLNB difficult to impossible. Larger radiocolloids such as unfiltered ^{99m}TcSC (mean particle size 200 nm; range 50-1000 nm) or colloidal human serum albumin (200-1000 nm particle size), while inferior to smaller colloids for elegant external imaging of a lymphatic basin, are much better suited for SLNB as they are avidly sequestered in SLNs for many hours, permitting GDP localization of SLNs in over 90% of cases.[2,6] As large radiocolloids do not migrate onward over time to non-SLNs in any significant quantity, SLNB poses no logistical challenges to surgical scheduling.[1,2,6]

^{99m}TcSC is the only radiocolloid approved for clinical use in the United States. Two reports[6,7] demonstrate superior SLN localization with unfiltered as compared to filtered sulfur colloid in breast cancer patients. Krag et al[6] successfully resected at least one SLN in 86% of patients injected with unfiltered ^{99m}TcSC as compared to 77% of those injected with filtered colloid. Linehan et al[7] reported SLN localization rates of 88% with unfiltered as compared to 66% with filtered ^{99m}TcSC (p < 0.01). Localization failure with filtered ^{99m}TcSC in this series was most often due to diffuse permeation of radioactivity throughout the soft tissues of the axilla, impeding probe identification of SLNs.

Preoperative Lymphoscintigraphy

Preoperative LS is indispensable in SLNB for cutaneous malignant melanoma and will undoubtedly prove essential in most other tumors. LS has conclusively shown that traditional concepts of cutaneous lymphatic drainage as elaborated by Sappey and others are often incorrect. Dynamic LS not only identifies which lymph node basin(s) a melanoma may drain to, but also demonstrates which node(s) is/are the first in line to receive lymph drainage from the tumor. Using orthogonal or parallax localization techniques, nuclear medicine physicians can usually mark the location of SLNs on the skin, axillary SLNs being more problematic in this regard than head and neck, inguinal or interval SLNs.

LS is more challenging in the breast cancer setting. Fewer lymph node basins are at risk and the overwhelming preponderance (> 80%) of sentinel lymphatic drainage is to the axilla. More often than not, with intraparenchymal injection of radiocolloid the radioactive diffusion zone at the injection site in the breast obscures much of the axilla and/or parasternal area, interfering with or precluding external imaging of SLNs in these areas even when both lateral and frontal images are acquired. Linehan et al[7] reported significantly lower localization rates of SLNs by external imaging as compared to probe-directed SLNB with both filtered ^{99m}TcSC (44% vs 66%; $p < 0.01$) and unfiltered ^{99m}TcSC (66% vs 88%; $p < 0.01$).

In addition, the disposition of the lymph nodes in the axilla is oblique with respect to the x, y and z axes. Marking the skin overlying an axillary SLN using conventional orthogonal methods is challenging, although this problem may be susceptible to clinical experience and technical refinements.[8]

However, there is little to be lost in acquiring external images preoperatively in breast cancer patients. LS can identify the presence of a nonaxillary or Berg's level III SLNs (see Figs. 18.6 and 18.7). When blue dye alone is used, preoperative LS is essential for identification of nonaxillary sentinel lymphatic drainage.

SLNB in Cutaneous Malignant Melanoma

The merits and drawbacks of immediate elective lymph node dissection (ELND) in cN-melanoma have been debated for years. While retrospective data suggest that ELND improves survival, four prospective randomized trials have failed to confirm this.

SLNB has rendered this issue of historical interest only, as the presence or absence of nonpalpable nodal metastases can now be determined with great accuracy and minimal morbidity. Patients with negative SLNs are spared the expense and complications of radical surgery while radical lymphedenectomy is clearly indicated in patients identified as pN+. The advent of efficacious adjuvant immunotherapy for pN+ melanoma provides an additional, compelling indication for SLNB.

Traditional concepts of cutaneous lymphatic drainage in the head and neck and truncal regions have been refined by LS in melanoma patients. A higher than expected incidence of bilateral drainage, "skip" drainage to a more distal node in a group than might be anticipated from the location of the primary melanoma, drainage to multiple lymph node groups, and other traditionally unorthodox patterns of lymphatic drainage to "interval" nodes have all been documented.

18

Norman et al[9] reported that ELND based on traditional anatomical guidelines rather than LS would not have removed all the lymphatics at risk for occult metastasis in 48 of 82 patients (59%) with cN-melanomas of the trunk and head and neck. Only one patient (1.2%) developed a metastasis in a lymphatic basin which was not predicted preoperatively by LS. Berman et al[10] documented discordance between LS and clinical prediction of regional lymphatic metastasis in 24 of 36 patients (67%) with head and neck melanomas. The primary lesion drained to a single cervical lymphatic zone or basin in only eight patients (22%); in nine patients (25%) the drainage was to three or more zones or basins. In a LS study of 97 patients with head and neck melanomas, O'Brien et al[11] found clinical-LS discordance in 34% of patients, multiple SLNs in 85%, and in 21 patients SLNs were found outside the parotid and the five nodal zones of the neck. Wells et al[12] reported discordance between LS and clinical prediction of lymphatic drainage in 21 of 25 patients with head and neck melanoma. Thus conventional surgical concepts of lymphatic patterns of melanoma metastasis are frequently inaccurate.

Sentinel drainage to unexpected or interval lymph node basins occurs in 20-25% of melanoma patients. Extremity melanomas may drain directly to epitrochlear, popliteal or supraclavicular nodes and truncal lesions to para-aortic, internal mammary, mediastinal, rectus sheath and intermuscular nodes (Fig. 18.2). Head and neck melanomas not infrequently drain to occipital, postauricular, facial artery and even axillary nodes.

Clearly, lymphatic mapping has enabled general surgeons, head and neck surgeons and surgical oncologists to revise their thinking about lymphatic drainage patterns in the management of melanoma and other cancers in which SLNB is being employed.

There are only three series[13-15] in which ELND was immediately performed after SLNB in melanoma patients.[2] These series included 383 patients in all, among whom 68 had at least one pN+ lymph node basin. SLNB was falsely negative in 4 of these (6%). Interestingly, Gershenwald et al[16] reported interval development of metastatic melanoma within nodal basins identified by SLNB as free of tumor in 4.1% of 243 patients at a median follow-up interval of 35 months, a figure not dissimilar to the FN rate for these three seminal series.

SLNB for melanoma has been performed using vital blue dye alone (with preoperative LS), radiocolloid and GDP alone (with LS) or both methods in combination. Overall, localization rates when radiocolloid and a GDP are used with or without blue dye (96-100%) are superior to those achieved with blue dye alone (70-90%).[2]

Completeness of resection of SLNs is essential to success, as highlighted by O'Brien et al[11] in head and neck melanoma. The authors performed SLNB using vital blue dye in 20 of 97 patients studied by LS, four of whom had positive SLNs. At least one SLN seen on LS was missed in 5 of the 20 patients and of the 16 patients with negative SLNs, 4 developed neck recurrences within 12 months. Three of these were among the 5 in whom not all SLNs were retrieved, and all relapsed in the same region of the neck in which the missed SLN was situated.

SLNB in Breast Cancer

Regional nodal status is the most significant prognostic variable in early breast cancer, and is often pivotal in selecting adjuvant systemic therapy. The current

18

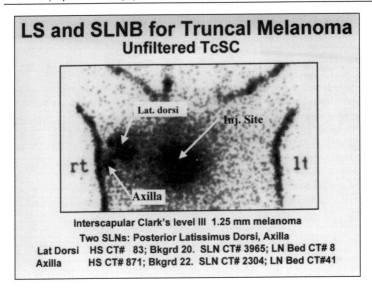

LS and SLNB for Truncal Melanoma
Unfiltered TcSC

Interscapular Clark's level III 1.25 mm melanoma
Two SLNs: Posterior Latissimus Dorsi, Axilla

Lat Dorsi	HS CT# 83; Bkgrd 20.	SLN CT# 3965; LN Bed CT# 8
Axilla	HS CT# 871; Bkgrd 22.	SLN CT# 2304; LN Bed CT#41

Fig. 18.2. LS of a patient with an intermediate thickness truncal melanoma who underwent SLNB one hour after this image was acquired. SLNs were located on the latissimus dorsi fascia and in the axilla. At SLNB, cutaneous hot spots were identified over the SLNs with 10-second GDP counts (HS CT#) and adjacent skin counts (Bkgrd) as shown. Note that SLN ex vivo counts (SLN CT#-taken with the SLN in direct contact with the tip of the GDP) were much higher than the hot spot counts; this is a function of the inverse square law, which states that the number of scintillations registered by a radiation detection device is inversely proportional to the square of the distance between the detector and the radiation source.

standard of care for nodal staging, axillary node dissection (AND), is performed by surgeons all over the world with reproducible results in prospective studies. AND eliminates the staging error inherent in selective use of AND. Surgical staging error is minimized or eliminated by resecting Berg's level I and II nodes or all three levels, respectively. Optimal control of axillary disease is achieved by AND, and while prospective randomized trials have not shown a survival benefit with this operation, some studies have suggested a small advantage.[17]

However, at present 70-80% of breast cancer patients in North America and Europe are pN- at the time of diagnosis. Routine AND cannot benefit these patients, yet incurs enormous expense because of the high incidence of this disease. Morbidity is significant. There is an inherent pathological understaging error associated with AND as only two sections are examined from each lymph node; metastatic disease is missed in 10-25% of patients. The cost and labor involved in serial sectioning and immunohistochemistry on large numbers of unselected nodes are prohibitive. Finally, AND gives no information on the status of nonaxillary regional nodal status. As the prognostic impact of metastases in internal mammary and other first-order nodes is essentially the same as that of axillary nodes, this represents an

18

additional source of understaging associated with routine AND.[17] In principle, there is much to recommend a paradigm shift to SLNB as the standard method of nodal staging in breast cancer.

SLNB Using Radiocolloid and a GDP without Blue Dye

SLN localization with radiocolloid/GDP alone is achievable in 84-99% of cases. In the University of Vermont technique,[5,18] the perimeter of the radioactive diffusion zone around the injection site is defined with the GDP and marked. Retraction of the breast during SLNB moves injection site radioactivity away from the regional nodal basins. The surgeon keeps the tip of the GDP angled away from the radiation "shine-through" from the injection site when mapping hot spots and performing SLNB.

The probe is moved radially over the breast and then in a grid over the upper rectus abdominis, parasternal, axillary and supraclavicular regions. Cutaneous "hot spots", defined as discrete areas of increased radioactivity (at least 25 counts per 10 seconds) on the skin, are identified and counted, and background counts of adjacent skin are taken. If no hot spots are found, a bolus of saline is injected peritumorally to increase interstitial tissue pressure and radiocolloid migration.

The hot spot is often close to or merges into the periphery of the injection site diffusion zone (Fig. 18.3). Hot spot identification in this circumstance is dependent on the surgeon's acquired skill in hearing a "valley" in the audio signal as the probe moves radially through the periphery of the diffusion zone (Fig. 18.3B). The distance of SLNs from the skin can be inferred from the size of the hot spot (Fig. 18.4).

An incision is made through the hotspot and dissection follows the "line of sight" established by the GDP to the SLN (Fig. 18.5), which is removed and counted ex vivo. The SLN bed is then resurveyed; if significant activity remains (> 10% of the ex vivo count of the hottest SLN), additional radiolabeled nodes are sought, removed and counted; all nodes with ex vivo counts of 10% or more of the hottest SLN are considered SLNs (see Figs. 18.6-18.8).

In patients with parasternal hot spots, an incision is made through the hot spot and the probe directs dissection through pectoralis major to the intercostals, which are taken off the adjacent inferior costochondral junction. The SLN is in the interspace or behind the sternum at its junction with the costal cartilage, and is retrieved with very low risk of pneumothorax.

SLNB is considerably more difficult in breast cancer than melanoma because of the proximity of at-risk lymphatic basins to injection site radioactivity. Even the most seasoned and adept surgeon is occasionally defeated by this problem. However, with practice, the surgeon becomes facile with the GDP, using its shielding and inherent directionality to best advantage. In the University of Vermont Multicenter Validation Trial,[5] 443 cN-breast cancer patients accrued by 11 surgeons underwent SLNB using 1 mCi unfiltered ^{99m}TcSC in 4 ml normal saline, followed immediately by axillary dissection. At least one cutaneous hot spot was localized in 413 patients (93.2%), and 114 patients proved to be pN+. SLNs were outside the axilla in 8% of cases and in 3% of all patients with positive SLNs, the involved SLNs were exclusively nonaxillary in location.

Among the 114 pN+ patients, all SLNs were negative in 13 for a false-negative (FN) rate of 11.4%. All 13 patients had laterally situated tumors (p = 0.004). The

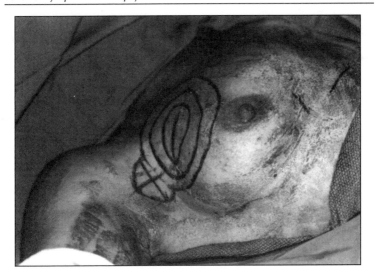

Fig. 18.3A. Injection site diffusion zone around an excisionally biopsied cancer in the upper outer quadrant, mapped at the 1X, 10X and 100X attenuations with the C-TRAK GDP (CareWise Medical, Morgan Hill, CA) (Fig. 18.3A). There is significant overlap with the axilla, and the hot spot merges with the periphery of the diffusion zone. The SLNs were deeply situated in the axilla (see Fig. 18.4).

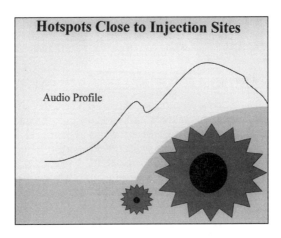

Fig. 18.3B. The cartoon in Figure 18.3B shows a cross-section of a breast with a radioactive injection site diffusion zone and a low axillary SLN (the gray bar to left represents patient's arm abducted 90° on an armboard). The audio profile represents the pitch of the GDP audio signal as the probe is moved over the diffusion zone and onto the axillary skin. As the periphery of the diffusion zone is approached, the pitch decreases only to increase again as the radioactivity from the nearby SLN begins to be picked up.

18

Hot Spot Size and Target Depth

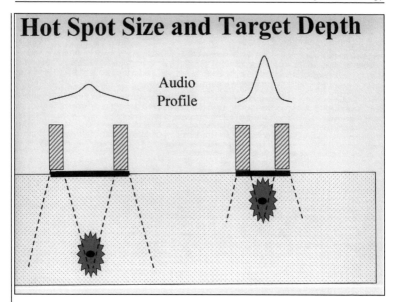

Fig. 18.4. Cartoon of hot spots from two equally radioactive SLNs, one superficial and the other deep within a lymph node basin. Hatched bars represent the GDP as it passes over the periphery of the hot spot, dotted lines represent the 20° field of view of the GDP, and the heavy horizontal lines represent the size of the hot spot on the skin. The GDP begins to detect the more deeply situated SLN further away from the center of the hot spot, but the signal is weaker because of the inverse square law. The size of the hot spot correlates with the depth of the SLN.

FN rates of individual surgeons ranged from 0-28.6%. The FN rate was most likely due to interference from injection site radioactivity.

These FNs occurred because proximity of the axilla to breast radioactivity interferes with probe localization of SLNs. As the diffusion zone cannot be eliminated, concomitant use of vital blue dye may reduce the FN rate. The findings of the Vermont study have resulted in the NSABP B-32 protocol requiring the use of isosulfan blue.

Among single-institution studies reported to date, SLN localization rates have ranged from 84-98% and FN rates from 0-14.3%.[17]

SLNB Using Blue Dye Without Radiocolloid

Five minutes after injection of 5 ml isosulfan blue dye into and around the primary cancer or biopsy cavity, an incision is made inferior to the hair-bearing axillary skin. Blue-stained lymphatic vessels and nodes are sought by blunt dissection. The dye-filled lymphatic is dissected to the first blue node(s), and is also dissected proximally to the tail of the breast to verify that the identified blue node(s) is/are the true (most proximal) SLNs.[20]

This technique has a longer learning curve than radiocolloid/GDP methods and SLN localization rates are significantly lower (64-85%). The amount of dissection is

18

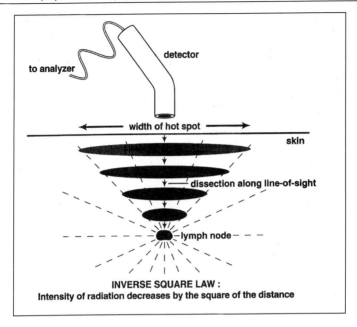

Fig. 18.5. Establishing the "line of sight" during SLNB with the GDP. As the SLN is approached during the dissection, the radioactivity becomes more intense and narrowly focused, consistent with the inverse square law and the phenomenon outlined in Figure 18.3. Reproduced with permission from Krag DN. Curr Probl Surg 1998; 35(11):951-1018.

more extensive, and localization of nonaxillary SLNs is not possible unless LS is performed preoperatively. FN rates range from 0-14.3%.[17]

SLNB Using Radiocolloid and Blue Dye in Combination

Albertini et al[21] studied 62 patients using isosulfan blue dye and filtered (200 nm filter) ^{99m}TcSC with a GDP. The radiocolloid was injected 2-4 hours preoperatively. Ten to 15 minutes prior to SLNB, 4-5 ml isosulfan blue dye was injected around the primary tumor. The GDP was used to identify cutaneous hot spots, and incisions were made through these.

SLNB was guided primarily by blue dye visualization. The GDP was used to confirm SLN identification but not to guide dissection unless blue lymphatics could not be seen with confidence. SLNs were defined as all blue-stained nodes and/or nodes with SLN/neighboring non-SLN activity ratios greater than 10. If, after removal of the SLN(s), the radioactivity of the SLN bed remained above 150% of background (established by GDP counts of axillary non-SLNs equidistant from the primary tumor), further SLNs were sought.

In an updated experience in 174 patients,[22] 136 of whom underwent immediate completion AND, SLN mapping was successful in 160 patients (92%) and 15 (9.4%) had internal mammary SLNs. Blue lymphatics and nodes were seen in only 50% of

18

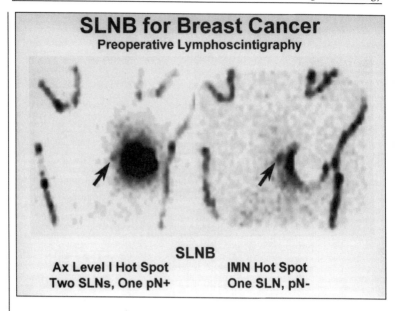

SLNB for Breast Cancer
Preoperative Lymphoscintigraphy

SLNB

Ax Level I Hot Spot   IMN Hot Spot
Two SLNs, One pN+   One SLN, pN-

Fig. 18.6. LS of a patient with a 1.3 cm breast cancer demonstrating an internal mammary hot spot, better seen on the right image in which a shield has been used to eliminate much of the injection site radioactivity. The axillary hot spot failed to image because the diffusion zone encompassed the entire superior half of the breast, and overlapped the axilla. The internal mammary hot spot was subtle (count 62 with adjacent skin count of 44), but the SLN had an ex vivo count of 349; the bed of the SLN was 3 (0.9% of the SLN ex vivo count), confirming removal of all radio-labeled nodes from this site. The axillary hot spot had a 10-second count of 198 and the adjacent skin count was 20. Two SLNs were removed, and the SLN bed count was 0.8% of the hotter axillary SLN.

cases; radiocolloid was essential to localization of the remaining 42%. Only one of the 36 pN+ patients had a FN SLNB (2.6%). The complementarity of the blue dye and radiocolloid methods was highlighted in 700 patients.[23] Of 1,348 SLNs removed, 374 were blue and hot (radioactive), 568 were hot only, and 406 were blue only.

Other series in which blue dye and radiocolloids were used in combination demonstrate a high rate of SLN localization (81-100%) and variable FN rates (0-15%).[17]

Intraoperative palpation of the axillary nodes during SLNB can reveal hard, tumor-replaced nodes which do not label with either radiocolloid or blue dye. Because these nodes are choked with tumor, lymphatic flow is diverted away from or around them. Thus, a FN SLNB is inevitable unless the surgeon is aware that the cN-breast cancer patient can harbor metastases which are palpable at surgery.[24]

Intradermal, Subdermal or Subareolar Injection of Radiocolloid in Breast Cancer

In most studies of SLNB in breast cancer, the radiocolloid has been injected into the peritumoral breast parenchyma. The resulting diffusion zone is often considerably

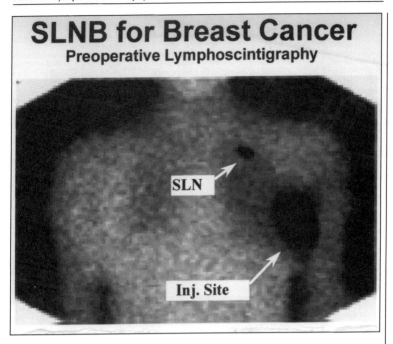

Fig. 18.7. A patient with an apical axillary (infraclavicular) hot spot with a count of 3204 (adjacent skin count 83). There were two SLNs at Halsted's ligament. An axillary hot spot (count 356) was also identified at SLNB, but found to be shine-through from the two apical SLNs (counts 6922 and 1795; SLN bed count 35).

larger than that produced by intradermal injection of radiocolloid around a primary melanoma. Breast radioactivity can be a very significant impediment to the surgeon in identifying hot spots and finding SLNs. However, the size of the diffusion zone created by a radiocolloid volume of 8-16 ml is not much larger than that created by 4 ml.

Borgstein et al[25] hypothesized that because both the breast parenchyma and skin are of ectodermal origin, they share a common lymphatic pathway to the same sentinel node. They injected 40 MBq (1.1 mCi) ^{99m}Tc-colloidal albumin peritumorally and blue dye intradermally over the primary tumor in 30 consecutive patients and showed that in each case the blue dye and radiocolloid had labeled the same lymph node(s), with no false-negatives.

Recent pilot studies suggest superior SLN localization rates with intradermal or subareolar injection techniques. Intradermal, subdermal or subareolar radiocolloid injection avoids inadvertent injection of radiocolloid into the biopsy cavity or into the pectoralis fascia or muscle. Nuclear medicine physicians are more comfortable with intradermal than intraparenchymal injection from their experience with LS and SLNB for melanoma. Finally, much of the radioactive diffusion zone around

18

SLNB for Breast Cancer
Preoperative Lymphoscintigraphy

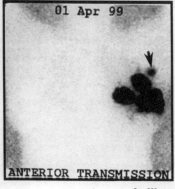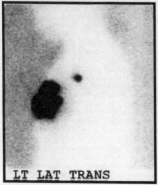

01 Apr 99

ANTERIOR TRANSMISSION

LT LAT TRANS

Axillary Hot Spot
Three level I SLNs, one pN+

Fig. 18.8. Two SLN specimens were found under an axillary hot spot (count 516 with adjacent skin count of 3). The ex vivo counts of the SLNs were 4723 and 1228; the second specimen was considered sentinel as its radioactivity was 26% that of the first, exceeding the 10% minimum required. The SLN bed count, at 13, was 0.27% that of the ex vivo count of the hotter SLN. SLNB was therefore deemed complete. The hotter SLN specimen contained two nodes, one of which was positive.

the site of intraparenchymal injection is eliminated, improving SLN localization and potentially reducing FN rates.

However, the intradermal/subdermal route of administration may underestimate the incidence of nonaxillary SLNs, as suggested by the rarity with which sentinel drainage to internal mammary and mediastinal nodes is seen from melanomas of the anterior chest wall. It has not as yet been established whether the intra/subdermal route of administration is superior or even equivalent to peritumoral injection for SLNB in breast cancer.

SLNB in Other Tumors

SLNB is now being applied in the management of other solid tumors, notably in squamous and Merkel cell carcinomas of the skin, thyroid cancer, mucosal squamous cell carcinoma of the head and neck, vulvar and endometrial carcinomas, and colorectal adenocarcinoma. Blue dye and radiocolloid/GDP techniques have been used alone and in combination in these studies. The pilot data published to date for each of these cancers are very encouraging, and suggest that SLNB will find a place in the management of many types of cancer in the future.[2,26]

18

Pathological Examination of SLNs

There is an incremental increase in yield of pN+ breast cancer by the use of serial section histopathology and immunohistochemistry of about 10-15% and 15-25%, respectively, in melanoma[2] and breast cancer.[2,17]

Reverse transcriptase-polymerase chain reaction (RT-PCR) analysis is receiving increasing attention as an ultrasensitive technique for the detection of occult metastases.[27-29] In theory, RT-PCR for detection of nodal metastases should be more sensitive than IHC by at least an order of magnitude. For melanoma, RT-PCR for tyrosinase, MART-1 and/or MAGE-3 greatly increases the yield of positive nodes as compared to routine histology or immunohistochemistry.[27,28,30-32] Moreover, RT-PCR positive histology-negative nodes are prognostically significant in this disease.[31,32]

Unlike melanoma, for which markers specific and reliable for detection of micrometastases have been identified, there is as yet no clearly defined RT-PCR marker or set of markers for breast cancer. RT-PCR is expensive, available only in a few academic medical centers, and prone to give false-positive and false-negative results. Further research and development is necessary for RT-PCR to become broadly applicable in clinical oncology.[17]

Intraoperative Frozen Sections and Touch Preparations of SLNs

In principle, intraoperative pathological evaluation of SLNs should permit immediate clarification of the nodal status of breast cancer patients. However, frozen section examination of SLNs in breast cancer patients is falsely negative in 27-32% of cases;[8,33] even when 60 frozen sections from each SLN are examined, the FN rate is still 5.5%.[23] In contrast, intraoperative SLN touch preparations yield a FN rate of only 0.8%.[34] This technique has been adopted for use in the NSABP B-32 prospective randomized trial of SLNB in breast cancer.

Radiation Safety Issues

[99mTc] is a cheap, ubiquitous, low to moderate energy gamma-emitting radioisotope with a half-life of 6 hours. Unlike β-particle emitters like radioactive colloidal gold it does not irradiate tissues at the site of injection to an unacceptable degree. Its radiation characteristics are safe for both patients and health care professionals and it has excellent imaging characteristics in a variety of nuclear medicine applications.

In the United States there are no limitations on bodily contact with individuals having a body burden of radioisotope of less than 1110 MBq (30 mCi).[1] The activity of [99mTc] administered in SLNB, 0.5 to 1.0 mCi, is a small fraction of that used in whole body skeletal scintigraphy (25 mCi) and is at most 3% of the threshold for imposing radiation restrictions. The patient's radioactivity is essentially nil by 60 hours postadministration, or 10 [99mTc] half-lives. The patient's personal radiation exposure is minimal, approximating that of a passenger on a nonstop transcontinental commercial flight at an altitude of 35,000 feet.

Even in worst-case scenarios the radiation exposure of hospital personnel does not exceed twice that received from natural background radiation.[2] The surgeon's hands receive an average of 10.2±5.8 mrem per SLNB in breast cancer patients.[35] The maximum safe annual dosage to extremity skin, as determined by the Nuclear Regulatory Commission, would be exceeded only if a surgeon performed more than 5,000 SLNBs per year.

18

The radiation dosage to anesthesia and nursing personnel during SLNB, standing on average one meter away from the surgeon, is just 5% of the surgeon's as predicted by the inverse square law.[2]

Veronesi et al[8] showed that radiation dose to the hands for surgeons and scrub nurses performing 30-50 breast cancer SLNBs per year is just 0.9% of the annual safe dose limits set by the International Commission on Radiological Protection. The dose to the surgeon's ocular lens is just 0.7% and total body dose 9% of the maximum allowable. The exposure of hands, lenses and total body for Pathology personnel handling radioactive tissue specimens is even smaller at 0.15%, 0.1% and 1.5% of the annual dose limits, respectively.

However, radiation safety is a zero tolerance issue from regulatory, political and labor relations standpoints. By law, radioisotopes can only be administered to patients in an approved area of the hospital (the Nuclear Medicine Department) where radioactive spills can be contained and monitored if they occur. All hospital personnel who come in contact with patients or tissue treated with radioisotopes must be formally instructed on radiation safety issues and made aware of which patients and tissues are radioactive. Female employees of child-bearing years must be informed of the use of radioactive materials so they can choose not to participate in SLNB if they have reason to believe they may be pregnant. That the radiation exposure involved is minuscule is no excuse for not adhering to these principles.

Prospective Multi-Center Randomized Trials of SLNB

The U.S. National Cancer Institute-sponsored Multicenter Selective Lymphadenectomy Trial is open for accrual of patients with cutaneous melanomas of greater than 1 mm thickness. All patients undergo preoperative LS and are randomized to either wide local excision plus observation or wide local excision plus SLNB. Patients in the latter arm in whom SLNB reveals pN-SLNs will be followed while those with positive SLNs will undergo therapeutic lymphadenectomy. This trial will determine whether there is a survival benefit with early lymphadenectomy, whether SLNB by itself has any therapeutic benefit, and what the FN rate actually is for SLNB in melanoma as reflected by the incidence of recurrence in the regional lymphatics after a negative SLNB.

The multi-institutional Sunbelt Melanoma Trial sponsored by Schering Corp. addresses the clinical significance of nodal metastases as established by conventional haematoxylin and eosin histology (H&E), immunohistochemistry (IHC) and RT-PCR using four markers (MAGE-3, MART-1, GP-100 and tyrosinase). The study will also address the question of benefit of adjuvant interferon-α_{2b} (IFN-α_{2b}) immunotherapy in patients with a low burden of nodal disease. Patients in whom SLNs are negative by all three methods are observed while those who are positive by RT-PCR only will be randomized to observation, therapeutic lymphadenectomy or lymphadenectomy plus IFN-α_{2b}. Those with one node positive by H&E or IHC will undergo therapeutic lymphadenectomy and then be randomized to observation or IFN-α_{2b} therapy.

There are two prospective randomized trials of SLNB in breast cancer in North America. In the NSABP B-32 trial, patients with unifocal invasive cN-breast cancer are randomly assigned to SLNB only or SLNB plus AND. The SLNs from those randomized to SLNB only will be evaluated intraoperatively by imprint cytology;

18

AND is performed if the SLNs are positive by imprint cytology or on subsequent paraffin section H&E histology. Patients will be followed for morbidity (serial volumetric assessments for arm lymphedema), goniometry for shoulder range of motion, and locoregional, distant disease-free and overall survival. Approximately 4,000 patients will be randomized.

The newly constituted American College of Surgeons Oncology Group (ACoSOG) has chosen a slightly different experimental design. All registered patients will undergo SLNB. Those with H&E-negative SLNs will be followed but their SLNs will undergo further evaluation by IHC; these patients will have no further surgery irrespective of the IHC status of their SLNs. Those with H&E-positive SLNs will be randomized to observation or AND.

Other prospective trials of SLNB in breast cancer are planned or underway in other parts of the world, notably in Great Britain (the Medical Research Council ALMANAC trial) and Italy.

Discussion

In the hands of well-trained, experienced surgeons this procedure is preferable to either wide local excision or wide excision plus ELND in cN-malignant melanoma. SLNB is also increasingly being applied by surgeons in the management of nonmelanoma skin cancers with metastatic potential. Gynecological oncologists are investigating this technique in patients with vulvar carcinoma as an alternative to routine inguinal dissection, a highly morbid and expensive procedure. So too, SLNB is being studied in the management of the N0 neck in patients with mucosal squamous cell cancers of the head and neck.

Radiocolloid/GDP SLNB methods localize SLNs in a larger proportion of patients than does the blue dye technique used without radiocolloid. GDP methods with or without blue dye are easier for surgeons to master and limit the amount of dissection required to find and excise SLNs.

The reliability of SLNB in patients who have had prior surgery or infections of the regional lymphatics and in those with cN+ tumors is questionable. Surgical disruption of lymphatic flow from a tumor to a lymph node basin or choking of lymphatics and effacement of regional nodes by gross tumor may adversely affect both localization and FN rates of SLNB because of altered lymph flow patterns. As noted above, in the small minority of cN-breast cancer patients in whom the surgeon appreciates palpable nodal disease only at the operating table, the involved nodes are neither hot nor blue while labeled nodes frequently prove to be pN-. For this reason, the NSABP B-32 protocol requires palpation of the axilla through the SLNB incision for palpable suspicious nodes which, if found, are removed and submitted as part of the SLNB procedure.

Of the tumors in which this procedure has thus far been studied, SLNB is likely to have its greatest impact on the management of breast cancer simply because of the high prevalence of this disease. The expense of routine AND, its morbidity, inherent pathological staging error and inability to evaluate nonaxillary regional nodes all contribute to a perception among some members of the public, medical profession and news media that a less invasive method of nodal staging should replace routine AND. In response to this a small number of surgeons are already performing SLNB in lieu of AND for nodal staging in breast cancer. A very few surgeons

18

and institutions are actively promoting their expertise in SLNB and representing this procedure as the new preferred standard.

Published data make clear that FN rates of SLNB are significantly more variable and higher in breast cancer than in melanoma. In the only multicenter prospective trial of SLNB in breast cancer thus far in print, the FN rate varied from 0-28.6% among the 11 participating surgeons, and was 11.4% overall.[5] While the addition of blue dye to radiocolloid/GDP has been mandated in NSABP B-32, the successor trial to the University of Vermont study, it is not at all clear that this will solve the problem of FN SLNBs in breast cancer. Single institution studies in which both methods were used report FN rates of up to 14.3%. Given this and the fact that only a tiny proportion of breast cancer surgeons are currently trained and experienced in SLNB, it is premature to conclude that this procedure should now supercede AND as the standard nodal staging operation in breast cancer.

In breast cancer, SLNB should be performed in addition to rather than instead of AND outside of a clinical trial. AND ensures that the patient will not suffer the consequences of a FN SLNB, which may amount to undertreatment with adjuvant systemic therapy and therefore being denied the overall and disease-free survival benefit which optimal systemic treatment might offer. SLNB allows the surgeon to identify for the pathologist the one, two or few SLNs which by definition are the nodes most at risk for harboring metastases. If routine H&E histology of all nodes is negative, the pathologist can then more closely scrutinize the SLNs by serial section histology and IHC. Thus the pN-status of the patient can be established with a greater degree of confidence or more appropriate adjuvant systemic therapy may be offered if occult pN+ disease is identified. Finally, with SLNB there is at least a 5-10% probability of identifying a nonaxillary SLN. The current standard of care disregards the prognostic importance of first-order regional nodes outside the axilla.

Technical expertise in SLNB is not easily acquired,[26] especially in the breast cancer setting. For all methods currently in use, it is essential that there be an excellent working relationship between the surgeon, operating room personnel, nuclear medicine physician and pathologist. Didactic instruction and proctoring of novice surgeons by more experienced colleagues are indispensable first steps in learning SLNB. It is also important that immediate completion lymphadenectomy be performed with SLNB during the learning phase[35] so that the surgeon can track his or her personal FN rate, although statistical interpretation of this important parameter is problematic.[17]

The number of SLNBs a surgeon must perform to become competent is controversial. Some authors contend that 25 to 30 operations are required to attain a SLN localization rate of over 90%. The 11 surgeons who participated in the University of Vermont Validation Trial in breast cancer completed 5 training cases before beginning to accrue patients to the study. The overall localization rate was 93%, and localization rates of individual surgeons correlated significantly with the number of cases each accrued to the trial.[5]

The NSABP B-32 and ACoSOG trials of SLNB in breast cancer have formalized training programs for participating surgeons which include didactic instruction and on-site proctoring by experienced surgeons. These surgical mentors then monitor the progress of their understudies through a series of training cases. Thus these trials provide an excellent educational opportunity for community surgeons to become technically proficient under supervision while at the same time contributing to important clinical research.

Conclusion

SLNB is one of the most exciting developments in surgical oncology of the last two decades, and is changing the management of a variety of solid tumors. That the sentinel node hypothesis is valid continues to be borne out in the peer-reviewed literature, now surpassing 250 publications. This technique promises to reduce the expense and morbidity of lymph node staging surgery while reducing both the inherent surgical and pathological staging errors of conventional lymphadenectomy.

To what degree if any that SLNB will replace formal lymphadenectomy in pN+ cancer patients is much less clear. There is well documented therapeutic benefit to radical node dissection from the standpoint of regional disease control in pN+ solid tumors of many types. Overenthusiastic use of SLNB for both staging and treatment of pN+ cancer may increase the risk of regional recurrence, which is a strong predictor of cancer mortality in most solid tumors.

Suggested Readings

1. Gulec SA, Moffat FL, Carroll RG. The expanding clinical role for intraoperative gamma probes. In: Freeman LM (ed.). Nuclear Medicine Annual 1997. Philadelphia: Lippincott-Raven 1997; 209-237.
2. Krag DN. Minimal access surgery for staging regional lymph nodes: the sentinel node concept. Curr Probl Surg 1998; 35:951-1018.
3. Krag DN, Meijer SJ, Weaver DL et al. Minimal-access surgery for staging of malignant melanoma. Arch Surg 1995; 130:654-658.
4. Glass LF, Messina JL, Cruse W et al. The use of intraoperative radio-lymphoscintigraphy for sentinel node biopsy in patients with malignant melanoma. Dermatol Surg 1996; 22:715-720.
5. Krag DN, Weaver D, Ashikaga T et al. The sentinel node in breast cancer. A multicenter validation study. New Engl J Med 1998; 339:941-946.
6. Krag DN, Ashikaga T, Harlow SP et al. Development of sentinel node targeting technique in breast cancer patients. Breast J 1998; 4:67-74.
7. Linehan DC, Hill ADK, Tran KN et al. Sentinel lymph node localization in breast cancer: Unfiltered radioisotope is superior to filtered. J Am Coll Surg 1999; 188:377-381.
8. Veronesi U, Paganelli G, Viale G et al. Sentinel lymph node biopsy and axillary dissection in breast cancer: results in a large series. J Natl Cancer Inst 1999; 91:368-373.
9. Norman J, Cruse CW, Espinosa C et al. Redefinition of cutaneous lymphatic drainage with the use of lymphoscintigraphy for malignant melanoma. Amer J Surg 1991; 162:432-437.
10. Berman CG, Norman J, Cruse CW et al. Lymphoscintigraphy in malignant melanoma. Ann Plast Surg 1992; 28:29-32.
11. O'Brien CJ, Uren RF, Thompson JF. Prediction of potential metastatic sites in cutaneous head and neck melanoma using lymphoscintigraphy. Amer J Surg 1995; 170:461-466.
12. Wells KE, Cruse CW, Daniels S et al. The use of lymphoscintigraphy in melanoma of the head and neck. Plast Reconstr Surg 1994; 93:757-761.
13. Morton DL, Wen D-R, Wong JH et al. Technical details of intraoperative lymphatic mapping for early stage melanoma. Arch Surg 1992, 27:392-399.

18

14. Thompson JF, McCarthy WH, Bosch CMJ. Sentinel node status as an indicator of the presence of metastatic melanoma in regional lymph nodes. Melanoma Res 1995; 5:255-260.

15. Reintgen DS, Cruse CW, Wells KE et al. The orderly progression of melanoma nodal metastases. Ann Surg 1994; 220:759-767.

16. Gershenwald JE, Colome MI, Lee JE et al. Patterns of recurrence following a negative sentinel lymph node biopsy in 243 patients with stage I or II melanoma. J Clin Oncol 1998; 16:2253-2260.

17. Moffat FL, Krag DN. Sentinel node biopsy for breast cancer: showtime or dress rehearsal? In Vivo 2000; 14(1):255-64.

18. Krag DN, Harlow SP, Weaver DL et al. Technique of selected resection of radiolabeled lymph nodes in breast cancer patients. Semin Breast Dis 1998; 1:111-116.

19. McMasters KM, Giuliano AE, Ross MI et al. Sentinel lymph node biopsy for breast cancer–not yet the standard of care. New Engl J Med 1998; 339:990-995.

20. Giuliano AE, Kirgan DM, Guenther JM, Morton DL. Lymphatic mapping and sentinel lymphadenectomy for breast cancer. Ann Surg 1994; 220:391-401.

21. Albertini JJ, Lyman GH, Cox C et al. Lymphatic mapping and sentinel node biopsy in the patients with breast cancer. J Amer Med Assoc 1996; 276:1818-1822.

22. Reintgen DS, Joseph E, Lyman GH et al. The role of selective lymphadenectomy in breast cancer. Cancer Control (MCC) 1997; 4:211-219.

23. Cox CE, Haddad F, Bass S et al. Lymphatic mapping in the treatment of breast cancer. Oncology 1998; 12:1283-1298.

24. Hill ADK, Tran KN, Akhurst T et al. Lessons learned from 500 cases of lymphatic mapping for breast cancer. Ann Surg 1999; 229:528-535.

25. Borgstein PJ, Meijer S, Pijpers R. Intradermal blue dye to identify sentinel lymph node in breast cancer. Lancet 1997; 349:1668-1669.

26. Morton DL, Ollila DW. Critical review of the sentinel node hypothesis. Surgery 1999; 126:815-819.

27. Smith B, Selby P, Southgate J et al. Detection of melanoma cells in peripheral blood by means of reverse transcriptase and polymerase chain reaction. Lancet 1991; 338:1227-1229.

28. Wang X, Heller R, VanVoorhis N et al. Detection of submicroscopic lymph node metastases with polymerase chain reaction in patients with malignant melanoma. Ann Surg 1994; 220:768-774.

29. Noguchi S, Aihara T, Nakamori S et al. The detection of breast carcinoma micrometastases in axillary lymph nodes by means of reverse transcriptase-polymerase chain reaction. Cancer 1994; 74:1595-1600.

30. Van der Velde-Zimmerman D, Roijers JFM, Bouwens-Rombouts A et al. Molecular test for the detection of tumor cells in blood and sentinel nodes of melanoma patients. Amer J Pathol 1996; 149:759-764.

31. Reintgen DS, Balch CM, Kirkwood J et al. Recent advances in the care of the patient with melanoma. Ann Surg 1997; 225:1-14.

32. Bostick PJ, Morton DL, Turner RR et al. Prognostic significance of occult metastases detected by sentinel lymphadenectomy and reverse transcriptase-polymerase chain reaction in early stage melanoma patients. J Clin Oncol 1999; 17:3238-3244.

33. Dixon JM, Mamman U, Thomas J et al. Accuracy of intraoperative frozen-section analysis of axillary nodes. Br J Surg 1999; 86:392-395.

34. Rubio IT, Korourian S, Cowan C et al. Use of touch preps for intraoperative diagnosis of sentinel lymph node metastases in breast cancer. Ann Surg Oncol 1998; 5:689-694.

35. Miner TJ, Shriver CD, Flicek PR et al. Guidelines for the safe use of radioactive materials during localization and resection of the sentinel lymph node. Ann Surg Oncol 1999; 6:75-82.

18

Index